METHODS IN MOLECULAR BIOLOGY™

Series Editor
John M. Walker
School of Life Sciences
University of Hertfordshire
Hatfield, Hertfordshire, AL10 9AB, UK

For other titles published in this series, go to
www.springer.com/series/7651

Heterologous Gene Expression in *E. coli*

Methods and Protocols

Edited by

Thomas C. Evans, Jr.

New England Biolabs, Inc., Ipswich, MA, USA

Ming-Qun Xu

New England Biolabs, Inc., Ipswich, MA, USA

Humana Press

Editors
Thomas C. Evans, Jr., PhD
DNA Enzymes Division Head
New England Biolabs, Inc.
Ipswich, MA 01938, USA

Ming-Qun Xu, PhD
Senior Scientist
New England Biolabs, Inc.
Ipswich, MA 01938, USA

ISSN 1064-3745
e-ISSN 1940-6029
ISBN 978-1-4939-5725-5
ISBN 978-1-61737-967-3 (eBook)
DOI 10.1007/978-1-61737-967-3
Springer New York Dordrecht Heidelberg London

Printed on acid-free paper

Humana Press is part of Springer Science+Business Media (www.springer.com)

Preface

As we move solidly into the 21st century, it is amazing to see that molecular biology and its related disciplines are advancing at an ever-increasing pace. Today, it is commonplace to use bioinformatic techniques to identify a gene encoding a protein of interest and then pay a company to synthesize it, codon optimized, and receive it already inserted into the expression vector of choice. This is all the more incredible when one considers that less than 100 years ago scientists were still trying to identify the substance responsible for inherited traits, a little over 50 years ago the first DNA polymerase was discovered, and only about 40 years ago the first restriction enzyme was purified.

It was the 1970s and 1980s that saw the expansive application of the prior basic research findings lead to the commonplace expression of proteins in a heterologous host. During this time recombinant DNA was first produced and shown to be stably maintained and replicated in *Escherichia coli*. The first recombinant protein, *E. coli* DNA polymerase I, was made commercially available from New England Biolabs, Inc., and a company, Genentech, was formed around the potential of recombinant DNA technology. It was also during this period that the first recombinant human protein, somatostatin, was produced.

Traditionally, *E. coli* has been the organism of choice to express proteins. Many proteins, however, were found to be insoluble or needed post-translational modifications that do not occur in *E. coli*. These issues are more likely to arise when eukaryotic proteins are expressed in *E. coli*. To address these problems, new expression hosts were developed to more effectively produce human proteins. These expression systems included insect and human cells. Although effective, these new expression systems were not as easy to manipulate or maintain as *E. coli*.

Currently, there is no perfect expression host. Membrane proteins constitute a significant percentage of the total cellular proteins but are very difficult to overexpress in a heterologous host. Furthermore, the ideal host would have the ability to express any protein, with relevant post-translational modifications, and still be as easy to work with as *E. coli*.

This volume is focused on this goal. The chapters herein describe methods, for example, to successfully express proteins in *E. coli* that would otherwise form aggregates in this host, to add post-translational modifications, to incorporate non-standard amino acid residues or moieties into *E. coli* expressed proteins, to identify binding partners, and to express membrane proteins. Although there is still no perfect expression host, and there may never be a perfect host, the work described herein moves *E. coli* closer to that ideal. In addition, a review of *E. coli* expression hosts is presented to help familiarize the researcher with the myriad of *E. coli* expression strains available. Finally, the strength of the *Methods in Molecular Biology* series is that the chapters are written in detail by scientists intimately familiar with the relevant techniques and protocols.

I hope that the reader finds the protocols described herein helpful to their research.

Thomas C. Evans, Jr.
Ming-Qun Xu

Preface

Contents

Contributors

EVELINA ANGOV • *Division of Malaria Vaccine Development, Walter Reed Army Institute of Research, Silver Spring, MD, USA*
QUAN BAI • *Key laboratory of Modern Separation Science in Shaanxi Province, Key Laboratory of Synthetic and Natural Functional Molecule Chemistry of Ministry of Education, Institute of Modern Separation Science, Northwest University, Xi'an, China*
JOHN BUSWELL • *New England BioLabs, Ipswich, MA, USA*
TAUSEEF R. BUTT • *LifeSensors Inc., Malvern, PA, USA*
NANCY CONSIDINE • *New England BioLabs, Ipswich, MA, USA*
MATTHEW P. DELISA • *School of Chemical and Biomolecular Engineering and Department of Biomedical Engineering, Cornell University, Ithaca, NY, USA*
THOMAS C. EVANS JR. • *DNA Enzymes Division, New England BioLabs, Ipswich, MA, USA*
ADAM C. FISHER • *School of Chemical and Biomolecular Engineering, Cornell University, Ithaca, NY, USA*
XINDU GENG • *Key laboratory of Modern Separation Science in Shaanxi Province, Key Laboratory of Synthetic and Natural Functional Molecule Chemistry of Ministry of Education, Institute of Modern Separation Science, Northwest University, Xi'an, China*
INCA GHOSH • *New England BioLabs, Ipswich, MA, USA*
LAN GUAN • *Department of Cell Physiology and Molecular Biophysics, Texas Tech University Health Sciences Center, Lubbock, TX, USA*
YOSHITAKA HARADA • *Laboratory of Microbial Technology, Division of Microbial Science and Technology, Department of Bioscience and Biotechnology, Graduate School, Kyushu University, Fukuoka 812-8581, Japan*
NORIO KAGAWA • *Office of Global COE, Nagoya University School of Medicine, Nagoya, 466-8550, Japan*
AKIRA KANNO • *Department of Chemistry, School of Science, The University of Tokyo, Tokyo, Japan; Research Institute of Pharmaceutical Sciences, Musashino University, Tokyo, Japan*
YUN CHEOL KIM • *Department of Microbiology, Immunology and Molecular Genetics, University of California Los Angeles, Los Angeles, CA, USA*
ORSOLYA KIRÁLY • *Department of Molecular and Cell Biology, Henry M. Goldman School of Dental Medicine, Boston University, Boston, MA, USA*
PATRICIA M. LEGLER • *Division of Biochemistry, Walter Reed Army Institute of Research, Silver Spring, MD, USA*
YURI Y. LONDER • *New England Biolabs, Inc., Ipswich, MA, USA*
DEREK MACMILLAN • *Department of Chemistry, Christopher Ingold Laboratories, University College London, London, UK*
ANDREA MAGGIONI • *Institute for Glycomics, Griffith University, Southport, QLD, Australia*
ARIO DE MARCO • *Biochemistry Unit, IFOM-IEO Campus for Oncogenomics, Cogentech, Milano, Italy*
ANASTASIYA MASHARINA • *New England BioLabs, Ipswich, MA, USA*

ELISSA MAUNUS • *New England BioLabs, Ipswich, MA, USA*
RYAN M. MEASE • *Division of Malaria Vaccine Development, Walter Reed Army Institute of Research, Silver Spring, MD, USA*
JUN-ICHI NAGAO • *Section of Infection Biology, Department of Functional Bioscience, Fukuoka Dental College, Fukuoka 814-0913, Japan*
JIRO NAKAYAMA • *Laboratory of Microbial Technology, Division of Microbial Science and Technology, Department of Bioscience and Biotechnology, Graduate School, Kyushu University, Fukuoka 812-8581, Japan*
CHRISTOPHER NOREN • *New England BioLabs, Ipswich, MA, USA*
KEN-ICHI OKUDA • *Laboratory of Microbial Technology, Division of Microbial Science and Technology, Department of Bioscience and Biotechnology, Graduate School, Kyushu University, Fukuoka 812-8581, Japan*
STEVEN J. ORCUTT • *LifeSensors Inc., Malvern, PA*
TAKEAKI OZAWA • *Department of Chemistry, School of Science, The University of Tokyo, Tokyo, Japan; Research Institute of Pharmaceutical Sciences, Musashino University, Tokyo, Japan*
RAYMOND J. PEROUTKA III • *LifeSensors Inc., Malvern, PA*
JONATHAN P. RICHARDSON • *Department of Chemistry, Christopher Ingold Laboratories, University College London, London, WC1H 0AJ, UK*
MARK A. ROCCO • *Department of Biomedical Engineering, Cornell University, Ithaca, NY, USA*
MIKLÓS SAHIN-TÓTH • *Department of Molecular and Cell Biology, Henry M. Goldman School of Dental Medicine, Boston University, Boston, MA, USA*
JAMES C. SAMUELSON • *New England Biolabs, Inc., Ipswich, MA, USA*
MARTIN SCHLAPSCHY • *Lehrstuhl für Biologische Chemie, Technische Universität München, Freising-Weihenstephan, Germany*
KOUKI SHIOYA • *Laboratory of Microbial Technology, Division of Microbial Science and Technology, Department of Bioscience and Biotechnology, Graduate School, Kyushu University, Fukuoka 812-8581, Japan*
ARNE SKERRA • *Lehrstuhl für Biologische Chemie, Technische Universität München, Freising-Weihenstephan, Germany*
KENJI SONOMOTO • *Laboratory of Microbial Technology, Division of Microbial Science and Technology, Department of Bioscience and Biotechnology, Graduate School, Kyushu University, Fukuoka 812-8581, Japan; Laboratory of Functional Food Design, Department of Functional Metabolic Design, Bio-Architecture Center, Kyushu University, Fukuoka 812-8581, Japan*
JAMES E. STRICKLER • *LifeSensors Inc., Malvern, PA, USA*
LUO SUN • *New England BioLabs, Ipswich, MA, USA*
PING SUN • *Macromolecular Crystallography Laboratory, Center for Cancer Research, National Cancer Institute at Frederick, Frederick, MD, USA*
JOE TIRALONGO • *Institute for Glycomics, Griffith University, Southport, QLD, Australia*
JOSEPH E. TROPEA • *Macromolecular Crystallography Laboratory, Center for Cancer Research, National Cancer Institute at Frederick, Frederick, MD, USA*
YOSHIO UMEZAWA • *Department of Chemistry, School of Science, The University of Tokyo, Tokyo, Japan; Research Institute of Pharmaceutical Sciences, Musashino University, Tokyo, Japan*

DAVID S. WAUGH • *Macromolecular Crystallography Laboratory, Center for Cancer Research, National Cancer Institute at Frederick, Frederick, MD, USA*
MING-QUN XU • *Chemical Biology Division, New England BioLabs, Ipswich, MA, USA*
TAKESHI ZENDO • *Laboratory of Microbial Technology, Division of Microbial Science and Technology, Department of Bioscience and Biotechnology, Graduate School, Kyushu University, Fukuoka 812-8581, Japan*
AIHUA ZHANG • *New England BioLabs, Ipswich, MA, USA*

Chapter 1

Adjustment of Codon Usage Frequencies by Codon Harmonization Improves Protein Expression and Folding

Evelina Angov, Patricia M. Legler, and Ryan M. Mease

Abstract

Over the past two decades, prokaryotic expression systems have been widely exploited for the bioproduction of many therapeutic proteins. Much of the success can be attributed to the implementation of basic principles of prokaryotic protein translation and protein folding to the problems of heterologous expression (e.g. codon usage substitutions, tRNA isoacceptor co-expression, chaperone co-expression); however, expression in a heterologous host still remains an empirical process. To improve heterologous protein expression further we have developed an algorithm termed "codon harmonization" that best approximates codon usage frequencies from the native host and adjusts these for use in the heterologous system. The success of this methodology may be due to improved protein folding during translation. Although so far exclusively applied to *Escherichia coli*, codon harmonization may provide a general strategy for improving the expression of soluble, functional proteins during heterologous host expression.

Key words: Codon harmonization, codon usage, codon optimization, expression, solubility, protein translation, protein structure, protein folding, circular dichroism.

1. Introduction

Biomedical research and biotechnological production processes rely on successful heterologous gene expression for the development of key reagents and biomolecules. Over-expression of target genes in *Escherichia coli* remains the preferred method due to the vast knowledge of *E. coli* genetics, the versatility in choice of vectors and expression host strains, the ease of use and low

The opinions or assertions contained herein are the private views of the authors and are not to be construed as official or as reflecting true views of the Department of the Army or the Department of Defense or the U.S.Army.

T.C. Evans, M.-Q. Xu (eds.), *Heterologous Gene Expression in E. coli*, Methods in Molecular Biology 705,
DOI 10.1007/978-1-61737-967-3_1,

cost, and often the high yields of protein produced. However, the primary reported causes for failures in efficient heterologous protein production include biased codon usage, gene product toxicity, solubility, mRNA secondary structure, and mRNA stability (reviewed in (1, 2)). The focus of this chapter is on one of the key factors leading to these failures, which are attributed to the degeneracy of the genetic code and the resultant species-specific codon usage frequencies (2). For *E. coli* as well as other organisms, synonymous codons are non-randomly utilized and a clear preference exists for particular codons which directly correspond to specific cognate isoacceptor tRNA molecule intracellular concentrations (3–6). Thus, for heterologous system expression, a high degree of discordance in codon usage between the species of the target gene and the heterologous expression host can lead to poor expression levels, solubility and product recovery, e.g. 80% AT bias in the structural genes from *Plasmodium falciparum* (7) compared with *E. coli*, which has a more balanced AT:GC genome content.

Various strategies have been used to minimize codon usage bias. One extensively used strategy has been generally termed as "codon optimization". In this method the codons most frequently used by the expression host are substituted at every position within the target gene sequence. The hypothesis for this strategy is that by introducing the most frequently used codons unilaterally throughout the length of the sequence; the resultant protein will be expressed at high levels primarily because the cognate isoacceptor tRNA molecules will not be rate limiting during protein synthesis (1, 2). This approach has led to successful production of some products (8); however, in some cases the very high expression levels have yielded insoluble protein localized to inclusion bodies, requiring denaturation and refolding strategies for purification (9, 10). Another strategy adjusts intracellular tRNA isoacceptor molecule concentrations by using specific plasmids which encode for rare tRNAs for that organism (reviewed in (1)). This approach resolves codon usage disparities in a very limited way and may lead to some, but not necessarily sufficient, amelioration in expression levels and product yield. In all cases the success or failure of these approaches used to optimize expression is sequence dependent and cannot be theoretically predicted.

Nascent polypeptide folding occurs co-translationally, which may be influenced by the kinetics of protein folding, the environment of the ribosomal tunnel, chaperone assistance, and the rate of tRNA binding (11–14). Less frequent codons within an RNA sequence may provide appropriate delays during protein synthesis, which enable longer periods for conformational sampling, thereby increasing the probability of finding the correct conformer (15,

4). On this premise, we developed an alternative approach for recoding target gene sequences for optimal heterologous expression in *E. coli*, which we refer to as "codon harmonization" (16). In this strategy, codon usage frequencies of the expression host are closely matched with the target gene's host codon usage frequencies. Particular attention is given to regions predicted to be slowly translated (i.e. "link/end" segments, defined as regions between domains or at the C-termini (17), where synonymous codons having usage frequencies either equal to or less than the respective codon usage frequency in the target gene's host are selected to ensure translational "pausing".

Using comparative analysis of *E. coli* gene sequences and their respective protein structures, Thanaraj and Argos (18) showed that sequences coded by higher usage frequency codons were mainly associated with secondary structural elements such as alpha helices, while sequences having one or more clusters of lower usage frequency codons tended to be associated with the linking residues between these elements. Furthermore, of the 20 possible amino acids, only 10 have a higher propensity of occurring within these link/end regions of *E. coli* proteins. These findings led to speculations that the positioning and clustering of low- and high-frequency usage codons were nonrandom and purposeful (19, 20). Clusters of low-frequency codons may have a specific role in slowing the rate of protein folding by providing the correct "pauses" in translation (5, 21). Thus, translation does not progress at a uniform rate but proceeds in pulses and the kinetics of protein folding is indirectly modulated by intracellular tRNA levels and codon usage frequencies (15, 22). Furthermore, control of nascent polypeptide synthesis and folding within the environment of the ribosomal tunnel would minimize the potential for partially folded polypeptides interacting with newly synthesized segments, thus minimizing deleterious aggregation and precipitation.

A flow chart of the steps from design to final product characterization is shown in **Fig. 1.1**. Protein expression profiles of native sequence compared to codon-harmonized *P. falciparum* 3D7 strain $MSP1_{42}$ indicate a significant overall increase in protein expression levels (*see* **Fig. 1.2**). The increased protein expression may be attributed to balancing the codon usage frequencies to the tRNA isoacceptor molecule availability found in the heterologous expression host. Furthermore, we hypothesized that adjusting the codon usage frequency of the target gene led to translation kinetics that more closely mimic the endogenous folding kinetics of the target. Evidence of protein fold and thermal denaturation differences between the native and codon-harmonized sequence products were analyzed by circular dichroism (*see* **Figs. 1.3 and 1.4**).

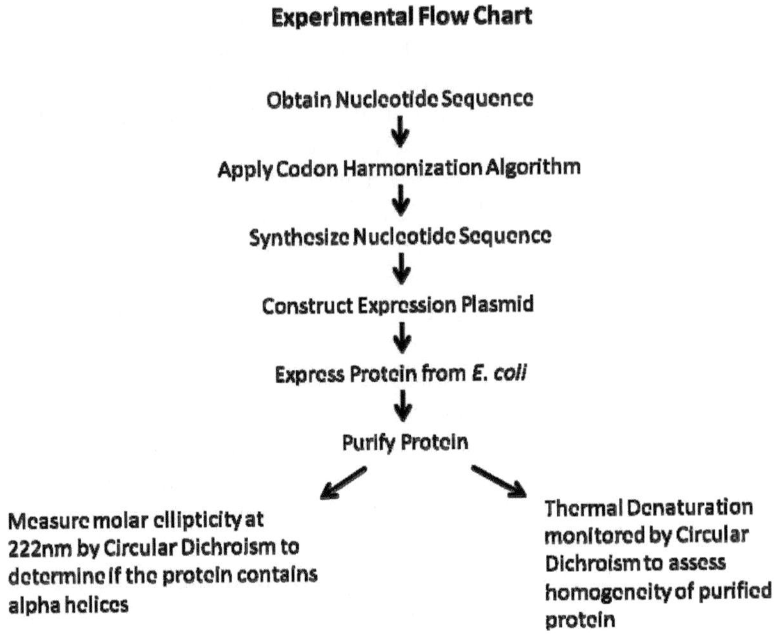

Fig. 1.1. Flow chart describing experimental plan.

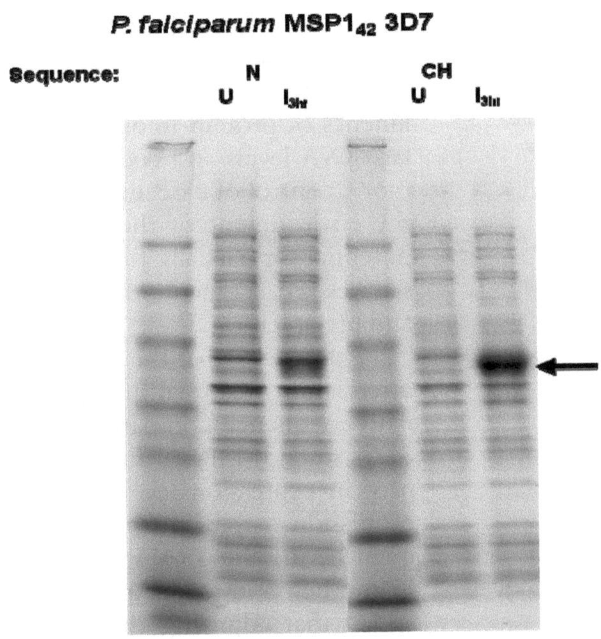

Fig. 1.2. Comparison of expression levels of native (N) and full gene codon-harmonized (CH) MSP1$_{42}$ gene fragment from 3D7 strain *Plasmodium falciparum* in *Escherichia coli*: U, cell lysates from uninduced cells; $I_{3\,h}$, lysate prepared 3 h post-induction (T3), respectively, after induction with 0.1 mM IPTG. Samples were separated by SDS-PAGE and stained with Coomassie Blue. The *arrow* indicates position of MSP1$_{42}$ migration.

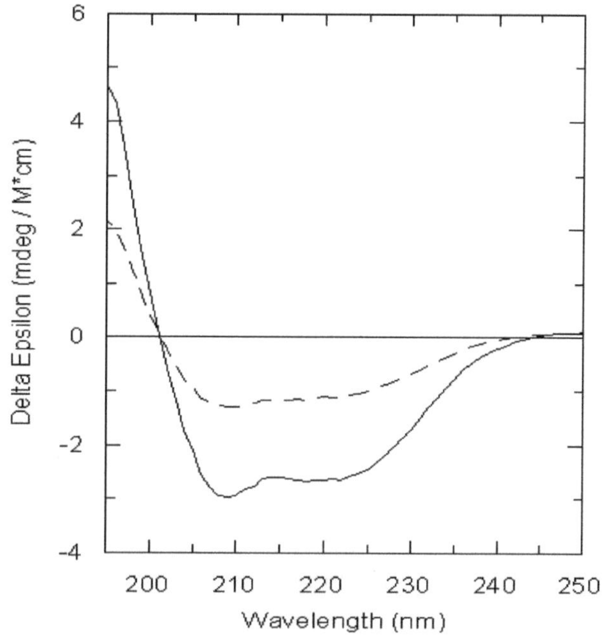

Fig. 1.3. Circular dichroism (CD) spectra of the *Plasmodium falciparum* $MSP1_{42}$ protein expressed from a codon-harmonized DNA sequence (—) versus the native DNA sequence (---) in 1X PBS pH 7.4 at 10°C. At 222 nm the $\Delta\varepsilon$ of protein produced from the native DNA sequence is 42% of the $\Delta\epsilon$ of the protein produced from the codon-harmonized DNA sequence, showing a higher percentage of alpha helical content in the protein expressed from the codon-harmonized DNA sequence.

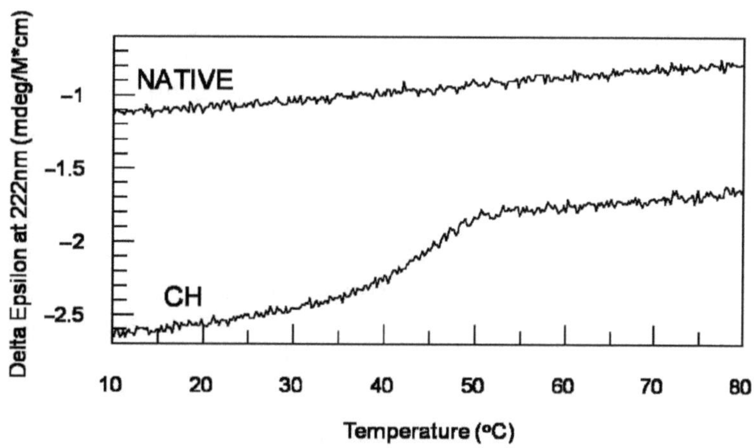

Fig. 1.4. Comparison of the irreversible thermal denaturation of $MSP1_{42}$ expressed from the native DNA sequence (*upper curve*) versus the codon-harmonized DNA sequence (*lower sigmoidal curve*) monitored by CD at 222 nm in 1X PBS pH 7.4 between 10 and 80°C. The broad transition and lower percentage of alpha helical content of the protein produced from the native DNA sequence suggests that the protein exists as a heterogeneous population of partially folded structures. The more cooperative transition and higher percentage of alpha helical content in the protein produced from the codon-harmonized DNA sequence suggests a more homogenous population of folded structures.

2. Materials

2.1. Target Gene Design and Synthesis

1. Target gene sequence (project specific)
2. Codon harmonization algorithm or alternative algorithm for recoding
3. Web-based sequence analysis software
4. Species-specific codon usage tables (http://www.kazusa.or.jp/codon/)

2.2. Protein Expression in E. coli

1. Bacterial culture media (e.g. Terrific Broth): Prepare 1 L by addition of 12 g tryptone, 24 g yeast extract, 4 mL glycerol in 800 mL H_2O. Autoclave to sterilize, cool to room temperature and then add 100 mL of filter-sterilized 0.17 M KH_2PO_4 and 0.72 M K_2HPO_4 solutions
2. Antibiotic (dependent on plasmid selected for expression) used to propagate plasmids in bacteria
3. IPTG: Isopropyl-β-D-thiogalactopyranoside. Used to induce expression of genes under control of the *lac* operon. Store solid at –20°C with desiccant
4. Amerex Instruments, Gyromax 767R Orbital Incubator shaker

2.3. Protein Analysis Reagents (SDS-PAGE)

1. Xcell SureLock™ MiniCell (Invitrogen)
2. Tris-glycine running buffer (containing SDS): 25 mM Tris-HCl, 192 mM glycine, 0.1% SDS, pH 8.3
3. Pre-cast minigels, 4–20% Tris-glycine (Invitrogen)
4. 2X loading sample buffer: 126 mM Tris-HCl, 20% glycerol, 4% SDS, 0.005% bromophenol blue, pH 6.8
5. SeeBlue protein marker (Invitrogen)
6. Gel staining solution [Coomassie Blue Stain]: 25% isopropanol, 10% acetic acid, 65% H_2O, 0.5 g Coomassie Brilliant Blue dye R-250. Dissolve and filter through 0.45 μm bottle top filter unit
7. Gel destaining solution [Coomassie Blue Destain]: 10% isopropanol, 10% acetic acid, 80% H_2O
8. DryEase Mini-Gel Drying System (Invitrogen)

2.4. Purification of Expressed Protein Product

1. Affinity chromatography using Ni^{2+}-NTA Sepharose (QIAGEN)

2.5. Protein Analysis by Circular Dichroism

1. 1X PBS: In 800 mL dissolve 9 g NaCl, 0.144 g KH_2PO_4, 0.795 g Na_2HPO_4. Adjust the pH to 7.4; add volume up to 1 L

2. PD-10 columns (GE Healthcare)
3. 1 mm quartz cuvette
4. Jasco J-815 Spectrometer fitted with Peltier temperature control unit

3. Methods

3.1. Design and Synthesis of Target Gene Sequence

1. Obtain the target gene nucleotide sequence for heterologous expression.
2. Apply the basic principles of the codon harmonization algorithm, which first matches the codon usage frequencies of the heterologous expression host with that of the species of the target gene. This can be done by evaluating the target nucleotide sequence codon usage preferences based on compiled species-specific codon usage tables for the native gene sequence and the heterologous expression host (codon usage tables found at http://www.kazusa.or.jp/codon/). Compilation of target sequences can be done electronically by the Codon Harmonization Algorithm through WRAIR investigator collaborations or can be performed manually by evaluating codon usage frequencies and usage tables for each, the target gene host and the expression host.
3. Identify putative link/end segments requiring slower translation progression by scanning for segments containing one or more of the 10 amino acid residues most frequently found in the intervening regions between domain structures in *E. coli* proteins (Trp, Tyr, His, Ile, Leu, Val, Ser, Thr, Cys, Pro) (17). If one or more of these amino acid residues is predicted to occur within a link/end segment then ensure that a lower frequency codon is substituted. However, in the absence of known crystal or NMR structures of the target protein these predictions are theoretical and may not necessarily reflect the native protein structure.
4. Determine if any "rare" codons (i.e. frequencies of <0.001) are introduced that would significantly limit translation kinetics leading to detrimental ribosomal stalling and/or early termination by the introduction of these extremely low-frequency tRNA isoacceptor molecules (*see* **Note 1**).
5. Design appropriate restriction sites on the 5′- and 3′-ends of the synthetic gene for subcloning into the desired expression vector. Ensure that no extraneous restriction sites that would interfere with subsequent cloning steps are included within the newly re-coded, harmonized sequence. To do this, perform restriction enzyme

mapping for the length of the codon-harmonized sequence (using http://tools.neb.com/NEBcutter2/index.php or other restriction site analysis methods).

6. Evaluate the codon-harmonized nucleotide sequence using protein translation tools to ensure that the reading frame has not been altered (e.g. using ExPASy Translate Tool at http://ca.expasy.org/tools/dna.html) and that all codon changes are silent synonymous substitutions.
7. For affinity chromatography using metal chelating chemistries, design N-terminal or C-terminal linker sequences that include six histidine residues.
8. Synthesize the codon-harmonized gene sequence. Currently, contracting the nucleotide synthesis step is competitively priced and cost-effective. Several commercial vendors are available, including Retrogen, Inc., Genescript, Inc., and Blue Heron Biotechnology's proprietary GeneMaker [R] platform (Invitrogen). Conversely, gene synthesis can be performed by a process of assembling overlapping oligonucleotides followed by amplification using PCR. The final assembled PCR product is subcloned into a suitable screening vector and verified by DNA sequencing (23).
9. Clone the re-coded target gene into the desired expression vector (i.e. the pET expression system, http://www.emdbiosciences.com/html/NVG/home.html). The pET expression plasmid is a popular bacterial plasmid designed especially for the production of high levels of the desired final protein product.

3.2. Evaluate Expression Levels

1. Standard molecular techniques are used to ligate the coding region into the expression vector and transform the plasmid DNA into bacterial cells (e.g. Molecular Cloning: A Laboratory Manual, fee-based website online laboratory manual at www.MolecularCloning.com).
2. Inoculate selected media using a bacterial cell glycerol stock (cryo-preserved cells). Incubate culture overnight under established optimal growth conditions (e.g. using rich culture media, the appropriate antibiotic for propagation of plasmid, at optimal temperature and agitation to promote maximum cell density) (Amerex Instruments, Gyromax 767R Orbital Incubator shaker) (*see* **Note 2**).
3. To evaluate expression levels of target protein, inoculate fresh culture media containing the appropriate antibiotic with cells grown overnight from **Step 2** (1% of the volume of fresh media). Allow the cells to grow at the established optimal temperature, usually in the range of 30–37°C with agitation (200 rpm) until a UV-vis absorbance reading at 600 nm

(OD_{600}) of between 0.6 and 1.0 is obtained. Reserve 1 mL of uninduced culture (adjust to an OD_{600} of 0.5) to prepare for polyacrylamide gel analysis. This sample is representative of the T0 or uninduced cell sample. Pellet by centrifugation at 5000 rpm for 5 min, remove supernatant and store the pellet at –20°C.

4. Induce expression of protein by adding isopropyl β-D-1-thiogalactopyranoside (IPTG) to the culture to obtain a final concentration of 0.1–2.0 mM IPTG (determined experimentally) (*see* **Note 3**).
5. Remove 1 mL of culture at each hour throughout the induction period for analysis of expression levels. To prepare samples for gel electrophoresis reserve 0.5 OD_{600} of cells from each time point (T0, T1, T2, T3, etc.) (being sure that the number of cells collected at each time point is identical so that equal amounts of protein are loaded per lane, e.g. 0.5 mL of culture with an absorbance of 1.0 OD_{600} equals 0.5 OD_{600} of cells), pellet the cells by centrifugation (5000 rpm for 5 min), decant media and add 50 μL H_2O, 50 μL formamide and 100 μl 2X sample buffer to the pellet. Vortex. Heat samples at 95°C for 2–5 min until fluid to prepare lysates. Store at 4°C until use.

3.3. Running Pre-cast Polyacrylamide Gels (SDS-PAGE)

1. For ease of handling and reproducibility, we prefer pre-cast minigel systems for analysis of protein expression and purity profiles, for example, 4–20% Tris-glycine gels from Invitrogen, although any SDS-PAGE standard format or other minigel systems can easily be substituted (24).
2. Select the appropriate percentage gel for the protein of interest and use a compatible buffer throughout the process. Remove gel from plastic pouch, take out comb and remove adhesive strip from foot of gel. Slide gels into Invitrogen Xcell SureLock Mini-cell with wells facing into central chamber. Close the seal and fill central chamber with the appropriate running buffer. Add running buffer to outside chamber until buffer is higher than the foot of the gel. Ensure that the seal is not leaking before loading samples. The wells of a 15 well × 1.5 mm gel can accommodate up to 15 μL of sample if loaded carefully with gel-loading pipette tips. Flush out any residual storage buffer from the wells with running buffer before loading samples. Load 5 μL SeeBlue pre-stained standards in lane 1. Load 5 μL of each lysate prepared. The marker serves as a protein size reference to compare with the expressed product. Load an aliquot of each sample from each time point (i.e. T0, T1, T2, T3, etc.) into consecutive wells (~5 μL each lysate). Separate proteins by running the gel system at 130 V (constant) for 90 min.

3. After completion of protein separation by SDS-PAGE, remove gel from pre-cast frame and place into the Coomassie Blue stain solution for 90 min with gentle rocking at room temperature to allow the dye to bind to the protein. Once staining is complete (approximately 1–2 h), place the gel in Coomassie Blue destain solution for 60 min with gentle rocking at room temperature. This solvent removes all dye in the gel that is not bound to protein. The expressed protein should be visible in the gel along with the protein marker. In an expression profile, the expressed protein can decrease over the period of the induction due to proteolysis or toxicity of the gene product (for an example of an expression profile *see* **Fig. 1.2**). The stained gel can be scanned for quantitative densitometry using, for example, BioRad VersaDoc 4000 Imager (which captures digital images from single and multicolor fluorescence, chemifluorescence, chemiluminescence and colorimetric samples).
4. Dry the gel using DryEase Mini-Gel Drying System (Invitrogen) for permanent storage.

3.4. Purification of Expressed Protein Product

1. For each target antigen, it is necessary to develop appropriate purification processes. We used Ni^{2+}-charged NTA sepharose affinity chromatography to purify the desired protein to homogeneity. See specific procedures for bacterial cell lysis and application of cleared lysates for affinity chromatography (http://www1.qiagen.com/literature/handbooks/literature.aspx?id=1000137).

3.5. Protein Secondary Structural Analysis by Circular Dichroism

1. Protein samples (1 mL of ~1 mg/mL) were loaded onto PD-10 gel filtration columns equilibrated with three column volumes of 1X PBS, pH 7.4.
2. Protein was eluted with the same buffer and 1 mL fractions were collected and kept on ice.
3. The absorbance at 280 nm was measured and the concentration of protein was determined using a calculated extinction coefficient (ProtParam Tool on http://ca.expasy.org/tools/protparam.html (25) and Beer's law ($A = \varepsilon cl$, where A is the absorbance at 280 nm, ε is the calculated extinction coefficient in M^{-1} cm^{-1}, c is the molar concentration of protein, l is the path length of the cuvette in centimetres). Protein was diluted with 1X PBS pH 7.4 until an absorbance at 280 nm of 0.2 was reached.
4. CD spectra were collected using a Jasco J-815 spectrometer fitted with a Peltier temperature control unit. A 1 mm quartz cuvette was used and three spectra were averaged. All spectra were collected at 10°C. Spectra were collected

between 190–250 nm. Analysis of proteins by CD reviewed in (26). Buffers other than PBS can interfere with measurements below 200 nm. A large negative minima corresponding to the random coil state occurs between ~196–198 nm and can be missed if data are not collected below 200 nm (27).

5. To make comparisons between protein samples for a given protein sequence, the ellipticity (in millidegrees (mdeg)) was converted into $\Delta\varepsilon$ using the following equation:

$$\Delta\varepsilon = \theta \frac{(0.1 \times \text{MRW})}{(P \times \text{CONC} \times 3298)}$$

where θ is in millidegrees, MRW is the mean residue weight (i.e. the molecular weight of the protein in daltons divided by the number of amino acid residues in the protein), P is the path length in centimetres and CONC is the protein concentration in milligram per millilitre determined using a calculated extinction coefficient and the molecular weight of the protein. The units of $\Delta\epsilon$ are mdeg M^{-1} cm^{-1}. Representative spectra are shown in **Fig. 1.3**. If the structure of the protein is known, the percentage of alpha helical content can be compared to a predicted value (28, 29) (e.g. http://www.embl-heidelberg.de/~andrade/k2d/) using these data.

6. Thermal denaturation curves were collected between 10 and 80°C; the temperature was increased at a rate of 2°C/min. The ellipticity (mdeg) at 222 nm was measured versus temperature to monitor the loss of alpha helical structure during the course of the melt. If denaturation is irreversible thermodynamic parameters cannot be calculated from these data. For comparisons the concentrations of protein should be within 10% of one another as the thermal denaturation curves also reflect the kinetics of aggregation which is concentration dependent. Data should be converted to $\Delta\varepsilon$ units to account for differences in concentration (*see* **Fig. 1.4**); points can be fit to a four-parameter logistic function to determine the melting temperature, T_m, using graphing software (Grafit 5.0.13 from Erithacus Software Limited).

4. Notes

1. The presence of codons that require rare isoacceptor tRNA molecules can lead to ribosomal pausing during translation and early termination of protein synthesis resulting in either low expression or a truncated product.

2. Cell culture growth conditions should be optimized for each clone before beginning an expression experiment. This includes temperature for growth (22–37°C), desired cell density for induction (0.6–2.0 OD_{600}), IPTG concentration (0.1–2 mM) and the duration of the induction period (2–4 h at 37°C or up to 20 h at 17°C). The concentration of antibiotic included must be high enough to allow for growth of only the desired clone, but not so high as to inhibit all bacterial growth.
3. Prior to induction, glucose can be added to cultures in concentrations ranging from 0 to 2%. For expression plasmids containing a T7 promoter, glucose can improve growth prior to induction by limiting basal expression of toxic gene products.

Acknowledgements

This work was supported by the United States Agency for International Development, Project Number 936–6001, Award Number AAG-P-00–98–00006, Award Number AAG-P-00-98-00005 and by the United States Army Medical Research and Materiel Command. The authors acknowledge the conceptual contributions of Drs. Jeffrey A. Lyon and Randall L. Kincaid for "codon harmonization". From the Division of Biochemistry, we thank Ms. Amy Michels for editorial assistance.

References

1. Makrides, S. C. (1996) Strategies for achieving high level expression of genes in *E. coli*. *Microbiol Rev* 60, 512–538.
2. Gustafsson, C., Govindarajan, S., Minshull, J. (2004) Codon bias and heterologous expression. *Trends Biotechnol* 22, 346.
3. Berisio, R., Schluenzen, F., Harms, J., Bashan, A., Auerbach, T., Baram, D., Yonath, A. (2003) Structural insight into the role of the ribosomal tunnel in cellular regulation. *Nat Struct Biol* 10, 366–370.
4. Purvis, I. J., Bettany, A. J., Santiago, T. C., Coggins, J. R., Duncan, K., Eason, R., Brown, A. J. (1987) The efficiency of folding of some proteins is increased by controlled rates of translation in vivo. A hypothesis. *J Mol Biol* 193, 413–417.
5. Wolin, S. L., Walter, P. (1988) Ribosome pausing and stacking during translation of a eukaryotic mRNA. *EMBO J* 7, 3559–3569.
6. Varenne, S., Buc, J., Lloubes, R., Lazdunski, C. (1984) Translation is a non-uniform process. Effect of tRNA availability on the rate of elongation of nascent polypeptide chains. *J Mol Biol* 180, 549–576.
7. Weber, J. L. (1987) Analysis of sequences from the extremely A + T-rich genome of *Plasmodium falciparum*. *Gene* 52, 103–109.
8. Itakura, K., Hirose, T., Crea, R., Riggs, A. D., Heyneker, H. L., Bolivar, F., Boyer, H. W. (1977) Expression in *Escherichia coli* of a chemically synthesized gene for the hormone somatostatin. *Science* 198, 1056–1063.
9. Pan, W., Ravot, E., Tolle, R., Frank, R., Mosbach, R., Turbachova, I., Bujard, H. (1999) Vaccine candidate MSP-1 from Plasmodium falciparum: a redesigned 4917 bp

polynucleotide enables synthesis and isolation of full-length protein from *Escherichia coli* and mammalian cells. *Nucleic Acids Res* 27, 1094–1103.
10. Singh, S., Kennedy, M. C., Long, C. A., Saul, A. J., Miller, L. H., Stowers, A. W. (2003) Biochemical and immunological characterization of bacterially expressed and refolded Plasmodium falciparum 42-kilodalton C-terminal merozoite surface protein 1. *Infect Immun* 71, 6766–6774.
11. Kleizen, B., van Vlijmen, T., de Jonge, H. R., Braakman, I. (2005) Folding of CFTR is predominantly cotranslational. *Mol Cell* 20, 277–287.
12. Kramer, G., Ramachandiran, V., Hardesty, B. (2001) Cotranslational folding – *omnia mea mecum porto? Int J Biochem Cell Biol* 33, 541–553.
13. Svetlov, M. S., Kommer, A., Kolb, V. A., Spirin, A. S. (2006) Effective cotranslational folding of firefly luciferase without chaperones of the Hsp70 family. *Protein Sci* 15, 242–247.
14. Etchells, S. A., Hartl, F. U. (2004) The dynamic tunnel. *Nat Struct Mol Biol* 11, 391–392.
15. Marin, M. (2008) Folding at the rhythm of the rare codon beat. *Biotechnol J.* 3, 1047–1057.
16. Angov, E., Hillier, C. J., Kincaid, R. L., Lyon, J. A. (2008) Heterologous protein expression is enhanced by harmoning the codon usage frequencies of the target gene with those of the expression host. *PLoS One* 3, e2189.
17. Thanaraj, T. A., Argos, P. (1996) Ribosome-mediated translational pause and protein domain organization. *Protein Sci* 5, 1594–1612.
18. Thanaraj, T. A., Argos, P. (1996) Protein secondary structural types are differentially coded on messenger RNA. *Protein Sci* 5, 1973–1983.
19. Ikemura, T. (1981) Correlation between the abundance of *Escherichia coli* transfer RNAs and the occurrence of the respective codons in its protein genes: a proposal for a synonymous codon choice that is optimal for the *E. coli* translational system. *J Mol Biol* 151, 389–409.
20. Bulmer, M. (1987) Coevolution of codon usage and transfer RNA abundance. *Nature* 325, 728–730.
21. Wen, J.-D., Lancaster, L., Hodges, C., Zeri, A.-C., Yoshimura, S. H., Noller, H. F., Bustamante, C., Tinoco, I., Jr. (2008) Following translation by single ribosomes one codon at a time. *Nature* 452, 598–603.
22. Baram, D., Yonath, A. (2005) From peptide-bond formation to cotranslational folding: dynamic, regulatory and evolutionary aspects. *FEBS Lett* 579, 948–954.
23. Moore, D. D. (2001) *Gene Synthesis: Assembly of Target Sequences Using Mutually Priming Long Oligonucleotides in Current Protocols in Molecular Biology.* Wiley, NJ, Unit 8.2B.
24. Laemmli, U. K. (1970) Cleavage of structural proteins during the assembly of the head of bacteriophage T4. *Nature* 227, 680–685.
25. Gasteiger, E., Hoogland, C., Gattiker, A., Duvaud, S., Wilkins, M. R., Appel, R. D., Bairoch, A. (2005) Protein identification and analysis tools on the ExPASy server. in (Walker, J. M. ed.), *The Proteomics Protocols Handbook* Humana Press, Totowa, NJ, pp. 571–607.
26. Kelly, S. M., Jess, T. J., Price, N. C. (2005) How to study proteins by circular dichroism. *Biochim Biophys Acta* 1751, 119–139.
27. Zubkov, V. A., Birshteen, T. M., Milevskaya, I. S., Volkenstein, M. V. (1971) Circular dichroism calculation for random-coil polypeptide chains. *Biopolymers* 10, 2051–2061
28. Andrade, M. A., Chacon, P., Merelo, J. J., Moran, F. (1993) Evaluation of secondary structure of proteins from UV circular dichroism spectra using an unsupervised learning neural network. *Protein Eng* 6, 383–390.
29. Merelo, J. J., Andrade, M. A., Prieto, A., Moran, F. (1994) Proteinotopic feature maps. *Neurocomputing* 6, 443–454.

Chapter 2

SUMO Fusion Technology for Enhanced Protein Expression and Purification in Prokaryotes and Eukaryotes

Raymond J. Peroutka III, Steven J. Orcutt, James E. Strickler, and Tauseef R. Butt

Abstract

The preparation of sufficient amounts of high-quality protein samples is the major bottleneck for structural proteomics. The use of recombinant proteins has increased significantly during the past decades. The most commonly used host, *Escherichia coli*, presents many challenges including protein misfolding, protein degradation, and low solubility. A novel SUMO fusion technology appears to enhance protein expression and solubility (www.lifesensors.com). Efficient removal of the SUMO tag by SUMO protease in vitro facilitates the generation of target protein with a native N-terminus. In addition to its physiological relevance in eukaryotes, SUMO can be used as a powerful biotechnology tool for enhanced functional protein expression in prokaryotes and eukaryotes.

Key words: SUMO, Smt3, SUMO protease1, protein expression, protein solubility, protein purification SUMOstar, SUMOstar protease.

1. Introduction

SUMO proteins are covalently attached to and removed from specific protein substrates in eukaryotic cells. SUMOylation as a reversible post-translational modification process has been shown to be involved in many cellular processes, such as nuclear-cytosolic transport (1), apoptosis (2), protein activation (3) and stability (4), response to stress (5), and progression through the cell cycle (6). A SUMO fusion system using yeast SUMO (*Saccharomyces cerevisiae* Smt3) as the N-terminal tag appears to enhance the expression and solubility of partner proteins and decrease proteolytic degradation (*see* **Fig. 2.1**) (7–13). After expression in *E. coli*, an N-terminal His 6 SUMO tag facilitates purification of the fusion protein. This tag can be efficiently removed by SUMO

T.C. Evans, M.-Q. Xu (eds.), *Heterologous Gene Expression in E. coli*, Methods in Molecular Biology 705, DOI 10.1007/978-1-61737-967-3_2, © Springer Science+Business Media, LLC 2011

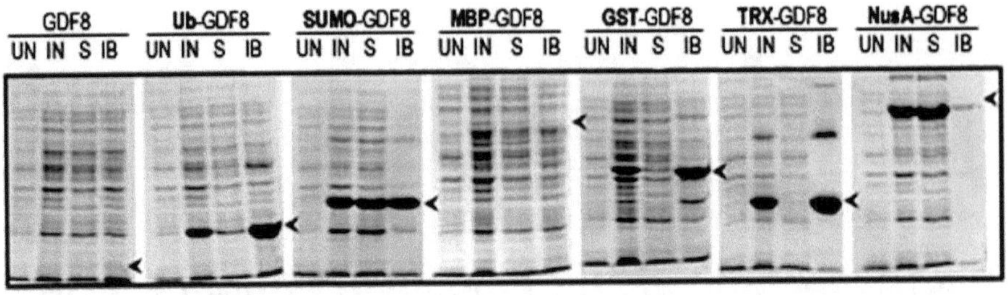

Fig. 2.1. Comparison of protein expression and solubility properties among GDF8 derivatives containing various N-terminal fusions (SUMO, GST, MBP, TRX, NusA, and Ub). All genes were expressed in a pET24 background. Equal amounts of protein from uninduced culture (UN), induced (IN), soluble fraction (S), and inclusion bodies (IB) were analyzed by 10% SDS-PAGE. Gels were stained with Coomassie blue. For details, please see the text. GDF8 fused with SUMO or NusA consistently showed higher amount of expression and solubility, whereas the GDF8 derivatives tagged with GST, MBP, or TRX were least effective in expression and solubility properties. GDF8 is expressed relatively poorly in *E. coli* as an unfused protein.

protease 1 (*S. cerevisiae* Ulp1), which recognizes the 3D structure of SUMO as well as the C-terminal sequence. Cleavage after the conserved C-terminal Gly–Gly motif of SUMO generates a partner protein with a native N-terminus and the capability of being re-purified and used for many biomedical or biopharmaceutical purposes.

SUMO fusion technology is an excellent tool for prokaryotic expression systems; however, the SUMO tag will be cleaved by SUMO proteases in a eukaryotic organism. LifeSensors, Inc. recently engineered a novel mutant SUMO tag, called SUMOstar, which is resistant to cleavage by SUMO protease in eukaryotic expression systems. This SUMOstar tag also maintains enhanced protein expression and solubility as shown in yeast (*S. cerevisiae* and *Pichia pastoris*, unpublished data), insect, and mammalian cells (14, 15). In addition, a novel SUMOstar-specific protease has been developed by LifeSensors, Inc. which is able to cleave the SUMOstar tag in vitro from the fusion protein. This novel SUMOstar fusion technology can be utilized for a variety of prokaryotic and eukaryotic expression systems.

Here, we will provide detailed protocols on how to construct a SUMO tag with a gene of interest, express this gene fusion in *E. coli* and purify the gene product from either the soluble fraction or inclusion bodies. We will also demonstrate how to cleave the SUMO tag in vitro using a SUMO-specific protease and purify the target protein with a native N-terminus.

2. Materials

2.1. Cloning into the pSUMO Vector

1. pSUMO vector (www.lifesensors.com) (*see* **Fig. 2.2**)
2. DNA containing the gene of interest

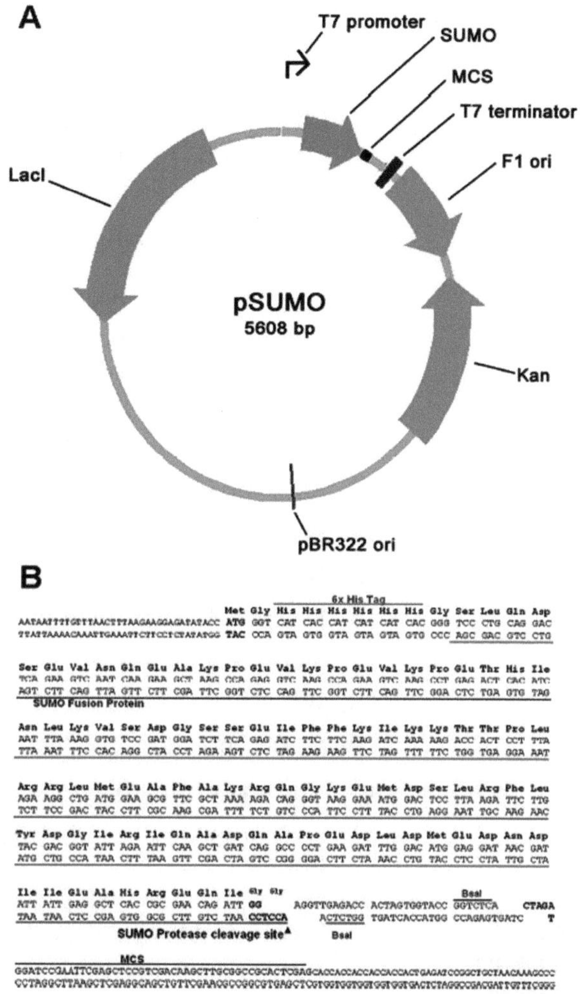

Fig. 2.2. The cloning vector pSUMO. (**a**) A feature map of the vector. Expression of a SUMO fusion protein is driven by an inducible T7 promoter. The pSUMO vector is a low copy plasmid with kanamycin selection. (**b**) The DNA sequence encoding the SUMO tag. The restriction digestion site of enzyme *Bsa*I is illustrated. Digestion with *Bsa*I results in two unique overhangs for directional cloning of the gene of interest.

3. PCR kit (High Fidelity DNA polymerase, 10x reaction buffer, dNTP nucleotide mix)
4. Appropriate DNA restriction enzymes and reaction buffers
5. PCR purification kit
6. 50°C water bath
7. DNA gel extraction kit
8. TAE buffer: 40 mM Tris, 1 mM EDTA, 20 mM acetic acid, pH 8.5

9. Agarose
10. T4 DNA ligase and reaction buffer
11. Competent *Escherichia coli* TOP10 cells
12. LB medium: 1% (w/v) bacto-tryptone, 0.5% (w/v) yeast extract, 1% (w/v) NaCl, pH 7.0
13. LB agar plates with appropriate antibiotic
14. Plasmid DNA miniprep kit

2.2. *E. coli* Transformation and Induction

1. SOC medium: 2% (w/v) bacto-tryptone, 0.5% (w/v) bacto-yeast extract, 0.05% (w/v) NaCl, 2.5 mM KCl, 20 mM glucose, pH 7.0
2. LB medium: 1% (w/v) bacto-tryptone, 0.5% (w/v) yeast extract, 1% (w/v) NaCl, pH 7.0
3. Appropriate antibiotics
4. Propane torch and flint striker
5. Bucket of ice
6. Shaking 37°C incubator
7. Sterile 2.5 l flasks
8. Water bath heated to 42°C
9. Sterile spreader
10. Automatic pipettor
11. Centrifuge with a rotor capable of holding 250 or 500 ml bottles

2.3. Preparation of Soluble and Inclusion Body (IB) Fractions

1. Sterile 35 ml centrifuge tubes
2. PBS: 2 mM KH_2PO_4, 8 mM Na_2HPO_4, 137 mM NaCl, 2.7 mM KCl, pH 8.0
3. Lysis buffer: 2 mM KH_2PO_4, 8 mM Na_2HPO_4, 287 mM NaCl, 2.7 mM KCl, 10 mM imidazole, 1% (v/v) Triton X-100, pH 8.0
4. IB wash buffer: 2 mM KH_2PO_4, 8 mM Na_2HPO_4, 287 mM NaCl, 2.7 mM KCl, 10 mM imidazole, 0.5% (v/v) Triton X-100, 1 mM EDTA, 1 M urea, pH 8.0
5. IB solubilization buffer: 50 mM CAPS, 0.3 M NaCl, 0.3% (w/v) *N*-laurylsarcosine, 1 mM DTT, pH 11
6. DNase: 50 mg/ml stock of deoxyribonuclease I from bovine pancreas (store at –20°C)
7. RNase: 50 mg/ml stock of ribonuclease A from bovine pancreas
8. PMSF: 1 M phenylmethylsulfonyl fluoride
9. IPTG: 1 M isopropyl β-D-thiogalactopyranoside

10. Dialysis buffer: 20 mM Tris, 150 mM NaCl, 10% (v/v) glycerol, pH 8.0

2.4. Affinity Purification

1. Ni-NTA resin
2. Lysis buffer: 2 mM KH_2PO_4, 8 mM Na_2HPO_4, 287 mM NaCl, 2.7 mM KCl, 10 mM imidazole, 1% (v/v) Triton X-100, pH 8.0
3. Wash buffer 1: 2 mM KH_2PO_4, 8 mM Na_2HPO_4, 287 mM NaCl, 2.7 mM KCl, 5 mM imidazole, and 1% (v/v) Triton X-100, pH 8.0
4. Wash buffer 2: 2 mM KH_2PO_4, 8 mM Na_2HPO_4, 287 mM NaCl, 2.7 mM KCl, 15 mM imidazole, pH 8.0
5. Elution buffer: 2 mM KH_2PO_4, 8 mM Na_2HPO_4, 287 mM NaCl, 2.7 mM KCl, 300 mM imidazole, pH 8.0
6. Strip buffer: 20 mM Tris, 100 mM EDTA, and 0.5 M NaCl, pH 7.9
7. IB solubilization buffer: 50 mM CAPS/KOH, 300 mM NaCl, and 0.3 % (w/v) *N*-laurylsarcosine, pH 11
8. Charge buffer: 50 mM $NiCl_2$
9. 20% (v/v) ethanol
10. Dialysis buffer: 20 mM Tris, 150 mM NaCl, 10% (v/v) glycerol, pH 8.0

2.5. SUMO Tag Cleavage

1. SUMO protease 1: 10 unit/μl (LifeSensors, Inc.)
2. PBS: 2 mM KH_2PO_4, 8 mM Na_2HPO_4, 137 mM NaCl, 2.7 mM KCl, pH 8.0
3. Refolding buffer: 20 mM Tris, 150 mM NaCl, 1% (w/v) CHAPS, 10% (v/v) glycerol, pH 8.0
4. 1 M DTT
5. 3.5 kDa MWCO (*m*olecular *w*eight *c*ut-*o*ff) dialysis tubing

3. Methods

3.1. Directional Cloning into pSUMO Vector Using BsaI

Here we describe the strategy of generating a SUMO tag fused gene of interest using the pSUMO vector and a single restriction enzyme for in-frame cloning (*see* **Notes 1** and **2**):

1. Design PCR primers to amplify the gene of interest. For the forward and reverse primers, use the sequences 5′-NNN **GGT CTC** NAG GTX XXX XXX XXX XXX XX–3′ and 5′-NNN **GGT CTC** *TCT AGA* TCA YYY YYY YYY YYY YYY-3′ as templates, respectively. Within these PCR

templates, "X" corresponds to nucleotides at the 5′-end of the gene of interest and "Y" to the reverse complement of nucleotides at its 3′-end. "N" is any nucleotide, the *Bsa*I site is in bold, and the *Xba*I site is in bold and italics (*see* **Note 3**).

2. Amplify the gene of interest with the designed primers in a PCR reaction using a thermostable high-fidelity DNA polymerase.
3. Clean up the PCR reaction using a PCR purification kit.
4. Digest the pSUMO vector and the generated PCR product separately in reaction tubes with *Bsa*I restriction endonuclease (10 U) for 1 h at 50°C (*see* **Notes 4** and **5**) and separate the reaction samples on a 1% (w/v) TAE agarose gel by running for 30 min at 10 V/cm.
5. Isolate (excise) both restricted DNA molecules, the pSUMO vector and the PCR fragment, from the agarose gel using a DNA gel extraction kit.
6. Mix the vector and the PCR fragment in the ratio 1:3 (mol:mol) and set up a 20 μl ligation reaction.
7. Incubate at room temperature for 2 h.
8. Use 5 μl of the ligation mixture to transform 50 μl of competent *E. coli* TOP10, DH5α, or other strains suitable for cloning according to the transformation protocol in **Section 3.2**.
9. Inoculate 3 ml culture with the positive colonies on LB-based selection plates and grow with shaking at 37°C overnight.
10. Spin down bacteria at 4000×*g* for 5 min and discard the supernatant.
11. Isolate the plasmid using a plasmid DNA miniprep kit. After sequence confirmation proceed to the transformation of the *E. coli* expression strain.

3.2. Transformation and Protein Expression

In this section, we describe how to express the SUMO fusion construct in *E. coli* BL21(DE3) cells. This strain is commonly used for inducible, T7 RNA polymerase-driven, high-level gene expression. Genes of two major proteases, OmpT and Lon, are deleted from the genome of this strain in order to decrease cellular degradation.

1. Gently thaw chemically competent cells on ice and keep the cells as cold as possible at all times. Always work aseptically when transforming and culturing *E. coli* cells.

2. Add 1 μl of DNA (approximately 0.1 μg) to 50 μl of chemically competent cells while on ice in a sterile 1.5 ml microfuge tube.
3. Mix gently with a pipette tip and let incubate on ice for 15–30 min.
4. Following incubation on ice, heat shock the cells by removing tube from the ice and immediately immersing in a 42°C water bath for 40 s. Place them back on ice after this heat shock period.
5. After 2 min on ice, add 200 μl of SOC pre-warmed to 37°C.
6. Leave the cells for recovery in the 37°C shaker for 1 h.
7. Aseptically transfer 0.1 ml of transformation culture to an LB plate containing 30 μg/ml kanamycin.
8. Spread transformation culture evenly with sterile spreader.
9. Place the plate into a 37°C incubator and incubate overnight.
10. Pipette 5 ml LB medium containing 30 μg/ml kanamycin into sterile, 15 ml snap cap tubes.
11. Inoculate the 5 ml LB with a single *E. coli* colony using an inoculation loop.
12. Incubate with shaking (250 rpm) at 37°C overnight.
13. Transfer 1 l of LB containing 30 μg/ml kanamycin to sterile 2.5 l flask.
14. Inoculate the 2.5 l flask with the 5 ml starter culture
15. Incubate with shaking (rpm = 250) at 37°C for approximately 3 h until the cell density reaches an $OD_{600\ nm}$ of 0.6–0.8.
16. When the $OD_{600\ nm}$ of 0.6–0.8 is reached, remove 1 ml of culture to serve as a negative induction control. Store this culture at –80°C.
17. Add IPTG to a final concentration of 1 mM. Incubate either at 37°C with shaking for 3 h or at 20°C for 16 h. Harvest cells by centrifugation at 4000×*g* for 15 min at 4°C. Store the pellet at –80°C. The typical wet weight of cultured cells is approximately 10–12 g^{-1}/l culture.

3.3. Cell Lysis and Protein Preparation

Overexpressed proteins in *E. coli* usually accumulate in the soluble fraction or form insoluble inclusion bodies. The SUMO fusion system significantly increases protein solubility and expression; therefore, SUMO-tagged proteins are usually present in the soluble fraction. However, in some cases, even SUMO-tagged proteins form inclusion bodies. Preparation of a soluble fraction and insoluble inclusion bodies from bacterial cells will be described here.

3.3.1. Preparation of Soluble Protein Fraction from E. coli Cells

1. Resuspend the cell pellet in lysis buffer (approximately 3 ml lysis buffer per gram cell paste).
2. Lyse cells by sonication (75% output for 10 × 15 s, with 30 s intervals between the pulse cycles) on wet ice (*see* **Note 8**).
3. Add DNase and RNase (20 μg/ml) to the lysates and incubate for 20 min on wet ice (*see* **Note 9**).
4. Add Triton X-100 to the sample to a final concentration of 1% (v/v) and incubate at 4°C for 1 h.
5. Centrifuge the sample (20,000×*g*, 30 min at 4°C) and pool the supernatant as the soluble protein fraction and keep the pellet for preparation of insoluble proteins (*see* **Section 3.3.2**).

3.3.2. Preparation of Insoluble Protein Fraction from Inclusion Bodies

1. Wash the pellet prepared above from the 1 l cultured cells with 30 ml IB wash buffer by resuspension and centrifugation at 10,000×*g* for 10 min at 4°C.
2. Discard the wash supernatant and repeat wash steps twice as described above.
3. Add 30 ml IB solubilization buffer to the pellet and incubate with shaking for 1 h at room temperature to extract insoluble proteins.
4. Centrifuge the sample at 15,000×*g* for 30 min at 4°C and collect the supernatant as the insoluble protein fraction.
5. Analyze the soluble and insoluble fractions prepared above using an SDS-PAGE according to the molecular weight of the gene product.
6. Use the protein samples immediately for purification or store them at 4°C for a short period of time (less than 10 days). For long-term storage keep the protein samples at –80°C and avoid repeated freeze–thaw cycles.

3.4. Affinity Purification Using Ni-IMAC Resin

The N-terminal His 6 tag of SUMO allows the affinity purification of SUMO fusion proteins using a Ni-IMAC resin (*see* **Fig. 2.3**). This method is an efficient and inexpensive way of purifying proteins from bacterial lysates. In this section, we will demonstrate the procedures of purifying SUMO fusion proteins from the soluble fraction and insoluble inclusion bodies.

3.4.1. Purification from a Soluble Extract

1. Pipette 25 ml Ni-IMAC resin into a column and allow to drain by gravity (here and in all subsequent purification steps allow the column to drain by gravity).
2. Wash the column with 5 column volumes (CV) of water.

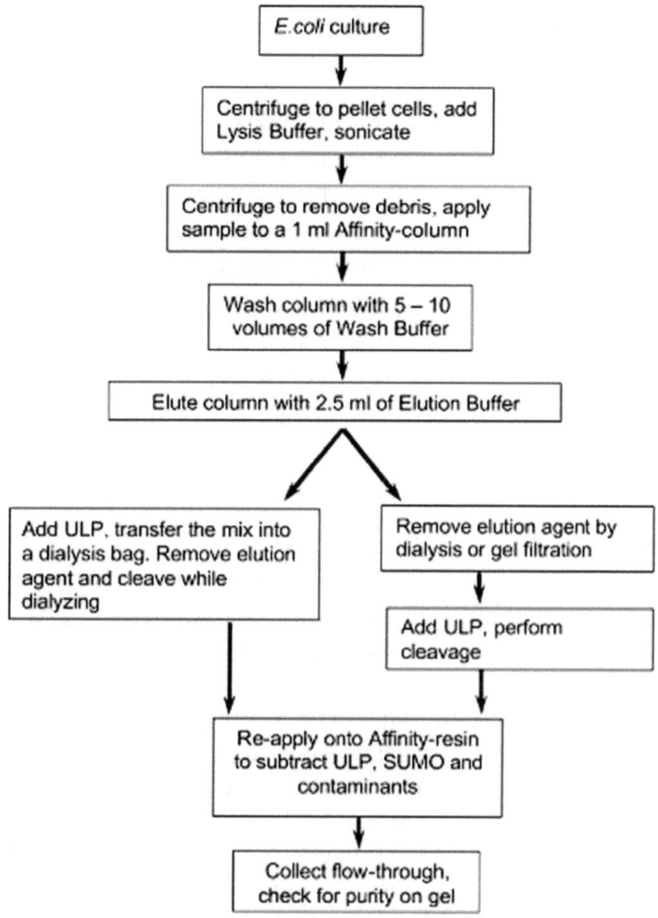

Fig. 2.3. Flow chart for purification and cleavage of His 6-SUMO-tagged proteins.

3. Charge the column with 5 CV charge buffer. If using a new Ni resin, the resin is pre-charged and this step may be omitted.
4. Equilibrate column with 10 CV of wash buffer 1.
5. Load sample onto Ni-IMAC resin. Make sure that the protein extract contains 5 mM imidazole. If not, add 1.7% (v/v) elution buffer to the protein sample and mix gently (*see* **Note 10**).
6. Wash column with 10 CV of wash buffer 1.
7. Wash with 10 CV of wash buffer 2.
8. Elute the bound SUMO fusion protein with 4 CV of elution buffer.
9. Strip column by adding 5 CV of strip buffer.
10. Wash column with 5 CV water.

11. Apply 5 CV 20% (v/v) ethanol to the column, allow ½ the volume to drain by gravity, and store the column at 4°C.

3.4.2. Purification from an Insoluble Extract

1. Equilibrate the pre-charged column with 10 CV of IB solubilization buffer containing 5 mM imidazole.
2. Load sample onto Ni-IMAC resin. Make sure the protein extract contains 5 mM imidazole as mentioned in **Section 3.4.1**.
3. Wash column with 10 CV IB solubilization buffer containing 5 mM imidazole.
4. Wash with 10 CV IB solubilization buffer containing 15 mM imidazole.
5. Elute with 4 CV IB solubilization buffer containing 300 mM imidazole.
6. Strip column by adding 5 CV strip buffer.
7. Wash column with 5 CV water.
8. Wash the column with 5 CV storage buffer as described in **Section 3.4.1**.

3.5. Removal of SUMO Tag

Unlike other proteases, SUMO protease 1 not only recognizes its specific amino acid sequence "x-Gly-Gly|x," but also the tertiary structure of the SUMO tag. SUMO proteases have been found to completely cleave a wide range (6–110 kDa) of proteins fused to SUMO and approximately 100 SUMO fusions have been cleaved without erroneous digestion (7, 9, 11). Variable conditions have also been tested on SUMO protease activity (7) (*see* **Fig. 2.4**). SUMO protease is capable of cleaving the SUMO-GFP fusion under a wide range of conditions such as in the presence of up to 2 M urea, 0.1 M guanidine-HCl, 300 mM imidazole, 1% Triton,

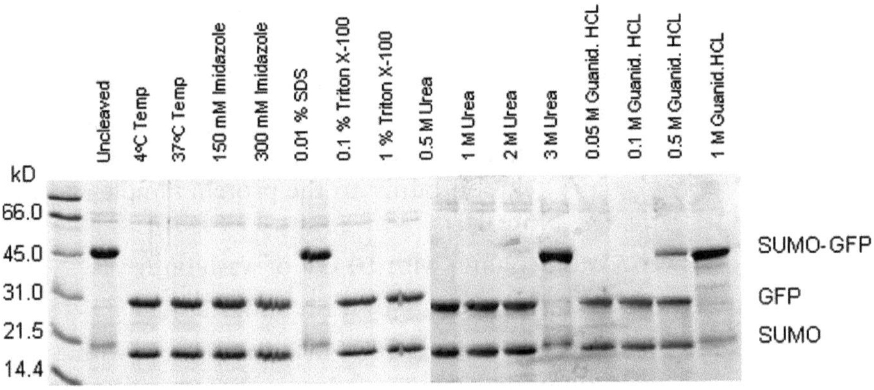

Fig. 2.4. Effect of temperature and various chemicals on SUMO protease 1 activity. Purified SUMO-GFP fusion (15 μg) and one unit of SUMO protease were combined in PBS buffer with indicated additives and the reactions were stopped after 20 min. Reaction products were resolved by SDS-PAGE and stained with Coomassie blue.

0.5 M NaCl, temperature from 4 to 37°C, and pH from 6 to 9 (7). Another important advantage of using SUMO as a fusion tag is the capability of generating the native N-terminus of a target protein which is particularly critical for producing proteins whose activity relies on their specific N-terminus (e.g. chemokines). The recombinant form of SUMO protease includes an N-terminal His 6 tag and so it can be easily removed using Ni-IMAC chromatography.

1. Dialyze a purified SUMO-fused protein using 3.5 kDa MWCO dialysis tubing against either 500 ml PBS for soluble fusions or 500 ml refolding buffer for insoluble SUMO-fused proteins for 24 h at 4°C with at least four fresh buffer exchanges.
2. Add SUMO protease 1 to the SUMO fusion protein sample in appropriate buffer (conditions are listed in **Table 2.1**) at a ratio of 1 unit enzyme per 100 μg substrate and incubate at 30°C for 1 h (*see* **Note 11**).

Table 2.1
Influence of various chemicals on the activity of SUMO protease 1

Chemical	Concentration	Percent of cleavage
Phosphate-buffered saline (PBS)	*See* **Section 2.3**	100
DTT or β-mercaptoethanol	20 mM	100
NaCl	150 mM	100
	500 mM	60
	1 M	30
Urea	1 M	100
	2 M	95
	3 M	5
Guanidine hydrochloride	500 mM	60
	1 M	0
Triton X-100	1%	100
Imidazole	300 mM	100
GSH (reduced glutathione)	20 mM	100
Maltose	20 mM	100
Glycerol	20% (v/v)	100
Ethylene glycol	20% (v/v)	100
Sucrose	20% (w/v)	100
Ethanol	10% (v/v)	100

3. Add DTT to the enzyme–substrate mixture to a final concentration of 2.0 mM. Do not exceed 2 mM if nickel affinity resin (Ni-IMAC) will be used for subtracting SUMO in the downstream purification, because high concentrations of DTT can disassociate the metal from the resin.
4. Incubate the mixture at 30°C for 1 h with slight shaking. Typically, >95% of SUMO fusions can be cleaved under these conditions. (To maximize cleavage, continue to incubate the mixture at 4°C overnight.)
5. Check the cleavage using SDS-PAGE. If the SUMO fusion is not approx. 95% cleaved, add more SUMO protease 1 and incubate for a longer time.
6. Dialyze the cleaved SUMO fusion with proper buffer for the next purification step and subtraction of SUMO and SUMO protease 1 for pure target proteins (*see* **Note 12**).
7. For the subtraction step, reapply the digestion mixture to the 5 ml Ni-IMAC column. Add two column volumes of PBS to the column. While the gene product of interest, which does not harbor a His 6-tag, flows through the column, both

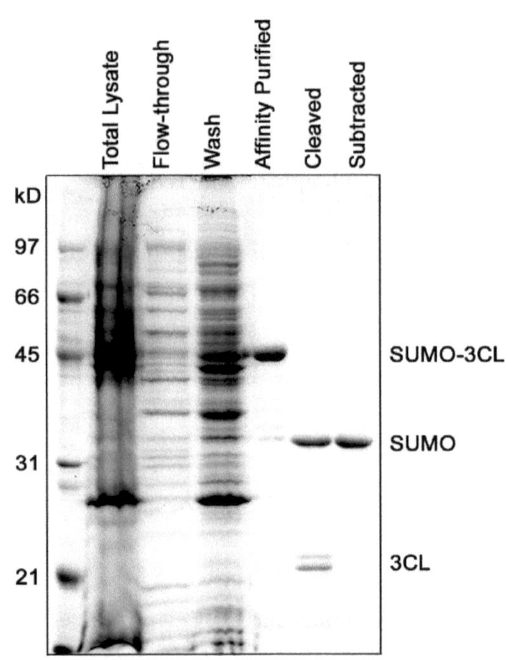

Fig. 2.5. Expression and purification of His 6-SUMO-3CL and protein. *Escherichia coli* grown in LB medium was induced at 20°C for 6 h. Proteins were purified on Ni-IMAC resin and eluted (lanes as described in **Section 3.4**). Cleavage and dialysis of 2 mg of each fusion were performed overnight at 4°C with 10 units of SUMO protease 1. Following dialysis, His 6-SUMO and SUMO protease 1 were subtracted by passing the proteins through a miniature Ni-IMAC column. Protein fractions were resolved by SDS-PAGE and stained with Coomassie blue.

the SUMO fusion and SUMO protease contain a His 6-tag and thus bind to the Ni-IMAC resin and become subtracted from the mixture.

8. Analyze the protein samples by SDS-PAGE (see a sample purification and cutting analysis in **Fig. 2.5**).

4. Notes

1. To take full advantage of SUMO fusion technology including the removal of the SUMO tag by SUMO protease, it is critical to make a fusion protein without any additional sequence between SUMO and the desired gene. SUMO protease recognizes the structure of the SUMO moiety and cleaves at the C-terminal end of a conserved – Gly–Gly sequence (16). Therefore, SUMO protease never cleaves inside the partner protein.
2. SUMO protease can cleave the SUMO moiety in any amino acid context except that in which a proline residue follows immediately after the last two glycines of SUMO.
3. In designing DNA primers for the amplification of the open reading frame, make sure to add at least two extra nucleotides at the 5′-end of the primer. Otherwise the restriction enzymes will not cleave PCR-amplified DNA efficiently.
4. If the gene of interest contains an inherent *Bsa*I site, one can use an alternate type IIS restriction endonuclease. Below are recommended ways to incorporate these enzymes into the DNA primers.

   *Aar*I: 5′ – **CACCTGC**NNNNAGGTXXXXXXXXXXXXXXX – 3′

   *Bbs*I: 5′ – **GAAGAC**NNAGGTXXXXXXXXXXXXXXX – 3′

   *Bbv*I: 5′ – **GCAGC**NNNNNNNNAGGTXXXXXXXXXXXXXXX – 3′

   *Bfu*AI: 5′ – **ACCTGC**NNNNAGGTXXXXXXXXXXXXXXX – 3′

   *Bsa*I: 5′ – **GGTCTC**NAGGTXXXXXXXXXXXXXXX – 3′

   *Bsm*AI: 5′ – **GTCTC**NAGGTXXXXXXXXXXXXXXX – 3′

   *Bsm*BI: 5′ – **CGTCTC**NAGGTXXXXXXXXXXXXXXX – 3′

   *Bsm*FI: 5′ – **GGGAC**NNNNNNNNNNAGGTXXXXXXXXXXXXXXX – 3′

*Btg*ZI: 5′ – **GCGATG**NNNNNNNNNNAGGTXXXXX XXXXXXXXXX – 3′

*Fok*I: 5′ – **GGATG**NNNNNNNNNNAGGTXXXXXXX XXXXXXXX – 3′

*Sfa*NI: 5′ – **GCATC**NNNNNAGGTXXXXXXXXXXXX XXX – 3′

where N represents any nucleotide and X represents the sequence of the gene of interest. The enzyme recognition sequence is underlined and in bold.

5. When performing the restriction digest, make sure not to use too much enzyme for an extended period of time. Over-digestion of the vector could result in inefficient cloning.
6. For the protein expression use *E. coli* expression strain freshly transformed with expression plasmid (<2 weeks old). Using older plates entails the risk of significant reduction in protein expression as compared with expression using freshly transformed plates.
7. When inducing the culture for protein expression it is important to know that even with the SUMO fusion tag some proteins are insoluble when expressed at 37°C. Lowering the temperature during the induction ensures a higher yield of soluble protein. Induction can be performed at 20°C overnight, in which case the preinduction cell density must be 0.8 OD instead of 0.5.
8. When sonicating, it is critical to not overheat the lysate. If a large bacterial pellet is being sonicated, it is wise to use a metal container for the best heat transfer from the lysate to the iced water.
9. The DNase I solution must be freshly prepared because freeze/thaw cycles significantly decrease DNase I enzymatic activity and the cell lysate might be too viscous after centrifugation.
10. During column purification it is critical to have imidazole in the lysis and wash buffers to ensure a clean protein preparation. However, if the protein starts to elute during the second wash, replace the wash buffer II with wash buffer I.
11. SUMO protease is a very robust enzyme and it can cleave in a variety of buffers and additives. **Figure 2.4** and **Table 2.1** summarize some tested conditions for the cleavage efficiency.
12. It is noteworthy that with SUMO the fusion partners are more soluble; however, after cleavage with SUMO protease the cleaved off protein might fall out of solution.

13. Due to the lack of endogenous SUMO protease in *E. coli* cells, SUMO fusion expression system is well established in prokaryotic cells (LifeSensors, Inc.). However, SUMO tag will be cleaved by endogenous SUMO protease after translation in eukaryotic cells (9). We recently engineered a novel SUMO tag, called SUMOstar, which is not removed from the fusion partner in eukaryotic cells. We have shown that the SUMOstar system enhances expression of proteins in *yeast, P. pastoris*, insect cells, and mammalian cells. In addition to enhancing intracellular expression, novel SUMOstar tags have been developed that enhance secretion of proteins in insect and mammalian cells. Thus, the SUMOstar fusion could be utilized for enhanced expression of functional proteins not only in prokaryotes but also in eukaryotic systems. After affinity purification of fusion proteins, SUMOstar tag can be cleaved in vitro by a specific SUMOstar protease (LifeSensors, Inc.).

References

1. Seeler, J. S., Bischof, O., Nacerddine, K., Dejean, A. (2007) SUMO, the three Rs and cancer. *Curr Top Microbiol Immunol* 313, 49–71.
2. Meinecke, I., Cinski, A., Baier, A., Peters, M. A., Dankbar, B., Wille, A., Drynda, A., Mendoza, H., Gay, R. E., Hay, R. T., Ink, B., Gay, S., Pap, T. (2007) Modification of nuclear PML protein by SUMO-1 regulates Fas-induced apoptosis in rheumatoid arthritis synovial fibroblasts. *Proc Natl Acad Sci USA* 104, 5073–5078.
3. Rajan, S., Plant, L. D., Rabin, M. L., Butler, M. H., Goldstein, S. A. (2005) Sumoylation silences the plasma membrane leak K+ channel K2P1. *Cell* 121, 37–47.
4. Martin, S., Nishimune, A., Mellor, J. R., Henley, J. M. (2007) SUMOylation regulates kainate-receptor-mediated synaptic transmission. *Nature* 447, 321–325.
5. Mabb, A. M., Wuerzberger-Davis, S. M., Miyamoto, S. (2006) PIASy mediates NEMO sumoylation and NF-kappaB activation in response to genotoxic stress. *Nat Cell Biol* 8, 986–993.
6. Li, S. J., Hochstrasser, M. (1999) A new protease required for cell-cycle progression in yeast. *Nature* 398, 246–251.
7. Malakhov, M. P., Mattern, M. R., Malakhova, O. A., Drinker, M., Weeks, S. D., Butt, T. R. (2004) SUMO fusions and SUMO-specific protease for efficient expression and purification of proteins. *J Struct Funct Genomics* 5, 75–86.
8. Marblestone, J. G., Edavettal, S. C., Lim, Y., Lim, P., Zuo, X., Butt, T. R. (2006) Comparison of SUMO fusion technology with traditional gene fusion systems: enhanced expression and solubility with SUMO. *Protein Sci* 15, 182–189.
9. Butt, T. R., Edavettal, S. C., Hall, J. P., Mattern, M. R. (2005) SUMO fusion technology for difficult-to-express proteins. *Protein Expr Purif* 43, 1–9.
10. Zuo, X., Li, S., Hall, J., Mattern, M. R., Tran, H., Shoo, J., Tan, R., Weiss, S. R., Butt, T. R. (2005) Enhanced expression and purification of membrane proteins by SUMO fusion in *Escherichia coli*. *J Struct Funct Genomics* 6, 103–11.
11. Zuo, X., Mattern, M. R., Tan, R., Li, S., Hall, J., Sterner, D. E., Shoo, J., Tran, H., Lim, P., Sarafianos, S. G., Kazi, L., Navas-Martin, S., Weiss, S. R., Butt, T. R. (2005) Expression and purification of SARS coronavirus proteins using SUMO-fusions. *Protein Expr Purif* 42, 100–110.
12. Guzzo, C. M., Yang, D. C. (2007) Systematic analysis of fusion and affinity tags using human aspartyl-tRNA synthetase expressed in *E. coli*. *Protein Expr Purif* 54, 166–175.
13. Dominy, J. E., Jr., Simmons, C. R., Hirschberger, L. L., Hwang, J., Coloso, R. M., Stipanuk, M. H. (2007) Discovery and characterization of a second mammalian thiol dioxygenase, cysteamine dioxygenase. *J Biol Chem* 282, 25189–25198.

14. Liu, L., Spurrier, J., Butt, T. R., Strickler, J. E. (2008) Enhanced protein expression in the baculovirus/insect cell system using engineered SUMO fusions. *Protein Expr Purif*, 62, 21–28.
15. Peroutka, R. J., Elshourbagy, N., Piech, T., Butt, T. R. (2008) Enhanced protein expression in mammalian cells using engineered SUMO fusions: secreted phospholipase A2. *Protein Sci* 17, 1586–1595.
16. Mossessova, E., Lima, C. D. (2000) Ulp1-SUMO crystal structure and genetic analysis reveal conserved interactions and a regulatory element essential for cell growth in yeast. *Mol Cell* 5, 865–876.

Chapter 3

Molecular and Chemical Chaperones for Improving the Yields of Soluble Recombinant Proteins

Ario de Marco

Abstract

Molecular chaperones and chemical compounds like amino acids and osmolytes share the capability to prevent protein aggregation and can contribute to rescue in vivo aggregated proteins. Therefore, both overexpression of the molecular folding machinery and induced accumulation of chemical chaperones are options to improve the correct folding of recombinantly expressed proteins. These two parameters may show synergistic effects, although success remains protein specific and, therefore, several combinations of molecular and chemical chaperones should be compared. However, proteins can fail to fold correctly even in optimized culture conditions. In this case, protein aggregates can be recovered and their refolding assisted by an osmolyte/chaperone-dependent system. The selection of aggregates with different degrees of complexity can be exploited to maximize the yields of native proteins at the end of the refolding process.

Key words: Molecular chaperones, osmolytes, protein refolding, protein aggregates, osmotic stress, heat-shock response, trehalose overproduction engineering.

1. Introduction

Any protein possesses the information necessary to reach its native structure coded into its linear amino acid sequence, as described by Anfinsen (1). However, in the absence of foldases and molecular chaperones, some proteins would be trapped in unproductive folding intermediates during their folding. When not prevented by cell proteolytic activity, the accumulation of unstable, partially unfolded intermediates leads to protein aggregation. The chance to form aggregates becomes higher in heterologous expression systems since unstable protein intermediates do not find suitable

T.C. Evans, M.-Q. Xu (eds.), *Heterologous Gene Expression in E. coli*, Methods in Molecular Biology 705, DOI 10.1007/978-1-61737-967-3_3, © Springer Science+Business Media, LLC 2011

folding machinery elements in the new expression host to complete their folding process.

The co-expression of recombinant chaperones can improve the folding efficiency of host cells and the yields of soluble target proteins, as initially demonstrated by Goloubinoff et al. (2). The accurate elucidation of the molecular background relative to the functional co-ordination of the chaperone network (3, 4) has recently allowed a systematic survey for identifying optimal conditions to exploit molecular chaperones for biotechnological applications (5, 6). This approach overcame the past empiric attempts of stimulating heat-shock protein accumulation by inducing metabolic stress using heat treatments and high salt concentrations or causing cell membrane perturbation by the addition of chemicals like ethanol and benzoic alcohol (7, 8).

Another attempt to improve protein stabilization during heterologous protein expression involved adding chemical chaperones to the cell culture media. These substances, such as osmolytes or amino acids, minimize unproductive interactions and have been successfully used both in vivo and in vitro to prevent protein aggregation and to stabilize protein intermediates during refolding (9). Furthermore, the enzymes involved in trehalose biosynthesis have been overexpressed in genetically engineered bacterial cells, enabling the induction of trehalose accumulation directly in the bacterial cytoplasm (10).

There are also examples of synergistic activities between molecular and chemical chaperones (7). Such observations suggest that molecular chaperones stabilize the protein substrates by interactions that are different or only partially overlapping with respect to the chemical chaperone mode of action and, therefore, the single contributions of both chaperone components can be exploited for improving the overall system functionality (*see* **Fig. 3.1**).

Nevertheless, there are still several cases in which none of the proposed approaches leads to substantial improvements in the recombinant protein folding. Nevertheless, protein aggregates can always be recovered and successively forced to refold in vitro under specific conditions.

We have previously shown that recombinant proteins form aggregates characterized by increasing levels of complexity (11) and that culture conditions, expression constructs, and molecular chaperone co-expression could contribute to the structural features of the aggregates (12). For instance, the same GST–GFP construct extracted from bacteria grown at different temperatures and variable DnaK availability formed soluble aggregates in shape of few nanometer rods, large nets of several hundreds of nanometer in diameter, and structured fibrils (*see* **Fig. 3.2**, (11, 12)). This information is relevant since both aggregate complexity and kinetics of (re)aggregation are crucial factors in the success of the

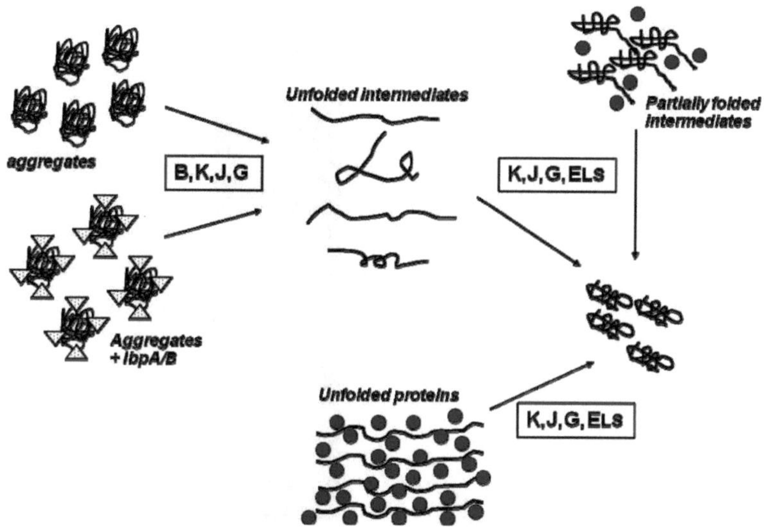

Fig. 3.1. Synergistic interactions between molecular and chemical chaperones. Large protein aggregates can co-precipitate with small chaperones IbpA and IbpB (IbpA/B) that prevent the formation of tight inclusion bodies. Such aggregates can be disassembled into soluble form by the disaggregation activity of ClpB (B) and DnaK (K) system (DnaK plus the regulative co-chaperones DnaJ [J] and GrpE [E]) and the protein intermediates can recover their native structure through the action of chaperonins GroEL (EL) and GroES (S), and the DnaK system. Protein intermediates can be stabilized by osmolytes that impair the interactions among still exposed hydrophobic regions. Osmolytes can also contribute to the unfolding of large aggregates and, consequently, in facilitating the molecular chaperone-dependent refolding route. Osmolytes are represented by *closed circles* and IbpA/B chaperones by *shaded triangles.*

refolding strategies (11, 12). The following protocols will focus on the possibility to use molecular chaperones and osmolytes, both alone and in combinations, to limit recombinant protein aggregation in bacteria and to improve the yields of productive refolding starting from precipitated proteins.

2. Materials

2.1. Vectors and Strains

1. Expression vectors pETM14, pETM22, pETM33, pETM44, pETM66, pETM80, and pETM82 with selective kanamycin resistance and pETM20 and pETM90 with selective ampicillin resistance (*see* **Note 1**).
2. BL21(DE3) bacterial cells (no resistance), Rosetta (DE3) strain (chloramphenicol resistance, Novagen) that enables the overexpression of tRNAs for rare codons, and Origami (DE3) strain (kanamycin resistance, Novagen) that allows the cytoplasmic formation of disulfide bonds (*see* **Note 2**).

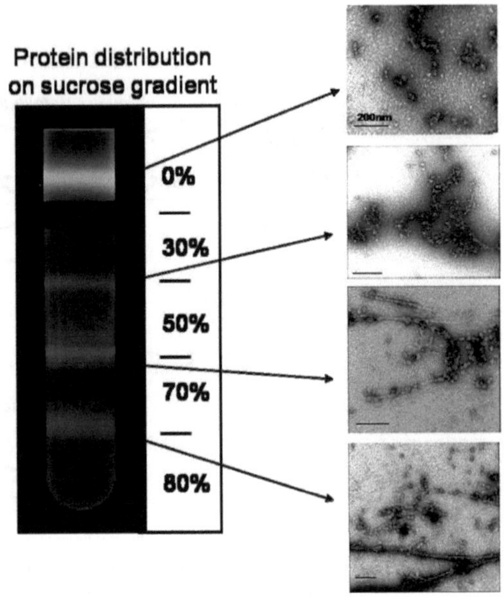

Fig. 3.2. Recombinant proteins produce aggregates of variable complexity. Bacteria overexpressing a GST–GFP fusion protein were cultured overnight at 20°C after IPTG induction and finally pelleted. Total lysate was loaded onto a sucrose gradient and separated by centrifugation. The protein fractions accumulated at the interfaces between solutions of different sucrose concentrations were recovered and analyzed by electron microscopy (11).

2.2. Chaperone Overexpression

1. Molecular chaperone vectors pBB528 (chloramphenicol), pBB530 (chloramphenicol), pBB535 (spectinomycin), pBB540 (chloramphenicol), pBB541 (spectinomycin), pBB542 (spectinomycin), pBB550 (spectinomycin), and pBB572 (ampicillin) (*see* **Fig. 3.3** and **Note 3**).
2. Antibiotic stock solutions (1000x) for chloramphenicol (10 mg/mL dissolved in ethanol), carbenicillin (100 mg/mL), kanamycin (30 mg/mL), and spectinomycin (50 mg/mL). Filter sterilize the solutions and store in single-use aliquots at –20°C.
3. Luria Bertani (LB) medium (10 g tryptone, 5 g yeast extract, 10 g NaCl × 1 L H_2O).
4. 100 mM isopropyl-beta-D-thiogalactopyranoside (IPTG) dissolved in H_2O; 1 mL aliquots are stored at –20°C.
5. SDS-PAGE running device.
6. Thermostatic orbital shaker with tube racks and bottle adaptors.

   Use the frozen stocks of bacteria harboring the suitable chaperone combinations to prepare chemically competent cells. Stocks remain competent for several months.

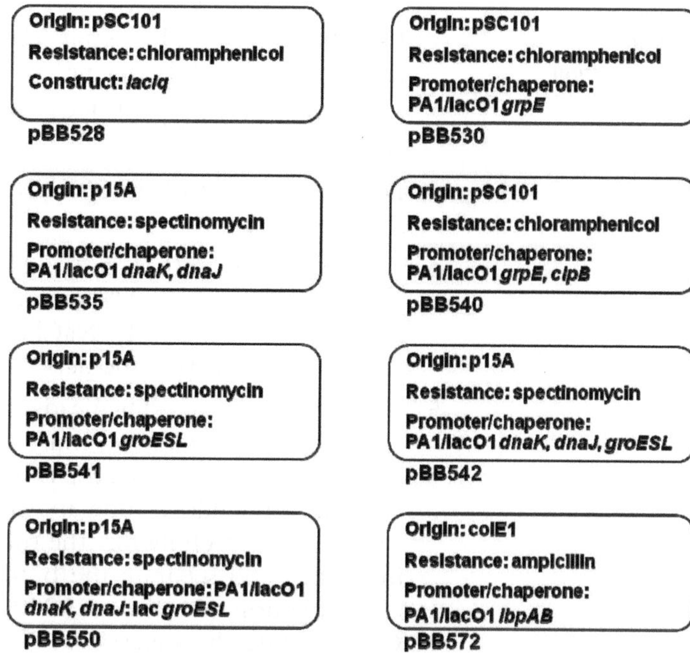

Fig. 3.3. Chaperone plasmid representation. Origin, promoters, resistance, and chaperone sequences present in each plasmid are reported.

2.3. Cell Culture

1. LB medium (10 g tryptone, 5 g yeast extract, 10 g NaCl × 1 L H_2O) and glycerol (70%)
2. 15 mL disposable tubes
3. 96-Well microtiter plates (Eppendorf) with permeable sealing membrane (Nunc)
4. Autoinducible growth medium prepared according to Studier (13)
5. Plastic inoculation tips (Sarstedt)
6. Thermostatic orbital shaker with tube racks (New Brawnswick)
7. Bench centrifuge (Eppendorf)
8. Trehalose (Sigma): Dry powder
9. Glycine betaine (Sigma): 1 M betaine in water. Store at –20°C
10. K-Glutamate (Sigma): 1 M glutamate in water. Store at –20°C
11. Hydroxyectoine (bitop AG): Dry powder
12. Di-myo-inositol 1,1′-phosphate (DIP) (bitop AG): Dry powder

2.4. Analytical Protein Purification

1. Lysis buffer: 50 mM Tris-HCl, pH 8.0, 0.5 M NaCl, 5 mM $MgCl_2$, 1 mg/mL lysozyme.
2. Lysozyme: 100 mg/mL lysozyme (L7651) in water. Store at –20°C.
3. DNase I: 1 mg/mL in 50% glycerol, 50 mM NaCl. Store aliquots at –20°C.
4. Washing buffer: 50 mM Tris-HCl, pH 8.0, 0.5 M NaCl, 15 mM imidazole, 0.02% triton X-100. Prepare stock solutions of Tris-HCl (1 M), NaCl (5 M), and imidazole (2 M). Store triton X-100 in the dark.
5. Ni Sepharose beads. The beads are stored at 4°C in the presence of preservatives and must be washed at room temperature before use. Pipet 25 μL of Ni Sepharose beads into a 1.5 mL Eppendorf tube in the presence of 300 μL of PBS buffer. Carefully mix the tube by repeated top/bottom inversions and separate the beads from the supernatant by centrifugation (2 min at 3300×*g* in a bench minifuge). Gently aspire the supernatant for preventing bead removal. Repeat the procedure twice and, finally, add 30 μL of fresh lysis buffer to cover the beads and protect them from drying.
6. SDS-sample buffer (2X): 100 mM Tris, pH 6.8, 4% SDS, 5 mM DTT, 20% glycerol, bromophenol blue (4.8 mg/100 mL). Aliquots can be stored at –20°C.
7. Rotating wheel (Intercontinental, Catania, I).
8. Colloidal blue dye: Instant Blue (Expedeon, Cambridge, UK) (*see* **Note 4**).
9. Water bath sonicator.

2.5. Inclusion Body Purification

1. Lysozyme: 100 mg/mL lysozyme (L7651) in water. Store at –20°C.
2. DNase I: 1 mg/mL in 50% glycerol, 50 mM NaCl. Store aliquots at –20°C.
3. Triton X-100 (Sigma, T8787).
4. $MgCl_2$: 1 M in water, stored at room temperature.
5. EDTA: 0.5 M disodium ethylenediaminetetraacetate, 50 mM Tris-HCl, pH 8.0.
6. Sucrose (Sigma).
7. Deoxycholate: 10% sodium deoxycholate in water. Store at 4°C in the dark.
8. Triton X-100: 10% triton X-100, 20 mM Tris-HCl, pH 8.0.
9. NaCl: 5 M NaCl in H_2O.

10. DTT: 1 M dithiothreitol, X mM dithioerythrol (DTT) 1 M in PBS, store at –20°C.
11. IPTG: 1 M isopropyl-beta-D-thiogalactopyranoside, X mM phosphate, pH X, X mM NaCl. Aliquots can be stored at –20°C.
12. 50 mL plastic tubes.
13. 2 L Erlenmeyer flasks.
14. Water bath sonicator.
15. Resuspension buffer: 50 mM Tris-HCl, 25% sucrose, 1 mM NaEDTA, 10 mM DTT, 100 μL lysozyme, 250 μL DNase I, 50 μL of $MgCl_2$.
16. Solubilization buffer: 50 mM Tris-HCl, 100 mM NaCl, 1% triton X-100, 1% Na deoxycholate, 10 mM DTT.
17. Washing buffer 1: 50 mM Tris-HCl, 200 mM NaCl, 0.5% triton X-100, 1 mM DTT.

2.6. Chaperone-Assisted Refolding

1. ATP: 100 mM ATP, 10 mM Tris-HCl, pH 7.5. Store aliquots at –80°C.
2. Phosphoenolpyruvate: 100 mM in water, store at –80°C.
3. Pyruvate kinase: 2 mg/mL pyruvate kinase, 10 mM Tris-HCl, pH 7.5, 10% glycerol. Store at –80°C.
4. Refolding buffer: 50 mM Tris-HCl, pH 7.5, 20 mM $MgCl_2$, 150 mM KCl, 2 mM DTT. Add DTT just before use.
5. Purified chaperones ClpB, DnaK, DnaJ, and GrpE (*see* **Section 3.1**).

2.7. Biophysical Characterization of the Purified Proteins

1. Superose 12 10/300 GL column (GE Healthcare).
2. SEC buffer: 25 mM Tris-HCl, pH 8.0, 250 mM NaCl.
3. ÄKTA-FPLC system (GE Healthcare).
4. Spectrofluorimeter (Jasco).
5. Cuvette for spectrofluorimetry: cat. no. 105.250 QS, light pass 10 mm (Hellma).

3. Methods

Both osmolytes and molecular chaperones can contribute to the accumulation of soluble and correctly folded recombinant proteins expressed in bacteria (7). However, the response is protein specific and, therefore, it is useful to compare several conditions in parallel, using small-volume systems. Here we focus on the advantage of exploiting chemical and molecular stabilizers rather than on describing high-throughput methods. However, the single proposed protocols can be combined together to give a large number of conditions to test. Ninety-six deep-well microplates are a suitable platform to grow bacteria for initial screening steps.

3.1. Recombinant Chaperone Production

1. His-tagged recombinant chaperones were expressed in BL21(DE3) bacteria and purified by ion metal affinity chromatography. For each chaperone, 10 mL of LB medium + 1% glycerol were inoculated with a stock solution of frozen-transformed cells and grown overnight at 30°C. The pre-culture was diluted into 1 L LB medium and the bacteria were grown at 30°C in an orbital shaker at 180 rpm until the OD_{600} reached 0.4. The shaker temperature was lowered to 20°C and the recombinant expression was induced after 30 min by adding 0.1 M IPTG. The overnight culture was pelleted 30 min at 12,000×*g* and the resulting pellet frozen in liquid nitrogen and stored at –80°C.
2. Pellets were re-suspended in 4 volumes of buffer (20 mM Tris-HCl buffer, pH 8.0, 500 mM NaCl, 10 mM imidazole, 5 mM $MgCl_2$, and 1 mM PMFS) and sonicated 5 min in a water bath (Diagenode). A volume of 1 mg/mL lysozyme and 1 μg/mL DNase I were added and the lysates were incubated with shaking at room temperature for 30 min. The samples were centrifuged at 90,000×*g* for 35 min, the supernatants filtered and loaded onto a pre-equilibrated Hi-Trap chelating column (GE Healthcare) charged with $CoCl_2$ connected to an FPLC system (GE Healthcare). The elution was performed using 20 mM Tris-HCl, pH 8.0, 250 Mm imidazole. The purified proteins were buffer exchanged into 50 mM PBS containing 10% glycerine using a HiTrap Desalting column (GE Healthcare), and the protein concentration was calculated after measurement of the absorbance at 280 nm.

3.2. Protein Expression in the Presence of Betaine

1. The template DNA is amplified by PCR using primers that insert the sequence recognized by *Nco*I at the 5′-site and another endonuclease sequence for any of the remaining restriction enzymes available in the multicloning site of pETM vectors at the 3′ end. The PCR product is ligated in parallel into all the pre-digested pETM vectors. After ligation, plasmid purification, and sequencing to verify the construct quality, the vectors are transformed into chemically competent BL21(DE3) and Rosetta(DE3) (*see* **Note 5**). Only plasmids conferring resistance to ampicillin can be effectively used with Origami(DE3) cells (Kan resistance).

2. Single colonies are used to infect growth medium. Autoinducible medium is more time-consuming to prepare than LB, but strongly simplifies the procedures during the expression phase of T7-dependent plasmids and enables more comparable yield data when many cultures are performed in parallel with clones/strains having different growth rates.

   Use no more than 3 mL of medium in a 15 mL tube (230–250 rpm in an orbital shaker), possibly using a rack oriented with a 30° angle. Aeration becomes the limiting step when using microtiter plates. The medium volumes are calibrated considering the maximal operative speed of the shakers. For instance, the microplate vortexer incubator (Glass-Col LLC) can reach 1000 rpm and it is compatible with 600 μL of medium loaded in each well. We suggest to start the cell culture at 37°C and to slow down the temperature to 20°C after 2 h. At the same moment, add betaine and NaCl to a final concentration of 5 mM and 0.4 M, respectively. Betaine can be substituted with potassium glutamate (*see* **Note 6**).
3. Bacteria grown in LB must be induced with IPTG (100 μM) when their OD_{600} reaches the value of 0.4–0.6, but growth in autoinducible medium obviates the need for IPTG addition. After protein expression, collect 1.5 mL of the overnight cultures into an Eppendorf tube, centrifuge 3 min at 16,000×*g*, and save the pellet.
4. The pellets from different samples corresponding to constructs with different tags expressed by the pETM vectors and grown into strains with specific features are used for the analytical protein purification as described at the **Section 3.7** to determine their yields (*see* **Note 7**).
5. The combination of strain and vector that results in maximal yields of soluble target protein will be used for large-scale production using 2 L flasks filled with 500 mL of culture medium. Purified protein must be analyzed for its monodispersity by measuring its aggregation index and by gel filtration analysis (described in **Section 3.8**) to confirm that the increased solubility was not merely the result of material transfer from precipitated protein to soluble aggregates.

3.3. Protein Expression in the Presence of Trehalose

1. The culture of bacteria harboring expression vectors in medium enriched in trehalose is performed similar to that previously described for betaine addition. However, it must be considered that trehalose is not very soluble and, therefore, it is not possible to use a concentrated stock solution. Add trehalose powder directly to the bacterial culture at a final concentration of 10 mM in the presence of 0.4 M NaCl.

3.4. Protein Expression Using Trehalose Overexpressing Proteins

1. Clone the cDNA sequences corresponding to *otsA* (trehalose-6-phosphate synthase) and *otsB* (trehalose-6-phosphate phosphatase) into a bicistronic expression vector under the control of an IPTG-inducible promoter and transform it into BL21(DE3), Rosetta (DE3), and Origami (DE3) *E. coli* strains. Grow the cells and make them competent for further transformation with the vector harboring the target protein (*see* **Note 5**). We used a pCDFDuet-1 vector (Novagen) (10) because it confers spectinomycin resistance and, consequently, is compatible with most of the expression vectors and strains that use ampicillin, kanamycin, and chloramphenicol for selection.
2. Doubly transformed bacteria are grown as described above. IPTG-dependent recombinant expression induces both trehalose and target protein accumulation, resulting in a lower degree of aggregation.

3.5. Protein Expression Using Molecular Chaperone Overexpression

1. Transform chemically competent wild-type BL21(DE3) cells with the suitable vectors to obtain bacteria overexpressing different chaperone combinations. Plate and grow overnight at 37°C and use single bacterial colonies for inoculating 3 mL of LB medium in a 15 mL tube. Culture inoculated media at 30°C until the OD_{600} reaches 0.4. Recover 1.4 mL from the cell suspension and supplement with 600 μL of 80% glycerol and prepare 4 × 500 μL aliquots. Freeze them in liquid nitrogen and store at –80°C.

   Induce recombinant chaperone expression in the remaining cultured bacteria by the addition of IPTG to 0.2 mM. Culture the bacteria for further 4 h at 30°C, pellet the cells by centrifugation (3 min at 16,000×g), and directly boil them in 100 μL of SDS loading buffer. Verify the chaperone accumulation by colloidal staining (Instant Blue, Novexin) of the proteins separated by SDS-PAGE to determine the transformation efficiency. GroEL, DnaK, and ClpB should be visible as sharp bands at 60, 70, and 90 kDa, respectively (*see* **Note 8**).

3.6. Osmolyte and Molecular Chaperone Combinations for Soluble Protein Accumulation

1. Transform competent BL21(DE3) cells overexpressing suitable molecular chaperone combinations (14) with a plasmid harboring the target protein. Note that it is crucial to use a plasmid encoding the target protein that confers a resistance different from the selective markers used by the chaperone vectors. Kanamycin is compatible with all the chaperone combinations, while ampicillin resistance limits the application possibilities. Molecular chaperone plasmids are IPTG inducible. Therefore, two strategies can be identified for the target protein expression.

2. *lac promoter*-dependent target protein expression. A single colony picked from the dish on which the transformation material has been plated is used to inoculate 3 mL of LB medium supplemented with 1% glucose and the appropriate antibiotics in a 15 mL tube. Grow the bacteria overnight at 30°C and use 100 μL to inoculate tubes filled with 3 mL of LB medium containing the requisite antibiotics. Grow at 37°C for 2 h, add 5 mM betaine (or other osmolyte) plus 0.4 M NaCl, and switch the temperature to 20°C. At OD_{600} of 0.8 induce both chaperone and target protein expression with 0.2 mM IPTG and incubate overnight using a rack with 30° angle in an orbital shaker (240 rpm). Recover the bacteria by centrifugation (3 min at 16,000×*g*), remove the supernatant, and freeze the pellet. Analytical small-scale purification is performed as indicated in **Section 3.7**.
3. Arabinose-dependent target protein induction. Overnight culture from transformation colonies is performed as above. Use 100 μL of the pre-culture to inoculate 3 mL of fresh medium in a 15 mL tube, grow bacteria at 37°C until the OD_{600} reaches the value of 0.4, and add 5 mM betaine (or other osmolyte) plus 0.4 M NaCl. Reduce the temperature to 20°C and, after 30 min (OD_{600} around 0.6), induce the molecular chaperone accumulation by the addition of 0.2 mM IPTG. After a further 20 min (OD_{600} around 0.8), add 1.5 mg/mL arabinose to induce the target protein expression. In such a way, the selected combination of chaperones will be already accumulated in the cells in suitable amounts before the target protein will start to be expressed. Output analysis is performed following the analytical purification protocol described in **Section 3.7**.

3.7. Small-Scale Screening for Evaluating Protein Expression

Resuspend the bacteria pellet in 400 μL of lysis buffer and incubate at room temperature under constant mixing using a wheel (20 rpm). Sonicate in a water bath for 5 min, add 50 μg/mL DNase I, incubate 20 min under constant mixing, and centrifuge 5 min at 16,000×*g*. Transfer the supernatant to an Eppendorf tube with pre-equilibrated Ni Sepharose. Incubate 30 min under constant rotation (30 rpm) at room temperature. Recover the beads by centrifugation (2 min at 3300×g), discard the supernatant, and resuspend the beads in 500 μL of washing buffer. Incubate and centrifuge using the same conditions as above, remove the supernatant, and repeat the complete washing procedure. Add 30 μL of SDS sample buffer to the beads recovered after the second washing step and boil them before running an SDS-PAGE.

3.8. Biophysical Characterization of the Purified Proteins

1. Protein polymerization and aggregation can be evaluated by size exclusion chromatography (SEC). Superose 12 10/300 GL column (GE Healthcare) has a suitable separation for most of the proteins and small complexes. Columns are stored in 20% ethanol at 4°C and must be carefully washed in water before beginning the measurements. Use initially low flow rates (0.20–0.25 mL/min) to avoid overpressure due to the resin volume changes. The flow rate can be later set at 0.5 mL/min for optimal peak separation. After washing with 5 column volumes of water, equilibrate the column with 5 column volumes of SEC buffer (25 mM Tris-HCl, pH 8.0, 250 mM NaCl). The accuracy of SEC depends on the reliability of the pumps controlling the liquid flux through the column. We use an ÄKTA-FPLC system (GE Healthcare).
2. Spectrofluorimeter (J-810, Jasco). The working temperature (25°C) is maintained by a water bath. Therefore, remind to switch on the equipment plus thermostated water bath at least 30 min before recording data. This will enable to perform the measurements in stable conditions. Fix the excitation at 280 nm, the scan rate at 3, and the range between 260 and 400 nm. The monochromators are set to 280 (excitation) and 340 (emission). The aggregation index is given by the ratio between the scattering signal at 280 nm and the emission at 340 nm. Values close to zero indicate negligible aggregation in the samples (15).

3.9. Inclusion Body Purification

Grow transformed bacteria in 1 L LB medium + suitable antibiotics at 37°C (500 mL medium for each flask) until OD_{600} reaches 0.6 and induce recombinant expression with 1 mM IPTG. After 3 h collect the bacteria yielding insoluble protein by centrifugation (15 min at 3000×*g*) at 4°C, resuspend the pellet in 40 mL PBS, load a 50 mL tube, and centrifuge again (10 min at 4500×*g*) at 4°C (*see* **Note 9**). Freeze the pellet in liquid nitrogen and store at –80°C.

Thaw the pellet in 12.5 mL of resuspension buffer (50 mM Tris-HCl, 25% sucrose, 1 mM NaEDTA, 10 mM DTT) on ice by alternating vortexing and sonication. Add 100 μL of lysozyme, 250 μL of DNase I, and 50 μL of $MgCl_2$. Mix gently by inversion and add 12.5 mL of 50 mM Tris-HCl, 100 mM NaCl, 1% triton X-100, 1% Na deoxycholate, 10 mM DTT. Mix gently and incubate 1 h at room temperature using a wheel for homogeneous rocking. Freeze in liquid nitrogen and thaw in water bath at 37°C. Repeat the freezing/thawing cycle, add 200 μL of $MgCl_2$, and incubate 1 h at room temperature with gentle rocking. No viscous phase would be visible in the suspension (*see* **Note 10**). Add 300 mL of NaEDTA and transfer on ice.

Centrifuge the suspension 10 min at 12,000×*g* at 4°C, remove the supernatant, and resuspend the pellet 10 min in 10 mL of 50 mM Tris-HCl, 200 mM NaCl, 0.5% triton X-100, 1 mM DTT by using a water bath sonicator. In case of poor resuspension, avoid intensive vortexing that will result in foam formation, but prefer a manual homogenizer with potter and piston. After a second step of centrifugation at the same conditions, resuspend the pellet in 20 mL of the same buffer without detergent. Centrifuge again and remove carefully the supernatant. At this point the pellet would be white and sandy and would contain mostly purified inclusion bodies. Dissolve the inclusion bodies in urea 9 M, pH 8.0, check a sample by SDS-PAGE, and store the remaining material at –20°C (*see* **Note 11**).

3.10. Chaperone-Assisted In Vitro Refolding of Unfolded Proteins

Proteins recalcitrant to in vivo solubility improvement through the employment of chemical and molecular chaperones can be recovered as inclusion bodies and solubilized in 9 M urea (*see* **Note 12**). Here we propose four variants of a refolding protocol based on the use of molecular chaperones as refolding helpers. The aggregate-binding chaperones ClpB and DnaK are made equimolar with respect to the substrate, while the regulative DnaJ and GrpE chaperones are present at lower concentrations. A common preliminary step is the removal of the chaotropic salt since at high concentrations it would denature the chaperones and prevent their activity. Unfolded proteins show a different aggregation dynamic when the chaotropic agent is progressively removed (*see* **Fig. 3.4**). A small-scale test with dilutions to 5, 2, and 1 M urea is sufficient to detect macroscopic precipitation and turbidity increase. Protocol number 1 is suitable only for unfolded proteins that do not precipitate when the urea concentration passes from 9 to 1 M, a concentration compatible with molecular chaperone activity (*see* **Note 13**).

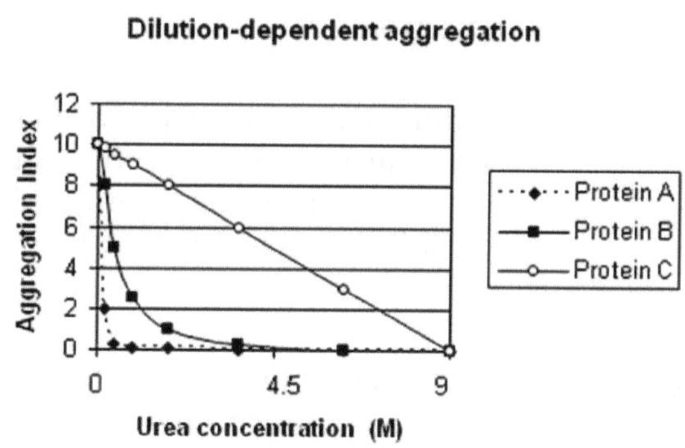

Fig. 3.4. Reaggregation kinetics of different protein constructs. Recombinant proteins were isolated as inclusion bodies and resuspended into 9 M urea to obtain their complete unfolding. The degree of monodispersity was calculated by means of the aggregation index (14) and monitored during the progressive dilution of the protein suspension in 50 mM Tris-HCl. A larger aggregation index corresponds to more aggregation.

3.10.1. Refolding Protocol 1

Dilute 4 aliquots of 1 μM unfolded protein resuspended in 9 M urea to 1 M urea using refolding buffer (50 mM Tris-HCl, pH 7.5, 20 mM $MgCl_2$, 150 mM KCl, 2 mM DTT), add the molecular chaperones (1 μM ClpB, 1 μM DnaK, 0.2 μM DnaK, and 0.1 μM GrpE), and the ATP recycling mixture (2 mM ATP, 3 mM phosphoenolpyruvate, 20 ng/μL pyruvate kinase) (*see* **Note 14**). Try to use a volume ratio >2 between protein suspension and chaperone solution to avoid further chaotropic dilution. The addition of trehalose (10 mM) or other osmolytes to the buffer can help stabilize partially unfolded intermediates and prevent non-productive folding. The presence in the buffer of reduced and oxidized glutathione (5 and 0.5 mM, respectively) is necessary to successfully complete the refolding of proteins stabilized by disulfide bonds. Incubate the four samples at 30°C for 1, 2, 4, and 16 h, respectively, before recovering and analyzing the folding of the target protein (*see* **Note 15**).

3.10.2. Refolding Protocol 2

Bind the 4 aliquots of 1 μM unfolded His-tagged protein resuspended in 9 M urea to a metal ion activated resin. Remember that DTT can interfere with the binding capacity and, therefore, it should be limited to 1 mM in the inclusion body resuspension medium (*see* **Note 16**). Resuspend the resin three times in refolding buffer plus osmolytes to wash it and then add the glutathione redox system, the molecular chaperones, and the ATP recycling mix. Incubate the four samples at 30°C for 1, 2, 4, and 16 h, respectively, pellet the resin (3 min at 3,300×*g*), and remove the supernatant. Wash the resin twice in 50 mM Tris-HCl, pH 8.0, 0.5 M NaCl, 20 mM imidazole, 1 mM DTT, and then elute the target protein in 50 mM Tris-HCl, pH 8.0, 0.5 M NaCl, 200 mM imidazole, 1 mM DTT. Elution can be performed in batch or after having collected the resin into an empty column. Analyze the recovered protein by appropriate activity and structural tests (**Section 3.7**).

3.10.3. Refolding Protocol 3

Dilute 4 aliquots of 1 μM unfolded protein resuspended in 9 M urea into refolding buffer until its precipitation. Recover the aggregates by centrifugation (3 min at 16,000×*g*), wash them in refolding buffer, and pellet them again. Resuspend in the refolding mixture and check the output as above.

3.10.4. Refolding Protocol 4

In the appropriate centrifuge tube, load 1 mL of unfolded protein resuspended (1 mg/mL) in 9 M urea onto a density step-gradient system composed of refolding buffer containing sucrose at four concentrations (80, 70, 50, and 30%, **Fig. 3.3**). Centrifuge 15 h at 180,000×*g* at 4°C using an SW40TI swinging bucket rotor and an ultracentrifuge (Beckman). Urea is diluted during centrifugation and protein samples composed of aggregates of increased complexity and density will separate at the interfaces between successive sucrose cushions. Recover the aggregates using a Pasteur pipette with the tip bent of 90° (use a flame to warm up the glass to make it workable). The different protein fractions can be diluted into the refolding mixture and analyzed as above (*see* **Note 17**).

4. Notes

1. The yield comparison of the same coding region fused to different tags improves the chance to identify a construct with enhanced solubility. Several strategies have been proposed to sub-clone or recombine a single PCR product into vectors with peculiar features (purification tags, fusion partners, protease cleavage sites differently organized at the N- and C-termini). The pETM collection offers a particularly large choice of expression vectors (16).
2. Redox conditions are crucial for the folding of several proteins. The bacterial cytoplasm is a strongly reducing environment and the formation of disulfide bonds that typically stabilize the structure of eukaryotic secreted proteins is prevented in these conditions. In contrast, the *E. coli* periplasm is sufficiently oxidizing and contains disulfide isomerases. Therefore, proteins sensitive to the reducing environment of the cytoplasm can be forced to accumulate in the periplasm by adding a leader peptide to the target sequence. This strategy is widely used, but not limited, to express recombinant antibodies. Otherwise, double mutant strains (*trxB*$^-$, *gor*$^-$) have been proposed that provide an oxidizing cytoplasm (17). Target proteins, especially when expressed as fusions to thioredoxin or DsbC, can form correct disulfide bridges in the cytoplasm of *trxB*$^-$, *gor*$^-$ *E. coli*. Nevertheless, these mutants grow at a significantly lower rate than the wild type and this aspect must be considered when evaluating the induction time of strains grown in parallel (and, in certain situations, the need for independent incubators at different temperatures).
3. The pBB chaperone vectors (**Fig. 3.3**) have been conceived to permit modular expression combinations and allow the maintenance of stoichiometric ratios among the chaperones that resemble the physiological accumulation during heat shock (6). The author will deliver the vectors to researchers working in non-profit institutes (ario.demarco@ifom-ieo-campus.it).

For-profit organizations should contact Dr. B. Bukau (bukau@zmbh.uni-heidelberg.de). As an alternative, cold-induced chaperones have been coexpressed to improve the folding efficiency of recombinant proteins at low temperatures (18).

4. Colloidal blue is more sensitive and rapid than conventional Coomassie staining. Furthermore, gel fixation is not requested and the washing is done in water (no methanol and acetic acid in the lab!).
5. Chemically competent cells are prepared by standard protocols. Online initiatives make them available and easy to improve for the user community. An example is reported at the following link: http://openwetware.org/wiki/Preparing_chemically_competent_cells.
6. The addition of osmolytes to bacterial cultures will result in their partial accumulation in cells. However, their cytoplasmic concentration can remain relatively low. In contrast, inducing osmotic stress will result in rapid and efficient accumulation of osmolytes inside the cell and the final cytoplasmic concentration can reach two orders of magnitude more than the external one (**Fig. 3.5a**). As an alternative to avoid metabolic rearrangement due to osmotic shock, it is possible to engineer the cell metabolic pathways to force the bacteria to produce higher amounts of osmolytes themselves (**Fig. 3.5b**).

   It must be remembered that osmotic shock can induce a detectable accumulation of molecular chaperones that

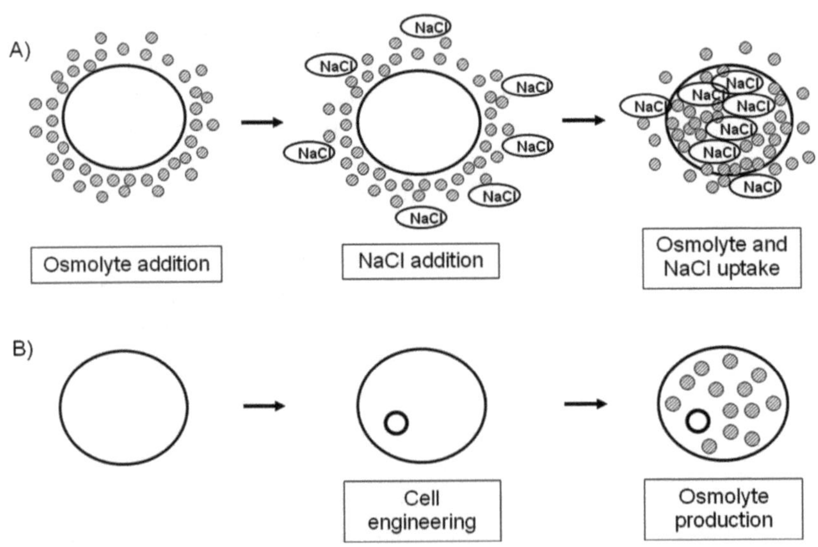

Fig. 3.5. Osmolyte accumulation into bacteria cells. (**a**) Osmolytes present in the culture medium slowly accumulate in the cell cytoplasm, but the addition of salt to the medium accelerates their uptake to counterbalance the osmotic shock. (**b**) Bacteria can be directly engineered to overproduce osmolytes that accumulate in the cytoplasm.

can partially contribute to the increased solubility of the overexpressed target proteins (7). Nevertheless, a specific osmolyte effect is supported by different independent observations. Salt alone and in combination with betaine or glutamate induced molecular chaperone accumulation to the same extent, but the effect on the target protein solubility varied in the three cases (7). In contrast, osmotic shock induced a more significant molecular chaperone accumulation, but had less positive effect on target protein solubility. Finally, there is a synergistic effect when both glutamate and betaine were added to the bacterial culture, indicating that the different chemical features of the two osmolytes could have only partially overlapping stabilizing effect (7).

Other osmolytes can be successfully used as chemical chaperones to improve recombinant protein solubility and folding, both in vivo and in vitro, among these, amino acids like arginine, glutamic acid, glycerol, or sucrose. Of particular interest are saccharides isolated from extremophilic organisms, like hydroxyectoine and di-myo-inositol 1,1′-phosphate (DIP), since they contribute to protein stability in vivo in very challenging environmental conditions.

7. It is useful to have the possibility to follow the development of protein aggregates during recombinant expression. The fusion of a fluorescent protein reporter at the target protein C-terminus has been proposed assuming that the lack of folding of the target protein at the N-terminus would prevent the fluorescent protein from reaching its native structure (19). As an alternative, we used a β-galactosidase reporter plasmid, the *IbpB* promoter of which is specifically activated by the presence of protein aggregates. Its enzymatic activity can be measured and used to infer the variation of aggregates during the culture and at different growth conditions (20).

8. The suitability of chaperones in helping substrate protein folding remains a matter of trial and error. We did not find any correlation between proteins with specific structural features and chaperone combinations and, therefore, so far it is impossible to predict them once known the protein sequence. Data reported in the literature do not contribute to identify consensus and the fact that they are very heterogeneous impairs to widen the pool of results to analyze for inferring general conclusions. However, we have observed that more complete chaperone combinations are in average more effective than combinations in which only one single chaperone group is overexpressed. A common annotation system for protein expression experiments might be particularly beneficial to discover more general rules (21).

9. Protein expression conditions can strongly influence the quality of the inclusion bodies in terms of dimension, composition, and interchain organization. Incubation temperature, IPTG concentration, and expression time are the main available parameters suitable for modulating the target protein aggregation. Specifically, high temperature (37–42°C) and saturating concentrations of IPTG (>100 μg/mL) favor significant inclusion body yields (>100 mg/L) in few hours of cell culture. Tags can also strongly vary the aggregation kinetics of recombinant proteins (12).
10. A common problem during inclusion body preparation is incomplete nucleic acid elimination. The presence of nucleic acids results in viscous structures in the suspension. Nucleic acids can be removed by enzymatic digestion (DNase I plus its co-factor $MgCl_2$) or by mechanical disruption using sonication or potter homogenizers.
11. The quality of the protein recovered from inclusion bodies depends on the solubilization conditions. The pellet can have a jelly-like consistency due to polysaccharide and nucleic acid deposition and these contaminants can be partially resuspended together with the proteins. Three to four freeze–thaw cycles followed by centrifugation (5 min at 16,000×*g*) can help in specifically precipitating debris and contaminants, while the monodispersed protein will remain in the soluble fraction.

    Guanidinium (8 M) is a more effective solubilization agent than urea, but it is not compatible with SDS-PAGE and, therefore, its use prevents the protein separation and identification on a gel. This analysis is useful to estimate the purity of the target protein since contaminants can interfere during refolding and, therefore, they should be removed. His-tagged proteins can be affinity purified even under denaturing conditions, but take into consideration that DTT used during resuspension to prevent disulfide bond formation can interfere with the metal ion affinity binding capacity. As an alternative, the differential solubility of target protein and contaminants at decreasing chaotropic salt concentrations can be exploited. The solution is progressively diluted and both the pelleted and soluble fractions after centrifugation (5 min at 16,000×g) are analyzed by SDS-PAGE to detect any specific partitioning between the two phases of the target protein and contaminants.

    Finally, inclusion body yields can largely vary and, therefore, the resuspension volumes must be optimized to avoid excessive dilutions. About 3 mL of chaotropic agent can be initially used and, in case of poor resuspension detectable by opalescence, further urea/guanidinium will be added. Strong detergents like sarcosyl (0.2%) can help in dissolving inclusion bodies, but are difficult to remove and can interfere during protein refolding.

12. In vivo protein refolding of precipitated proteins mediated by chaperones is a physiological mechanism to recover protein functionality after environmental stress. This peculiarity has been exploited to improve the solubility of recombinant proteins in combination with molecular chaperone overexpression (5, 6).

    In vitro foldase-assisted protein refolding has also been proposed in a version that uses the chaperone system bound to a matrix while the misfolded protein is present in the mobile phase (22).

13. Protein refolding is performed by progressive dilution of unfolded protein to change the concentration of chaotropic agent in the refolding buffer. The most sensitive refolding step is the one in which there is a folding intermediate with exposed hydrophobic patches because it is prone to aggregate with similar molecules. There are several proposed method variants that use either dialysis or dilution strategies. Furthermore, the unfolded proteins can be bound to a substrate to minimize non-productive interactions before the buffer is exchanged to induce refolding.

    However, less studied is the reaggregation process of each construct during chaotropic salt dilution and its influence on productive folding. We have previously reported that the same cloning region fused to different tags or expressed under variable conditions showed reaggregation patterns resembling those schematically shown in **Fig. 3.4** (12). Some constructs start aggregating at minimal chaotropic salt dilutions while others remain monodisperse even at extremely elevated dilutions. The knowledge of such a parameter is useful for the choice of both the material to use for refolding and the strategy to apply. For instance, a preliminary dilution of the denatured protein without incurring aggregation is compatible with the addition of foldases that would themselves denature in buffers containing high urea concentration.

14. As an alternative, the ATP necessary for chaperone activity can be generated using a mix of ATP (5 mM, pH 7.0), phosphocreatine (10 mM), and phosphocreatine kinase (100 μg/mL).

15. The efficiency of refolding protocols can be easily evaluated when the native protein has a measurable activity. Otherwise, biophysical analyses on purified proteins are necessary to determine the folding state of proteins for which

no functional test is available. Since the target protein is only one among several proteins present in the reaction medium, it is beneficial to have it labeled with an exclusive small tag such as hexa-histidine, FLAG, myc, or HA to facilitate purification and analysis. We have produced the recombinant chaperones in pETM vectors and removed the tags by 3C protease digestion after purification to avoid interferences.

16. Nowadays several resins are available on the market and some of them would maintain their binding capacity for His-tag at DTT concentrations up to 4–5 mM.
17. A preliminary isolation of inclusion bodies is not strictly necessary. It is possible to selectively recover aggregates with a wide range of structural complexity also from total lysate (**Fig. 3.2**) and to start refolding with fractions enriched in aggregate subsets of different mass. This approach resulted in more efficient disaggregation when low-density aggregates were used, probably because of the weaker inter-chain bonds in such precipitates (11).

Acknowledgments

The author would like to thank B. Bukau, E. Deuerling, P. Goloubinoff, and A. Mogk for the stimulating discussions and for having provided the chaperone vectors.

References

1. Anfinsen, C. B. (1972) The formation and stabilization of protein structure. *Biochem J* 128, 737–749.
2. Goloubinoff, P., Gatenby, A. A., Lorimer, G. H. (1989) GroE heat shock proteins promote assembly of foreign prokaryotic ribulose biphosphate carboxylase oligomers in *Escherichia coli*. *Nature* 337, 44–47.
3. Mogk, A., Tomoyasu, T., Goloubinoff, P., Rüdiger, S., Röder, D., Langen, H., Bukau, B. (1999) Identification of thermolabile *E. coli* proteins: prevention and reversion of aggregation by DnaK and ClpB. *EMBO J* 18, 6934–6949.
4. Goloubinoff, P., Mogk, A., Ben Zvi, A. P., Tomoyasu, T., Bukau B. (1999) Sequential mechanism of solubilization and refolding of stable protein aggregates by a bichaperone network. *Proc Natl Acad Sci USA* 96, 13732–13737.
5. de Marco, A., De Marco, V. (2004) Bacteria co-transformed with recombinant proteins and chaperones cloned in independent plasmids are suitable for expression tuning. *J Biotechnol* 109, 45–52.
6. de Marco, A., Deuerling, E., Mogk, A., Tomoyasu, T., Bukau, B. (2007) Chaperone-based procedure to increase yields of soluble recombinant proteins produced in *E. coli*. *BMC Biotechnol* 7, 32.
7. de Marco, A., Vigh, L., Diamant, S., Goloubinoff, P. (2005) Native folding of aggregation-prone recombinant proteins in *Escherichia coli* by osmolytes, plasmid- or benzyl alcohol-overexpressed molecular chaperones. *Cell Stress Chaperones* 10, 329–339.
8. Balogh, G., Horvath, I., Nagy, E., Hoyk, Z., Benko, S., Bensaude, O., Vigh, L. (2005) The hyperfluidization of mammalian cell

membranes acts as a signal to initiate the heat shock protein response. *FEBS J* 272, 6077–6086.
9. Roberts, M. F. (2005) Organic compatible solutes of halotolerant and halophilic microorganisms. *Saline Syst* 1, 5.
10. Schultz, T., Liu, J., Capasso, P., de Marco, A. (2007) The solubility of recombinant proteins expressed in *Escherichia coli* is increased by *otsA* and *otsB* co-transformation. *Biochem Biophys Res Commun* 355, 234–239.
11. Schrödel, A., de Marco, A. (2005) Characterization of the aggregates formed during recombinant protein expression in bacteria. *BMC Biochem* 6, 10.
12. Schrödel, A., Volz, J., de Marco, A. (2005) Fusion tags and chaperone co-expression modulate both the solubility and the inclusion body features of the recombinant CLIPB14 serine protease. *J Biotechnol* 120, 2–10.
13. Studier, F. W. (2005) Protein production by auto-induction in high-density shaking cultures. *Protein Expr Purif* 41, 207–234.
14. de Marco, A. (2007) Protocol for preparing proteins with improved solubility by co-expressing with molecular chaperones in *Escherichia coli*. *Nat Protocols* 2, 2632–2639.
15. Nominé, Y., Ristriani, T., Laurent, C., Lefevre, J.-F., Weiss, E., Travé, G. (2001) A strategy for optimizing the monodispersity of fusion proteins: application to purification of recombinant HPV E6 oncoprotein. *Protein Eng* 14, 297–305.
16. Dümmler, A., Lawrence, A.-M., de Marco, A. (2005) Simplified screening for the detection of soluble constructs expressed in *E. coli* using a modular set of vectors. *Microb Cell Fact* 4, 34.
17. Prinz, W. A., Åslund, F., Holmgren, A., Blackwith, J. (1997) The role of thioredoxin and glutaredoxin pathways in reducing protein disulfide bonds in the *Escherichia coli* cytoplasm. *J Biol Chem* 272, 15661–15667.
18. Qing, G., Ma, L. C., Khorchid, A., Swapna, G. V., Mal, T. K., Takayama, M. M., Xia, B., Phadtare, S., Ke, H., Acton, T., Montelione, G. T., Ikura, M., Inouye, M. (2004) Cold-shock induced high-yield protein production in *Escherichia coli*. *Nat Biotechnol* 22, 877–882.
19. Waldo, G. S., Standish, B. M., Berendzen, J., Terwilliger, T. C. (1999) Rapid protein-folding assay using green fluorescent protein. *Nat Biotechnol* 17, 691–695.
20. Schultz, T., Martinez, L., de Marco, A. (2006) The evaluation of the factors that cause aggregation during recombinant expression in *E. coli* is simplified by the employment of an aggregation-sensitive reporter. *Microb Cell Fact* 5, 28.
21. de Marco, A. (2008) Minimal Information: an urgent need to assess the functional reliability of recombinant proteins used in biological experiments. *Microb Cell Fact* 7, 20.
22. Altamirano, M. M., Golbik, R., Zahn, R., Buckle A. M., Fersht, A. R. (1997) Refolding chromatography with immobilized minichaperones. *Proc Natl Acad Sci USA* 94, 3576–3578.

Chapter 4

Genetic Selection of Solubility-Enhanced Proteins Using the Twin-Arginine Translocation System

Adam C. Fisher, Mark A. Rocco, and Matthew P. DeLisa

Abstract

The expression of heterologous proteins in robust hosts such as *Escherichia coli* is often plagued by the tendency of the protein of interest to misfold and aggregate. To engineer and improve the folding properties of virtually any protein of interest, the quality control process inherent to the bacterial twin-arginine translocation (Tat) pathway can be exploited. The Tat pathway preferentially transports folded substrates across the inner membrane of *E. coli* with remarkable quality control that can provide selection pressure for protein folding and solubility. By fusing desired proteins to the N-terminus of mature TEM-1 β-lactamase and using an N-terminal signal peptide to target the fusion to the Tat pathway, it is possible to perform genetic selections for folded, soluble proteins. Here, we present a method for exploiting the folding quality control process associated with the Tat pathway for engineering folding-enhanced proteins.

Key words: Protein engineering, protein folding, Tat pathway, solubility, selection.

1. Introduction

The twin-arginine translocation (Tat) pathway in Gram-negative bacteria is primarily responsible for secreting cofactor containing proteins and protein complexes into the periplasm (1, 2). Genetic and biochemical studies have revealed that the translocation machinery (translocon) for this pathway is comprised minimally of the Tat(A/E)BC proteins (3, 4), which together promote export of folded proteins from the cytoplasm. The existence of this pathway was originally hypothesized by Ben Berks after observing that several periplasmic enzymes required cytoplasmic cofactors (e.g., iron–sulfur clusters, molybdopterin

T.C. Evans, M.-Q. Xu (eds.), *Heterologous Gene Expression in E. coli*, Methods in Molecular Biology 705,
DOI 10.1007/978-1-61737-967-3_4,

cofactors) to function properly (5). Closer inspection of these precursor proteins revealed N-terminal signal peptides that preserved the structural organization of export signals specific for the general secretory (Sec) pathway, but contained a characteristic n-region sequence motif (S/TRR*X*FLK) that included two consecutive and invariant arginine residues. It was later observed that the export of cofactor-containing proteins by the bacterial Tat pathway was hindered when the incorporation of the cofactor was blocked in the cytoplasm (6). Since cofactor attachment occurs at a relatively late stage of the folding process, it was hypothesized that precursors were folded in the cytoplasm and subsequently exported in a folded or at least partially folded conformation (7). In support of this notion, several recent reports have demonstrated that only proteins exhibiting native-like characteristics are exported by the Tat pathway (8–11). Further, a technique was developed for genetic selection of soluble proteins from combinatorial libraries exploiting this folding quality control mechanism of the Tat pathway (12). The growing body of literature on mechanistic and applied aspects of the Tat system is answering several fundamental questions regarding the Tat system's unique ability to export folded proteins. First, how can the translocon transport fully folded proteins without disruptive leakage of ions across the inner membrane? Second, what are the specific folding requirements for Tat pathway transit? Third, what cellular components are responsible for proofreading and quality control of precursors of the Tat pathway?

The structure of the translocon complex provides some clues as to how fully folded proteins can transit the inner membrane without disrupting the ionic gradient or the structural integrity of the cell envelope. In *E. coli*, the integral membrane proteins TatA (and its paralog TatE), TatB, and TatC form the necessary components of the Tat translocon. The TatB and TatC proteins form a stable unit that appears to act as the receptor element of the translocation pathway (13, 14). In particular, TatC has been shown to contain the primary binding site for the signal peptide (15). TatA forms a separate oligomeric ring-like structure thought to be the protein-conducting channel (16) that assembles with TatBC following the binding of a Tat substrate (17). TatA channel complexes exhibit variable diameters with central pores that can reach up to 70 Å in diameter (16). It is notable that the range of variation in channel diameter between individual TatA complexes (30–70 Å) matches the range of diameters found in native *E. coli* Tat substrates (18). Therefore, the channel may change diameter to allow TatA to pack tightly around substrates of differing sizes while preventing ion leakage during transport. Further, purified TatA complexes contained a 'lid' region that might operate as a gate, preventing solute access and regulating substrate entry (16).

Ever since the supposition that the Tat system exports folded proteins (7), an intriguing question is whether the Tat pathway is able to accept both folded and unfolded polypeptides as substrates or if only proteins that have reached a substantially native state in the cytoplasm are competent for translocation. Because initial assembly of many native Tat substrates requires careful coordination of cofactor synthesis and insertion, subunit recruitment, and protein targeting, it is critical that Tat transport not take place before cofactor insertion. Indeed, it was initially shown that certain native Tat substrates were transport incompetent when lacking necessary cofactors (6). This observation led to the hypothesis that there is a substrate 'proofreading' mechanism inherent to the transport process whereby correct assembly (e.g., cofactor/subunit attachment) is 'sensed' prior to export. This process of proofreading is in part achieved by dedicated molecular chaperones (e.g., *E. coli* DmsD and TorD) that coordinate cofactor insertion with protein export by binding twin-arginine signal peptides and preventing premature export (19). While the chaperone-mediated 'proofreading' of substrate proteins conveniently explains how transport of cofactor-containing enzymes is regulated, it does not account for the fact that proteins without cofactors are also transported by the Tat pathway (5). For many of these proteins, it has been shown that export via the Tat system also requires some degree of folding. In one notable case, it was shown that heterologous proteins that require disulfide bonds for proper folding (e.g., *E. coli* alkaline phosphatase) were transport incompetent in *E. coli* strains deficient in disulfide bond formation (9). This study demonstrated that only proteins that have attained native conformations in vivo are exported by the Tat translocon, suggesting that a general 'folding quality control' mechanism is intrinsic to the export process. But how would such discrimination between folded and misfolded proteins be accomplished?

It is likely that a quality control mechanism that accepts or rejects substrates based on fundamental folding (9) would be distinct from or overlap with the 'proofreading' mechanism. Thus, we expect that the Tat 'folding quality control' system would be in operation for all substrates of the Tat pathway, while the 'proofreading' system provides an additional checkpoint in the biosynthesis of complex cofactor-containing proteins and protein complexes (19). Moreover, since the Tat translocon is necessary and sufficient for protein export, and because certain Tat substrates do not contain cofactors, the folding quality control might reside with the membrane Tat(A/E)BC components (9). Of course, at least some 'folding quality control' may occur *prior* to the interaction between the substrate and TatBC (e.g., proteolytic degradation, chaperone interaction). Nonetheless, it was observed that unfolded alkaline phosphatase (PhoA) fused to a Tat-specific signal peptide (ssTorA) was targeted to the TatBC complex in vivo but transport could not be completed (20). More recent in vitro experiments showed that even though both folded and unfolded ssTorA-PhoA were found to physically associate with the Tat translocase, a perturbed interaction was observed for the signal sequence of unfolded ssTorA-PhoA with the TatBC receptor site as compared to that of folded ssTorA-PhoA (21). These results are consistent with some degree of quality control by TatBC. Further evidence for the direct involvement of the translocon in quality control was observed when two FeS-containing substrates that receive cofactors in the cytoplasm were mutated in their FeS binding (10). As expected, transport was completely blocked. However, these mutants were rapidly degraded *only* via interaction with the Tat translocon, suggesting that the Tat apparatus directly initiates the turnover of rejected substrates. Further, turnover of the mutant substrates was completely dependent on the TatA/E subunits that are involved in the later stages of translocation. One explanation for this behavior is that cytoplasmic 'proofreading' factors remain associated with native substrates and inhibit translocation once bound to the Tat components. In this scenario, translocation may render the substrate protease sensitive. Alternatively, the substrate may become trapped in the translocon in a transport-incompetent state and the system may recruit necessary degradation machinery. This would represent a more active role for the Tat system in disposing of misfolded substrates.

More recently, it was shown that folding speed and surface hydrophobicity also govern the transport of Tat substrates, as faster folding substrates (22) and substrates with buried hydrophobic residues are preferred (11). In the latter case, an artificial Tat substrate of 10 repeats of a 19-residue, unstructured, flexible peptide appended with 6 consecutive histidine residues could be transported through the pathway. Though transport became less efficient with increasing size, it was concluded in this instance that translocation of small, soluble, hydrophilic polypeptides is possible despite the lack of folded structure. Even in this unnatural case, the absence of an exposed hydrophobic core motif was the critical factor in determining whether a protein could be translocated by the Tat system. Since an exposed hydrophobic core is the typical hallmark of misfolded, aggregation-prone proteins, one possible explanation for how the Tat system might monitor the folding state of its substrates is via hydrophobic interactions between the Tat components and the substrate itself.

Taken together, the results of these prior studies support the use of the Tat pathway as a tool to isolate correctly folded, aggregation-resistant proteins. Here, we present a detailed method for utilizing the Tat pathway and its intrinsic 'folding

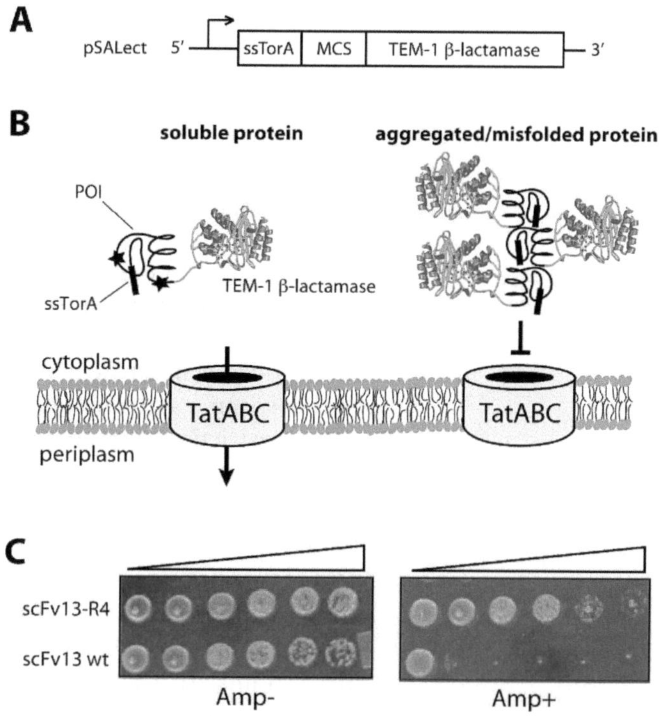

Fig. 4.1. Genetic selection for folded proteins using the Tat pathway's 'folding quality control' mechanism. (**a**) Schematic of pSALect showing the ssTorA signal peptide followed by a mini-multiple cloning site (MCS) and finally the TEM-1 Bla sequence. The *arrow* indicates the IPTG-inducible promoter. The plasmid carries the *cat* gene for chloramphenicol resistance. (**b**) Schematic showing the basis for the folding selection where ssTorA is the Tat-specific signal peptide from the *E. coli* trimethylamine-*N*-oxide reductase TorA enzyme and POI is the protein of interest. (**c**) The Amp resistance phenotype of cells expressing a well-folded single-chain Fv antibody variant, scFv13-R4, specific for β-galactosidase, and the poorly folded scFv13wt. Each was cloned in the selection vector and plated on selective (Amp+) and non-selective (Amp-) media. Cells were serially diluted 10-fold (indicated by *white triangle*) from an overnight culture such that the most dilute (right-most spot in each image) is a 10^6 dilution of overnight culture.

quality control' mechanism to engineer folding-enhanced proteins. This method relies on C-terminal TEM-1 β-lactamase (Bla) as a selectable marker for substrate transport to the periplasm mediated by the Tat pathway (**Fig. 4.1a** and **b**). We have used this method, in conjunction with laboratory protein evolution, to improve the in vivo solubility of the aggregation-prone Alzheimer's Aβ42 peptide (12) as well as a poorly folded single-chain Fv antibody fragment (23) (**Fig. 4.1c**). We have laid out the method that has proven to be most effective for engineering solubility-enhanced proteins and we have provided notes to guide the application of this selection strategy to any future protein of interest (POI).

2. Materials

One of the benefits of using the Tat pathway to perform in vivo selections in *E. coli* is that standard microbiology techniques and very common materials are used.

2.1. Determining Selection Conditions

1. Parental selection plasmid pSALect encoding the POI and a selection control plasmid (pSALect) encoding a well-folded protein such as the *E. coli* maltose-binding protein (MBP) or green fluorescent protein (GFP). Briefly, pSALect encodes the Tat-specific signal peptide ssTorA and TEM-1 Bla with several restriction sites between (**Fig. 4.1a**). The plasmid carries the *cat* gene for chloramphenicol (Cm) resistance. The pSALect plasmid is described in more detail elsewhere (12) and available on request.
2. MC4100 (or any other strain of *E. coli* commonly used for protein expression such as BL21, TG1) and a Tat-deficient isogenic mutant of that strain (e.g., B1LK0; Δ*tatC* (24)). These strains are available on request.
3. Luria–Bertani (LB) agar medium (~12.5 mL for each 100-mm plate).
4. Sterile 100-mm Petri dishes (one for each antibiotic concentration tested).
5. LB liquid medium (approximately 2 mL for each strain tested plus sufficient volume for serial dilutions).
6. 100-mg/mL ampicillin (Amp) in sterile water (carbenicillin (Carb) is an alternative) (*see* **Note 1**).
7. 25-mg/mL chloramphenicol in ethanol.
8. Sterile 96-well plate (centrifuge tubes could substitute).
9. Multichannel or standard pipettor (with accuracy down to 5 μL).

2.2. Library Transformation and Selection

1. Plasmid library under investigation (several micrograms) and the parental plasmid (~100 ng) (both are pSALect derivatives).
2. Electrocompetent MC4100 (or comparable *E. coli* strain) in 270-μL aliquots stored at –80°C.
3. Electroporation cuvettes with a 2-mm gap (VWR Scientific).
4. Pre-warmed (37°C) SOC liquid medium (3 mL for each transformation).
5. LB liquid medium (with 0.2% glucose) (250 mL for each transformation).

6. LB agar medium (~25 mL for each 150-mm plate plus ~12.5 mL for each 100-mm dilution plate).
7. Sterile 150-mm Petri dishes (sufficient for antibiotic concentrations used during selections).

2.3. Cell Culture and Counter-Screening

1. Sterile 96-well plates (sufficient for number of colonies under investigation)
2. LB liquid medium (~55 mL for each 96-well plate)
3. LB agar medium (~25 mL for each 150-mm plate)
4. Sterile 150-mm Petri dishes (minimum 3 required for each 96-well plate)
5. Multichannel pipettor (with accuracy down to 5 μL)

3. Methods

The methods described herein are based on a genetic selection for protein folding and solubility that exploits the quality control feature of the *E. coli* Tat pathway (12). This selection employs a tripartite sandwich fusion of a POI with an N-terminal Tat-specific signal peptide and C-terminal TEM-1 Bla, thereby coupling antibiotic resistance with Tat pathway export. If the POI is correctly folded, the ssTorA-POI-Bla chimera is localized in the periplasm where the Bla moiety confers resistance to β-lactam antibiotics such as Amp or Carb. In contrast, if the POI is misfolded, the ssTorA-POI-Bla chimera is incapable of Tat export and cells are sensitive to β-lactam antibiotics. Here, we will focus on effective methods for using the Tat pathway to improve the folding and solubility of heterologous proteins in *E. coli.* The selection for soluble, folded proteins from recombinant libraries seems – and in many ways is – relatively straightforward (*see* **Fig. 4.2**). Here, we outline the specific steps critical to the success of this particular form of selection (*see* **Fig. 4.3**).

3.1. Determining Selection Conditions

1. The gene encoding a POI should be cloned into pSALect (25) between the regions encoding the Tat-specific signal peptide (ssTorA) and mature TEM-1 Bla (*see* **Fig. 4.1a**). For comparison, a plasmid should be cloned encoding a well-folded protein (e.g., *E. coli* maltose-binding protein (12), but preferably similar in nature to the POI) in this same position.
2. These plasmids should be transformed into cells with a functioning Tat pathway (e.g., MC4100) and isogenic mutants lacking a functioning Tat pathway (e.g., B1LK0).

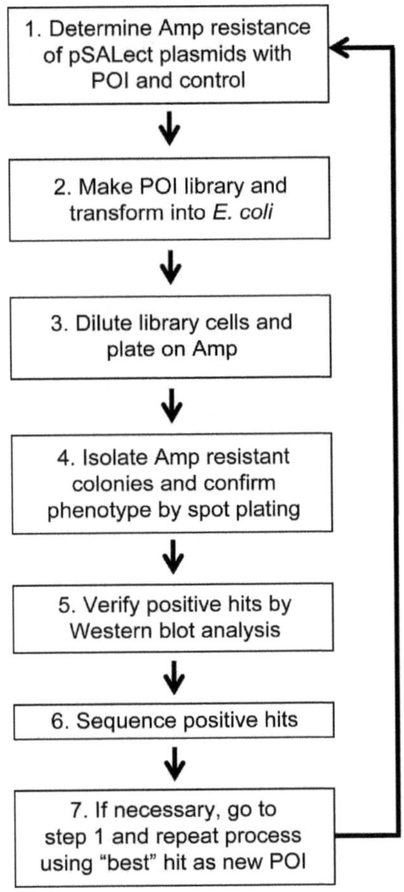

Fig. 4.2. Flowchart for the selection of folded proteins using the Tat folding reporter.

3. Each strain of transformed cells should be plated on an LB agar plate containing chloramphenicol and incubated at 37°C overnight.
4. Colonies from these plates should be picked and cultured overnight at 37°C in Luria–Bertani (LB) medium in the presence of antibiotics selective for the plasmid (25-μg/mL chloramphenicol, Cm).
5. An array of 100-mm LB agar plates should be prepared with Amp concentrations ranging from ~25 μg/mL to 1 mg/mL (*see* **Note 2**). Additionally, an LB agar plate with 25 μg/mL Cm should be prepared. Each plate should be marked with a distinctive symbol indicating the 'top' (12 o'clock position) of the plate. These can be prepared the next morning or prepared and stored overnight at 4°C.
6. The next morning aliquots of the culture should be spun down and washed in fresh equivolume LB medium with

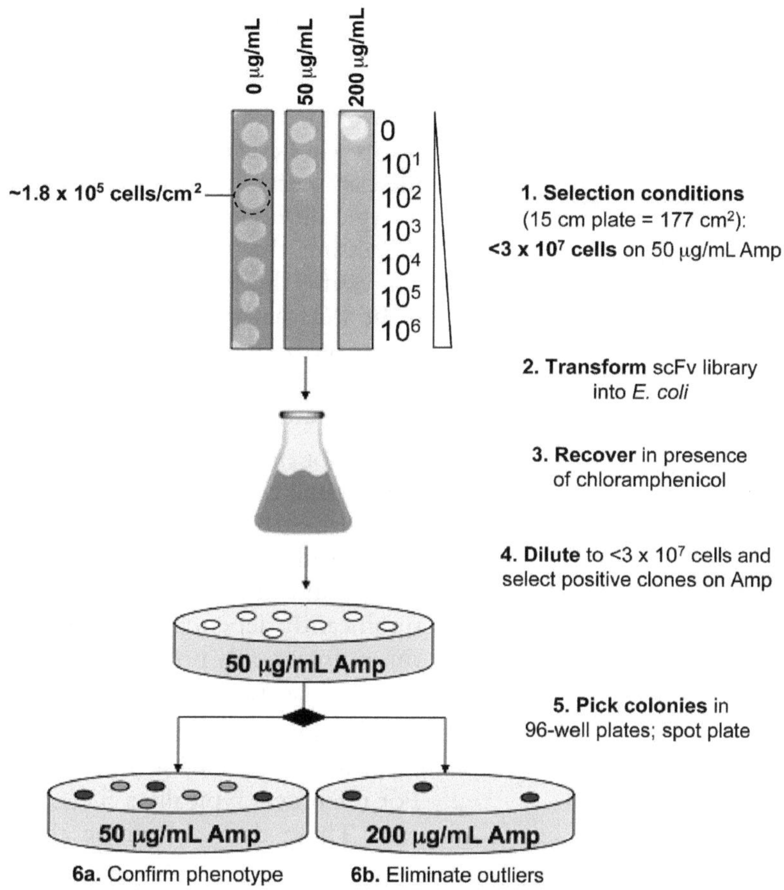

Fig. 4.3. Detailed strategy for the isolation of folded proteins (e.g., scFv antibodies) from recombinant libraries using the Tat folding reporter. Cells were serially diluted 10-fold (indicated by *white triangle*) from an overnight culture such that the most dilute cells (*bottom spots* in each image) are a 10^6 dilution of overnight culture.

25 μg/mL Cm. Take an OD_{600} measurement of the overnight culture for reference.

7. In a sterile 96-well plate (or multiple sterile tubes) prepare a dilution series of the cultures in fresh LB medium with 25 μg/mL Cm. Seven consecutive 10-fold dilutions (to a dilution factor of 10^7) are recommended. About 5 μL of each dilution is needed for each Amp concentration tested.
8. In a hood or over a flame, use a multichannel pipettor (or a standard pipettor) to place 5-μL droplets of each dilution of each sample on each LB agar plate. Be sure the cells are properly suspended before application to the plates. By starting with the most dilute sample and working up, one set of tips can be used for the entire spot plating experiment. *See* **Fig. 4.1c** for an example of what the spotting will look like after growth.

9. Allow the spot plates to dry by a flame until surface droplets are no longer visible (approximately 15 min). Place the plates upside down in a 30°C incubator and observe cell growth between 12 and 24 h. The growth on the spot plates will be used to confirm Tat specificity and to specify the parameters of selection. There should be a clear phenotypic difference between (a) MC4100 and B1LK0 cells expressing the POI (*see* **Note 3**) and (b) MC4100 cells expressing the POI and the well-folded control, especially if the POI has poor solubility. The difference in antibiotic resistance between cells expressing the POI and well-folded control represents the 'distance' that may be covered during laboratory evolution.
10. Each spot on the plates is representative of a certain number of total cells. By using the approximation that in 1 mL there are ~10^9 cells/OD_{600} unit, an estimate of the number of cells in each spot can be obtained. Note that this is only an estimate as the actual number of cells may vary significantly depending on a number of variables. However, this estimate only provides the *basis* for future dilutions and therefore accuracy is not necessary.
11. Provided that the number of cells per surface area is kept constant, the behavior on Amp plates scales up (**Fig. 4.4**). For reference, typically a 5-μL droplet is ~6 mm in diameter. These data determine the approximate size of the

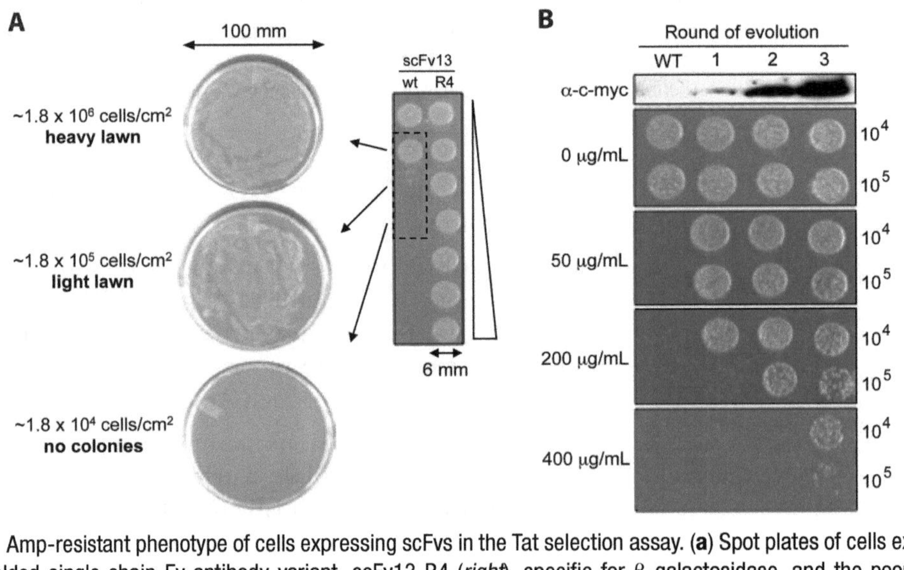

Fig. 4.4. Amp-resistant phenotype of cells expressing scFvs in the Tat selection assay. (**a**) Spot plates of cells expressing a well-folded single-chain Fv antibody variant, scFv13-R4 (*right*), specific for β-galactosidase, and the poorly folded scFv13-wt (*left*). Scale-up is evidenced by similar growth phenotypes of equivalent cells/cm^2 on LB agar with 50 μg/mL Amp grown on 100-mm plates (*left*) versus 5-μL spots (*right*). (**b**) Three rounds of scFv mutagenesis and selection and the resultant Western blot of the soluble fraction (*top*) and the Amp-resistant phenotype of cells on spot plates (*bottom*). Western blot was probed with anti-c-myc antibodies that recognize affinity epitopes on the scFvs.

library that can be selected on each concentration of Amp. For an example, *see* **Fig. 4.3**.

3.2. Selecting for Enhanced Folding

1. The next step is the construction of a diverse mutant library from which to select solubility-enhanced proteins. There are numerous methods of library synthesis which are beyond the scope of this protocol, although we prefer to generate diversity using random mutagenesis PCR (26) or with the GeneMorphII Random Mutagenesis Kit (Stratagene) (*see* **Note 4**).
2. Once a library of sufficient diversity (*see* **Note 5**) has been constructed and isolated in solution at a concentration >50 ng/μL, frozen electrocompetent cells (e.g., MC4100) (*see* **Note 6**) in glycerol should be thawed on ice. About 270 μL of electrocompetent cells can be transformed with 30 μL of DNA plasmid library solution in 2-mm gap electroporation cuvettes. In parallel, 45 μL of electrocompetent cells should be transformed with up to 5 μL of the parental POI plasmid in 1-mm gap electroporation cuvettes.
3. Following electroshock, recover the contents of each 2-mm gap cuvette in a glass tube containing 3 mL of pre-warmed SOC medium. The contents of the 1-mm gap cuvettes can be recovered in 1 mL of SOC medium. Shake at 37°C for 1 h.
4. Following recovery, plate a series of 10-fold dilutions on LB agar plates supplemented with 25 μg/mL Cm. For example, take 20 μL of recovered cells and dilute in 180 μL of fresh LB, take 20 μL of this diluent and dilute further into 180 μL LB; repeat at least six times. This will define the total number of transformants in the tube (*see* **Note 7**).
5. With the remaining library transformants, inoculate 250 mL of LB containing 25 μg/mL Cm and 0.2% glucose for each tube recovered. About 50-mL cultures can be used to do the same for the control transformants. Shake at 37°C overnight (*see* **Note 8**).
6. An array of 150-mm LB agar plates should be prepared with Amp concentrations ranging above and below the ideal selection concentration. These can be prepared the next morning or prepared and stored overnight at 4°C.
7. Count colonies on dilution plates to ensure that sufficient library diversity has been conserved during transformation in light of (a) the diversity of the constructed library and (b) the number of cells that were calculated to plate on selective Amp plates. If insufficient, retry the transformation.

8. Take OD_{600} measurements of overnight cultures; typically ~3 units. Use this as a basis for diluting to the correct cell concentration for plating.
9. Plate the appropriate number of cells on 150-mm LB plates supplemented with the appropriate Amp concentrations. This should be done for the library and the control (parental) culture. Incubate plates at 30°C overnight.
10. Compare the antibiotic resistance of cells expressing the library to the antibiotic resistance of cells expressing the parental POI. To avoid false positives, colonies should be picked from plates that (a) have the least Amp, to ensure selection along a continuum of antibiotic resistance (*see* Ref. (23) for further details); and (b) show a significant difference from the parental resistance phenotype. Colonies from this plate can be picked as desired with sterile toothpicks and used to inoculate 200 μL of LB supplemented with Cm in 96-well plates. Shake the plates at 37°C overnight.

3.3. Selective Spot Plating and Counter-Screening

1. The purpose of spot plating is to (a) confirm that selected colonies display the desired resistance phenotype and (b) ensure that false positives beyond the continuum distribution are excluded (*see* **Note 9**). To do this, begin by diluting each well from step 10 of **Section 3.2** by 10^4; 10^2 dilutions can be performed twice in 96-well plates to accomplish this. Place 5-μL drops of each diluted sample onto 15-cm LB agar plates in the same orientation on each plate. Again, a mark should be used to indicate the top (12 o'clock) position on the plate. Minimally, plates should be as follows: plate A: LB with 25 μg/mL Cm; plate B: LB with the Amp concentration used for selection; and plate C: LB with double the Amp concentration used for selection. Allow droplets to evaporate over a flame and incubate at 30°C overnight.
2. The next morning spots should be selected that grow on plates A and B, but not plate C. These spots represent candidates to pick, culture, miniprep, and sequence to identify mutations.
3. More than one round of laboratory evolution may be needed to meet the desired folding objective(s) (*see* **Note 10**). However, before beginning a new round of laboratory evolution the following should be performed:
   a. Transform isolated plasmids back into MC4100 and B1LK0 cells. Confirm that the Amp resistance phenotype is specific to the Tat pathway (*see* **Note 3**).
   b. Sequence the entire mutated gene to confirm presence of mutations and absence of frameshifts or other abnormalities.

4. Notes

1. We have noticed little difference in the assay whether using Amp or Carb, overnight incubation of plates is at 30°C and relatively short. Although we have never found it necessary, longer incubation times may require Carb due to its greater stability. We suggest making one concentrated stock of Amp and storing aliquots at –20°C until needed. Aliquots should be thawed on ice only once.
2. There is no clear guideline for the range of Amp concentrations that need to be tested. In our experience, cells expressing poorly folded targets typically do not confer significant resistance above ~50 μg/mL Amp. It is important to figure out the appropriate conditions for library selection, so err on the side of testing excess Amp concentrations.
3. There are two possibilities that we have observed. In one case, the general growth phenotype (i.e., on the Cm plate) of Tat mutant cells will be attenuated by expressing the POI. In the other case, Tat mutants display a normal growth phenotype, but sensitivity to Amp. In both cases, the growth on selective plates is dependent on a functional Tat pathway and justifies moving forward.
4. In our experience, random mutagenesis PCR is simple, but it is difficult to control the error rate and there are inherent biases in the types of errors that are generated. The GeneMorphII Random Mutagenesis Kit from Stratagene is the least biased method of generating errors. It is also simple, but more expensive.
5. There is no way to determine a priori the diversity needed for a directed evolution experiment. However, we generally strive for libraries of $>10^6$ clones with ~1% error rate at the amino acid level.
6. Briefly, a 1-L culture of MC4100 in YenB medium should be iced and spun down between OD_{600} 0.5 and 0.8. At 4°C, cells should be resuspended and washed three times with sterile water and finally concentrated in <5-mL sterile 15% glycerol. Aliquots should be stored at –80°C and thawed on ice just prior to use. When transforming, the volume of plasmid solution should not exceed 10% of the total volume.
7. These dilution plates will indicate how many cells were transformed with plasmid. For example, if 10 colonies resulted from plating 100 μL of a 10^6 dilution of a 1 mL culture, one could estimate that there were 10^8 cells

transformed with plasmid: $10 \times (1\ mL/100\ \mu L) \times 10^6$. Multiple-fold coverage of each library clone is possible if the number of transformed cells is higher than the diversity of the constructed library. If it is lower, then full coverage of the library is not possible. Regardless, the number of cells to be plated has already been calculated and should not be changed.

8. This is an enrichment that selects for transformants. Do not plate the transformants on Cm plates and then scrape the plates to collect the library. For libraries above 10^5 in diversity, plating is grossly insufficient for selecting transformants.
9. Selections will provide surprising false positives, consistent with the adage 'you get what you select for' (27). By eliminating events that far exceed the selection conditions, many if not most false positives can be avoided.
10. It is possible to continue this directed evolution until there are no longer observable, authentic gains in antibiotic resistance. The appropriate endpoint will be determined by the user and might include surpassing a threshold of soluble cellular expression (g/L) or achieving observable crystallization.

References

1. Halbig, D., Wiegert, T., Blaudeck, N., Freudl, R., Sprenger, G. A. (1999) The efficient export of NADP-containing glucose-fructose oxidoreductase to the periplasm of *Zymomonas mobilis* depends both on an intact twin-arginine motif in the signal peptide and on the generation of a structural export signal induced by cofactor binding. *Eur J Biochem* 263, 543–551.
2. Rodrigue, A., Chanal, A., Beck, K., Muller, M., Wu, L. F. (1999) Co-translocation of a periplasmic enzyme complex by a hitchhiker mechanism through the bacterial tat pathway. *J Biol Chem* 274, 13223–13228.
3. Settles, A. M., Yonetani, A., Baron, A., Bush, D. R., Cline, K., Martienssen, R. (1997) Sec-independent protein translocation by the maize Hcf106 protein. *Science* 278, 1467–1470.
4. Weiner, J. H., Bilous, P. T., Shaw, G. M., Lubitz, S. P., Frost, L., Thomas, G. H., Cole, J. A., Turner, R. J. (1998) A novel and ubiquitous system for membrane targeting and secretion of cofactor-containing proteins. *Cell* 93, 93–101.
5. Berks, B. C. (1996) A common export pathway for proteins binding complex redox cofactors? *Mol Microbiol* 22, 393–404.
6. Sanders, C., Wethkamp, N., Lill, H. (2001) Transport of cytochrome c derivatives by the bacterial Tat protein translocation system. *Mol Microbiol* 41, 241–246.
7. Berks, B. C., Sargent, F., Palmer, T. (2000) The Tat protein export pathway. *Mol Microbiol* 35, 260–274.
8. Bruser, T., Yano, T., Brune, D. C., Daldal, F. (2003) Membrane targeting of a folded and cofactor-containing protein. *Eur J Biochem* 270, 1211–1221.
9. DeLisa, M. P., Tullman, D., Georgiou, G. (2003) Folding quality control in the export of proteins by the bacterial twin-arginine translocation pathway. *Proc Natl Acad Sci USA* 100, 6115–6120.
10. Matos, C. F., Robinson, C., Di Cola, A. (2008) The Tat system proofreads FeS protein substrates and directly initiates the disposal of rejected molecules. *EMBO J* 27, 2055–2063.
11. Richter, S., Lindenstrauss, U., Lucke, C., Bayliss, R., Bruser, T. (2007) Functional Tat transport of unstructured, small, hydrophilic proteins. *J Biol Chem* 282, 33257–33264.

12. Fisher, A. C., Kim, W., DeLisa, M. P. (2006) Genetic selection for protein solubility enabled by the folding quality control feature of the twin-arginine translocation pathway. *Protein Sci* 15, 449–458.
13. Cline, K., Mori, H. (2001) Thylakoid Delta pH-dependent precursor proteins bind to a cpTatC-Hcf106 complex before Tha4-dependent transport. *J Cell Biol* 154, 719–729.
14. de Leeuw, E., Granjon, T., Porcelli, I., Alami, M., Carr, S. B., Muller, M., Sargent, F., Palmer, T., Berks, B. C. (2002) Oligomeric properties and signal peptide binding by *Escherichia coli* Tat protein transport complexes. *J Mol Biol* 322, 1135–1146.
15. Alami, M., Luke, I., Deitermann, S., Eisner, G., Koch, H. G., Brunner, J., Muller, M. (2003) Differential interactions between a twin-arginine signal peptide and its translocase in *Escherichia coli*. *Mol Cell* 12, 937–946.
16. Gohlke, U., Pullan, L., McDevitt, C. A., Porcelli, I., de Leeuw, E., Palmer, T., Saibil, H. R., Berks, B. C. (2005) The TatA component of the twin-arginine protein transport system forms channel complexes of variable diameter. *Proc Natl Acad Sci USA* 102, 10482–10486.
17. Mori, H., Cline, K. (2002) A twin arginine signal peptide and the pH gradient trigger reversible assembly of the thylakoid [Delta]pH/Tat translocase. *J Cell Biol* 157, 205–210.
18. Sargent, F., Berks, B. C., Palmer, T. (2002) Assembly of membrane-bound respiratory complexes by the Tat protein-transport system. *Arch Microbiol* 178, 77–84.
19. Jack, R. L., Buchanan, G., Dubini, A., Hatzixanthis, K., Palmer, T., Sargent, F. (2004) Coordinating assembly and export of complex bacterial proteins. *EMBO J* 23, 3962–3972.
20. Richter, S., Bruser, T. (2005) Targeting of unfolded PhoA to the TAT translocon of *Escherichia coli*. *J Biol Chem* 280, 42723–42730.
21. Panahandeh, S., Maurer, C., Moser, M., Delisa, M. P., Muller, M. (2008) Following the path of a twin-arginine precursor along the TatABC translocase of *Escherichia coli*. *J Biol Chem* 283, 33267–33275.
22. Ribnicky, B., Van Blarcom, T., Georgiou, G. (2007) A scFv antibody mutant isolated in a genetic screen for improved export via the twin arginine transporter pathway exhibits faster folding. *J Mol Biol* 369, 631–639.
23. Fisher, A. C., DeLisa, M. P. (2009) Efficient isolation of soluble intracellular single-chain antibodies using the twin-arginine translocation machinery. *J Mol Biol* 385, 299–311.
24. Bogsch, E. G., Sargent, F., Stanley, N. R., Berks, B. C., Robinson, C., Palmer, T. (1998) An essential component of a novel bacterial protein export system with homologues in plastids and mitochondria. *J Biol Chem* 273, 18003–18006.
25. Lutz, S., Fast, W., Benkovic, S. J. (2002) A universal, vector-based system for nucleic acid reading-frame selection. *Protein Eng* 15, 1025–1030.
26. Fromant, M., Blanquet, S., Plateau, P. (1995) Direct random mutagenesis of gene-sized DNA fragments using polymerase chain reaction. *Anal Biochem* 224, 347–353.
27. Zhao, H., Arnold, F. H. (1997) Combinatorial protein design: strategies for screening protein libraries. *Curr Opin Struct Biol* 7, 480–485.

Chapter 5

Protein Folding Liquid Chromatography

Quan Bai and Xindu Geng

Abstract

A method for carrying out protein folding with simultaneous separation by protein folding liquid chromatography (PFLC) is described herein. Furthermore, a two-dimensional chromatographic column, termed a 2D column, which can be independently employed for accomplishing PFLC in either weak cation exchange mode or hydrophobic interaction chromatography mode is reported. The content of this chapter describes the most commonly employed methods and operations of PFLC, such as the use of urea or guanidine hydrochloride as a denaturant with the protein in either the reduced or oxidized state and solving problems caused by the formation of the precipitates during protein folding. The PFLC can be performed using conventional chromatographic columns and a new chromatographic cake. A protocol for fast renaturation with simultaneous purification of inclusion body protein of the recombinant human interferon-gamma to obtain purity ≥95% and high specific bioactivity in a single step and in 1 h is introduced.

Key words: Protein folding liquid chromatography, proinsulin, granulocyte colony-stimulating factor, protein refolding.

1. Introduction

Protein folding is vital not only for theoretical studies in molecular biology but also for lowering the cost of therapeutic proteins in industry. Many kinds of recombinant proteins have been produced in different hosts, especially *E. coli*, but the overexpression of heterologous proteins in bacteria often results in the accumulation of the protein product in inactive and insoluble deposits inside the cells called the inclusion bodies. A challenge is how to convert the insoluble inclusion body proteins into soluble and bioactive ones, or proteins in the native state (N state),

T.C. Evans, M.-Q. Xu (eds.), *Heterologous Gene Expression in E. coli*, Methods in Molecular Biology 705,
DOI 10.1007/978-1-61737-967-3_5, © Springer Science+Business Media, LLC 2011

i.e., solving the problem of protein refolding (renaturation) (1). Commonly employed methods for the renaturation of inclusion body proteins include dilution and dialysis; however, these methods usually result in very low recoveries, typically 5–20%, of the expected total mass or bioactivity.

Fifteen years ago, one of the authors first reported high-performance hydrophobic interaction chromatography (HPHIC) to be a powerful tool for protein refolding (2). Since then many kinds of liquid chromatography (LC) have been employed for protein folding (3–5). When LC is used for protein folding, the recovery of bioactivity and mass increases significantly and permits the rapid assessment of protein folding efficacy. This is a new field in both LC and molecular biology and was termed protein folding liquid chromatography (PFLC) (5–7). It is defined as "a kind of liquid chromatography, incorporating various kinds of biochemical and/or physicochemical processes originally accomplished in solution, which can result in either raising the efficiency or shortening the time of protein folding" (5–7). Hydrophobic interaction chromatography (HIC), ion exchange chromatography (IEC), size exclusion chromatography (SEC), and affinity chromatography (AFC) have been employed for PFLC at laboratory or industrial scales (5, 6, 8).

An ideal PFLC would have the following four characteristics as shown in **Fig. 5.1**: (a) complete removal of the detergents, (b) renaturation of several proteins, (c) separation of the target protein from impure proteins, and (d) easily recycling waste denaturing reagent (3). The principle of protein folding by PFLC was explained from chemical equilibria to be valid for each kind of LC (5, 6, 8) and molecular mechanism only valid for HIC (3, 6, 8).

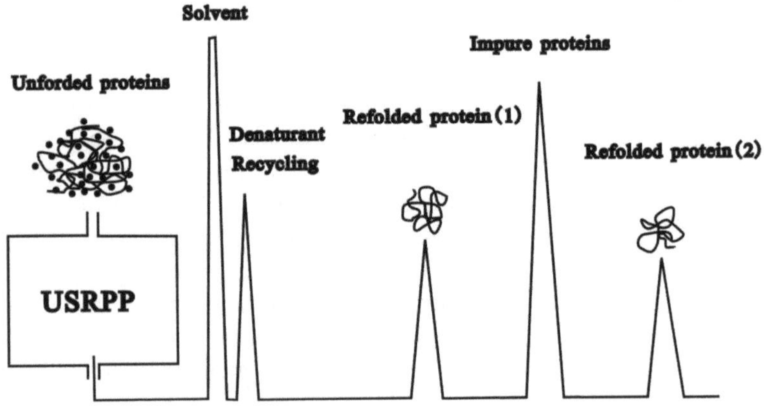

Fig. 5.1. Scheme of an ideal protein folding liquid chromatography having four functions simultaneously (3, 5, 6). Four functions: removal of denaturants, refolding of target proteins, separation from contaminant proteins including misfolded intermediates of the target protein, and easy recovery of denaturants.

Different aspects of PFLC have been the subject of a number of recent reviews (5, 8–12).

Although PFLC has so many advantages, several problems still exist. If proteins have strong hydrophobicity, some aggregates may form when loading the denatured protein solutions onto an LC column, resulting in increased back pressure of the employed column, perhaps even blocking it. Additionally, mass and bioactivity recoveries of the target protein would decrease. These problems are much more serious in large-scale protein renaturation. To alleviate this issue, a series of units for simultaneous renaturation and purification of proteins (USRPP), also called chromatographic cake (*see* **Fig. 5.2**), were manufactured and employed to replace the conventional column (3, 13). This "cake" has a very short length ranging only 1.0–5.0 cm in size, but a very large diameter ranging from 2.0 to 50 cm. Due to the large cross-sectional area of the "cake," the back pressure increases insignificantly when some protein precipitates accumulate on the filter or on top of the column bed. It was reported that this unit has very good resolution for protein separation at both laboratory and large scales and has been applied successfully for the renaturation with simultaneous purification of some recombinant proteins produced by *E. coli*, such as recombinant human interferon-gamma (rhINF-γ) (14), recombinant human proinsulin (15), recombinant human granulocyte colony-stimulating factor (rhG-CSF) (16), and recombinant human stem cell factor (rhSCF) (17). Compared to conventional methods, the recovery of bioactivity of these renatured recombinant proteins was increased two- to threefold with a purity of $\geq$95% with only one step. With milligram amounts of target protein the whole process to purify and refold the protein can be accomplished in 1 h.

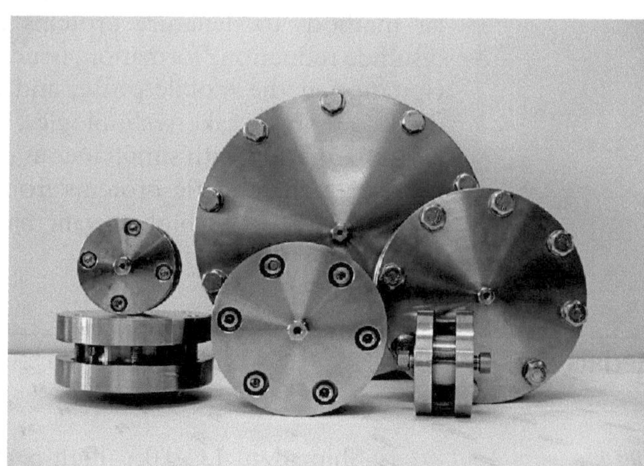

Fig. 5.2. Photo of a set of USRPP or chromatographic cakes in laboratory and preparative scales (3, 13).

The high efficiency of the renaturation with simultaneous separation of unfolded proteins depends on the optimization of PFLC conditions including the selection of the kind of LC, stationary phase, mobile phase, flow rate, and pH value (14, 18). The first important thing is to select the kinds of LC. Generally speaking, any kind of LC can be employed for carrying out protein folding; thus various kinds of chromatographic columns are required to storage for this selection. To satisfy the simultaneous renaturation and separation of proteins, reverse phase liquid chromatography (RPLC) cannot be used for most proteins, because the RPLC conditions denature most proteins. Affinity chromatography (AFC) is only valid for a subset of proteins, limiting its broad application. SEC, IEC, and HIC are popularly employed for this purpose, although SEC is the worst resolution in LC.

It was reported that a single column having two orthogonal retention mechanisms of HIC and weak cation exchange (WCX) was first employed for fast protein separation separately with the two modes (19, 20) and this type of column is termed a 2D column. If the 2D (HIC, WCX) column can be employed for protein refolding also with the two modes of HIC and IEC, respectively, this column should be the first choice for accomplishing PFLC (*see* **Note 1**). In this chapter, the 2D column is initially employed for carrying out both the renaturation of the urea reduced/denatured lysozyme (Lys) by WCX mode alone and establishing a new methodology for the renaturation with simultaneous separation of a pair protein of Lys and ribonuclease A (RNase A) together by HIC mode (*see* **Notes 2** and **3**). A real recombinant human protein drug, recombinant human interferon-gamma (rh-IFN-γ), by PFLC is also introduced.

In addition to PFLC, this chapter will touch upon popular methods to denature proteins using chemical denaturants, disulfide reduction/formation, issues stemming from protein precipitation at the mobile phase, and chromatography using standard column or cake technologies. Furthermore, a protocol for fast renaturation with simultaneous purification of heterologously expressed, therapeutic proteins from inclusion bodies to obtain a purity ≥95%, and high specific bioactivity in one step and less than an hour is described.

2. Materials

2.1. Liquid Chromatography

1. Shimadzu LC-10A high-performance liquid chromatograph (Shimadzu, Japan) including two pumps, a gradient elution system, a Rheodyne 7725i manual sample injector, and an SPD-10Avp UV detector.

2. 2D (HIC, WCX) chromatography column (100 × 4.6 mm ID) was purchased from Xi'an Aolan Technology Development Co. Ltd (Xi'an, China, http://www.aulast.com). The particle diameter and average pore diameter are 7 μm and 30 nm, respectively.
3. HPHIC-USRPP (10 × 50 mm ID) was purchased from Xi'an Aolan Technology Development Co. Ltd (Xi'an, China). The particle diameter and average pore size is 7 μm and 30 nm, respectively, and the end group is PEG600.
4. Solution A: 10 mM potassium dihydrogen phosphate (PBS), 1 mM EDTA, 3.0 mM reduced glutathione (GSH, Sigma, USA), 0.6 mM oxidized glutathione (GSSG, Sigma, USA), 2 M urea, pH 7.0.
5. Solution B: 10 mM PBS, 1.0 M NaCl, 1 mM EDTA, 3.0 mM GSH, 0.6 mM GSSG, and 2 M urea, pH 7.0.
6. Solution C: 3.0 M $(NH_4)_2SO_4$, 50 mM PBS, pH 7.0.
7. Solution D: 50 mM PBS, pH 7.0.
8. Denaturation buffer: 7.0 M guanidine hydrochloride.
9. Reduced/denatured buffer: 8.0 M urea, 0.1 mM Tris-HCl, 1 mM EDTA, 30 mM dithiothreitol, pH 8.5.
10. *E. coli* cleaning buffer: 20 mM PBS, 1 mM EDTA, 0.2 mg/mL lysozyme (chicken egg white, Sigma), pH 7.4.
11. Inclusion body cleaning buffer I: 20 mM PBS, 1 mM EDTA, 2 M urea, 1 M NaCl, pH 7.4;

    inclusion body cleaning buffer II: 20 mM PBS, 1 mM EDTA, 0.5% Triton X-100, 1 M NaCl, pH 7.4;

    inclusion body cleaning buffer III: 20 mM PBS, 1 mM EDTA, 1 M NaCl, pH 7.4.

2.2. Enzyme Assays

1. Lysozyme assay buffer: 20 mM sodium phosphate, pH 7.0.
2. RNase A assay buffer: 0.1 M sodium acetate, pH 5.0.
3. 0.5% yeast solution: 0.05 g yeast ribonucleic acid (source) dissolved in 100 mL RNase A assay buffer.

2.3. Bradford Dye-Binding Assay

1. Bradford reagent: dissolve 100 mg Coomassie Blue G-250 (Fluka, MO, USA) in 50 mL 95% ethanol. Add 100 mL 85% (w/v) phosphoric acid to this solution and dilute the mixture to 1 L with water. Filter with filter paper and store at 4°C.
2. BSA standard: 0.2 mg/mL bovine serum albumin.

3. Methods

3.1. Simultaneous Refolding and Separation of Proteins with PFLC with WCX Mode

1. Dissolve 5 mg lysozyme (Lys, chicken egg white, Sigma, USA) and 5 mg RNase A, separately in 1.0 mL denaturation solution containing 7.0 mol/L GuaHCl and incubated at room temperature for 24 h.
2. Dissolve 5 mg Lys in 1.0 mL reduced/denatured sample solution, then add DTT to 30 mM, and incubate at 40°C for 3 h. Store this solution at 4°C. This DTT step reduces disulfide bonds to facilitate complete denaturation.
3. Inject 20 μL of the urea reduced/denatured Lys solution directly into the 2D (HIC, WCX) column equilibrated in solution A for 20 column volumes. Eluted with a linear gradient from 100% of solution A to 100% of solution B within 30 min and with a 10 min delay. Use a flow rate of 1.0 mL/min and a detection wavelength of 280 nm. The obtained chromatogram is shown (*see* **Fig. 5.3b**).
4. Inject 20 μL native Lys (N state) into the 2D (HIC, WCX) column and run with the same conditions as described in Step 3. The obtained chromatogram is shown (*see* **Fig. 5.3a**). It can be seen that peak 1 in **Fig. 5.3a** contains only solvent, but peak 1′ in **Fig. 5.3b** is much higher than that of peak 1, because peak 1′ contains both solvent and Lys in the unfolded state (U state) and/or dimer. Peak 2 in **Fig. 5.3** is Lys only in the N state.

Based on the retention of the renatured proteins, the fraction containing Lys corresponding to peak 2′ of **Fig. 5.3** should be the refolded Lys. From peaks 1′ and 2′ of the chromatogram "b," the reduced/denatured Lys is not only renatured but also separated from its U state completely; the chromatogram obtained from the 2D column, therefore, is an assay to determine the efficiency of protein refolding. Interestingly, compared to the peak height and shape of peak 2 shown in chromatogram "a," that of the refolded Lys corresponding to the peak 2′ is lower and broader. It is easy to understand that the mass loss of the refolded Lys can be attributed to a fraction of Lys that converts to its U state. The broader peak of peak 2′ is due to the fact that Lys only partially carries out refolding due to the dynamic request for the oxidation of the two –SH. Based on the enzymatic activity experiment, it, at least, takes 8 h to incubate at room temperature after adjusting the fraction solution pH to 8.0 (*see* **Notes 4–8**).

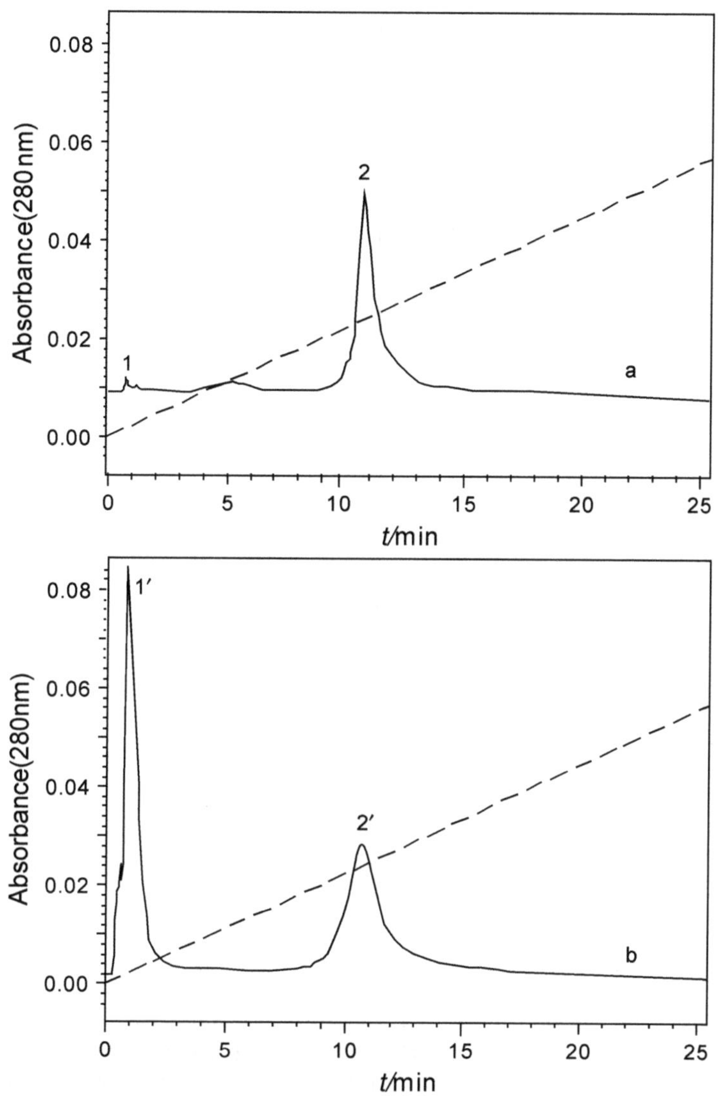

Fig. 5.3. Simultaneous refolding and separation of reduced/denatured lysozyme with WCX. Conditions: initial concentration: 5.0 mg/ml; linear gradient elution (*dash line*), 100% solution A to 100% solution B, 30-min linear gradient and with a 10-min delay; flow rate, 1.0 mL/min; chromatogram "a," Lys in N state; sample size: 20 μL; chromatogram "b," the WCX mode-renatured Lys which was originally reduced–denatured by 8.0 M urea solution; sample size: 20 μL; peak 1: solvent, peak 1′ solvent and Lys in U states; peak 2: Lys (N state), peak 2′ refolded Lys.

3.2. Renaturation with Simultaneous Separation of Denatured Lysozyme and RNase A by HIC Mode

1. Inject 20 μL Lys and 30 μL RNase A denatured by 7.0 M GuaHCl solution directly into the 2D (HIC, WCX) column equilibrated with solution C and renatured with simultaneous separation by HIC mode. Elute with solution D by a linear gradient elution 30 min with a 10-min delay. The obtained chromatogram "b" is shown (*see* **Fig. 5.4b**) and

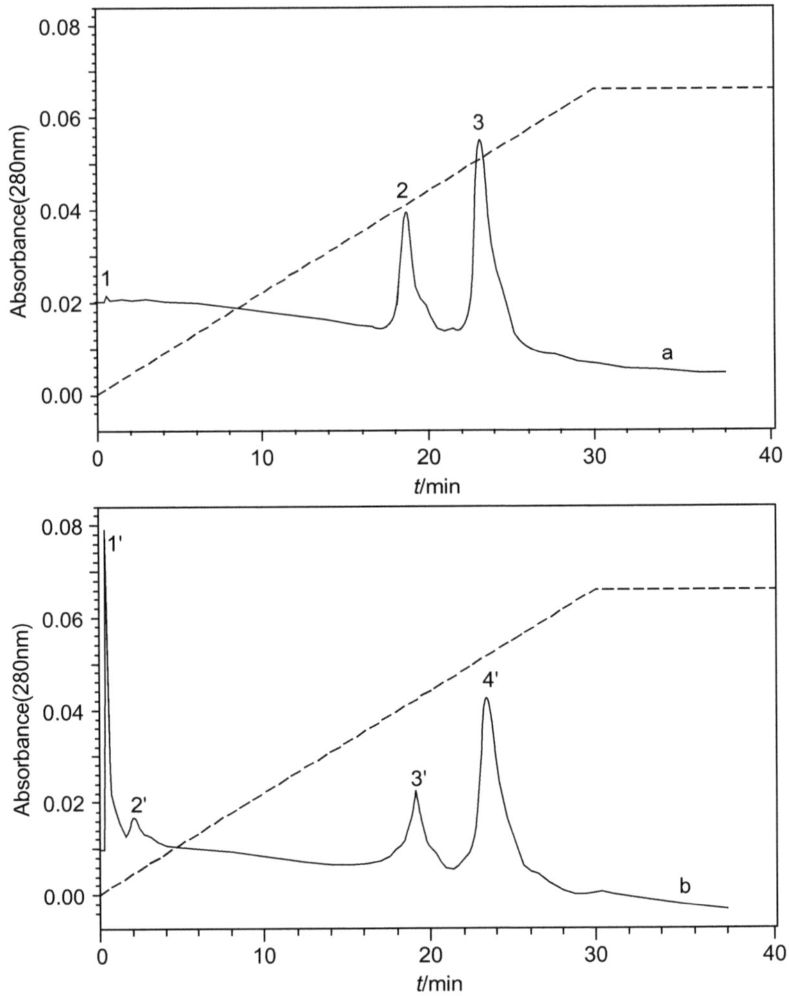

Fig. 5.4. Simultaneous refolding and separation of denatured lysozyme and RNase A with HIC. Conditions: initial concentration: 5.0 mg/mL; linear gradient elution (*dash line*): 100% solution C to 100% solution D for 30 min with a 10-min delay; flow rate, 1.0 mL/min; chromatogram "a," native RNase A and Lys; chromatogram "b," the HIC mode-renatured RNase A and Lys which was originally denatured by 7.0 M GuaHCl solution; peaks, 1, solvent; 2, native RNase A; 3, native Lys; 1′ and 2′, solvent and unfolded RNase A and Lys; 3′, refolded RNase A; 4′, refolded Lys.

the two proteins in the N state were also run under the same chromatographic condition as in the U state. The obtained chromatogram "a" is also shown (*see* **Fig. 5.4a**). It can be seen that three peaks of 1–3 in **Fig. 5.4a** and four peaks of 1′–4′ in **Fig. 5.4b** were obtained.

2. From **Fig. 5.4**, the peak 1 is solvent and 1′ and 2′ are solvent and unretained of the two proteins in U state; 2, 3′ separately denote RNase A in N state and the refolded RNase A; 3, 4′ represent Lys in N state and the refolded one, respectively. This fact indicates that the denatured Lys and RNase

A can not only be renatured but also be completely separated with each other and also separated from their unfolded states, respectively.

3. The fractions 3′ and 4′ were separately collected and subsequently were tested by the assay of their enzymatic activities (21).
4. The mass and the bioactivity recoveries of the renatured Lys and RNase A with this manner of HIC mode can reach to 85%, respectively.

3.3. Lys Bioactivity Assay

1. The measurement of the enzymatic activity of the refolded Lys was routinely done at 20°C by following the decrease in absorbance at 450 nm of a *Micrococcus lysodecticus* cell wall suspension in 20 mM sodium phosphate (pH 7.0) (22).
2. Dissolve the *M. lysodecticus* cell wall in 20 mM sodium phosphate (pH 7.0).
3. Adjust OD value of the buffer about 0.5–0.7.
4. Pipette 3.0 mL of the *M. lysodecticus* cell wall suspension buffer in the cuvette and then pipette 100 μL of the HPLC fraction solution containing lysozyme and mix it well rapidly.
5. With distilled water as the reference buffer, monitor the decrease of OD value at 450 nm for 3 min, and note the data per 20 s.
6. When measuring the time dependence of the recovery of activity, aliquots were taken and directly analyzed for activity.
7. Activities were normalized to a native Lys solution of known concentration.

3.4. RNase A Bioactivity Assay

1. The measurement of the enzymatic activity of the refolded RNase A was routinely carried out at 20°C by following the decrease in absorbance at 300 nm of 0.05% yeast solution in 0.1 M sodium acetate (pH 5.0).
2. Pipette 2.5 mL of the 0.05% ribonucleic acid from yeast buffer in the cuvette and then pipette 100 μL of the HPLC fraction solution containing RNase A, and mix it well rapidly.
3. With distilled water as the reference buffer, monitor the decrease of OD value at 300 nm for 3 min, and note the data A_t per 30 s.
4. After 30 min, measure the OD value at 300 nm again, and note the data A_f. A_f is the final OD value at 300 nm.
5. When measuring the time dependence of the recovery of activity, aliquots were taken and directly analyzed for activity.
6. Activities were normalized to a native RNase A solution of known concentration.

7. With the linear relationship of $\log(A_t - A_f)$, the slope (S) can be obtained from the line.
8. Based on the slope S, calculate the recovery of enzymatic activity.
9. Following the equation, calculate the enzymatic activity ratio:

$$\text{Unit/mg} = S(-2.3) \times 4/W$$

where W is the amount of RNase A in the sample.

3.5. Determination of Protein Concentration

The concentration of Lys and RNase A in both U and N states was done at 280 nm ($E_{1\%} = 2.63$ and 2.37, respectively).

3.6. Renaturation of rhIFN-γ with Simultaneous Purification by USRPP

3.6.1. Disruption of E. coli Cells Containing rhIFN-γ by Sonication

1. rhIFN-γ was expressed in *E. coli* (pBV220/DH5α). About 30 g wet *E. coli* cells containing rhIFN-γ was added in 300 mL cleaning buffer (1:10, W/V) containing 20 mM PBS (pH 7.4), 1 mM EDTA, 0.2 mg/mL lysozyme. After cleaning, the cells were centrifuged at 18,000 rpm for 20 min at 4°C.
2. Remove the supernatant and clean the cells again as in Step 1.
3. Recover the cells and store at –20°C overnight.
4. After unfreezing the cells, 30 g cells were disrupted by sonication with intermissive mode in 300 mL cleaning buffer (1:10, W/V) in the ice-water bath.
5. Until the dope become like rheumy and no babble appears, the rhIFN-γ inclusion bodies was recovered by centrifugation with 18,000 rpm at 4°C.
6. To clean the inclusion bodies with the cleaning buffer (*see* **Section 2.1.10**) again, the rhIFN-γ inclusion bodies were stored at –20°C.

3.6.2. Preparation of Crude rhIFN-γ Extract (see Notes 9 and 10)

1. The rhIFN-γ inclusion bodies were washed twice using buffer I (*see* **Section 2.1**, Step 11). After stirring gently at 4°C for 1 h, centrifuge it at 18,000 rpm for 20 min and removed the supernatant.
2. Wash the inclusion bodies twice with buffer II (*see* **Section 2.1**, Step 11), and then centrifuge it and remove the supernatant.

3. Wash the inclusion bodies twice with buffer III (*see* **Section 2.1**, Step 11).
4. The cleaned inclusion bodies were recovered by centrifugation at 18,000 rpm for 15 min.
5. About 10 g of the cleaned inclusion bodies were solubilized in 100 mL 7 M GuaHCl (1:10, W/V).
6. Adjust pH value to 8.0 with 0.1 M HCl and stirr the solution gently at 4°C overnight.
7. The supernatant after centrifugation at 18,000 rpm was crude rhIFN-γ extract.

3.6.3. Renaturation with Simultaneous Purification of rhIFN-γ with "Cake" (see Notes *11–16)*

1. The chromatographic cake ("cake") has very short length and very large diameter. **Figure 5.5** shows the resolution of seven standard proteins with an HIC "cake" of 1.0 cm in thickness and 5.0 cm in diameter under flow rate of 8.0 mL/min. It is seen that the protein resolution is excellent.

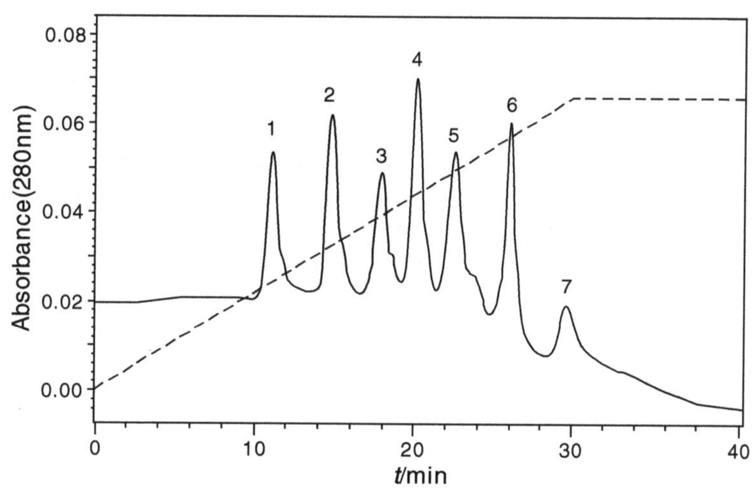

Fig. 5.5. Chromatogram of standard proteins by chromatographic cake (10 × 50 mm). Stationary phase: PEG-600; linear gradient elution (*dash line*): 100% solution C to 100% solution D for 30 min with a 10-min delay; flow rate, 8.0 mL/min; peak: 1, cytochrome-C; 2, myoglobin; 3, RNase A; 4, lysozyme; 5, α-chymotrypsin; 6, α-amylase; 7, insulin. Total mass of proteins, 2.1 mg.

2. About 500 μL rhIFN-γ extract solution (500 μL containing about 500 μg total proteins) was loaded on the "cake" initially equilibrated with buffer solution A by means of a cumulative manner, for example, 100 μL for each injection and totally five times. Subsequently a 35-min non-linear gradient elution of 0–100% buffer solution B (0–15 min, 0–50% B; 15 min, curve –2; 15–23 min, 50–40%B; 23 min, curve 2; 23–35 min, 40–100%B) was performed at a flow rate of 5.0 mL/min. The chromatogram is shown (*see* **Fig. 5.6**). The eluted fractions of aim protein were collected for the

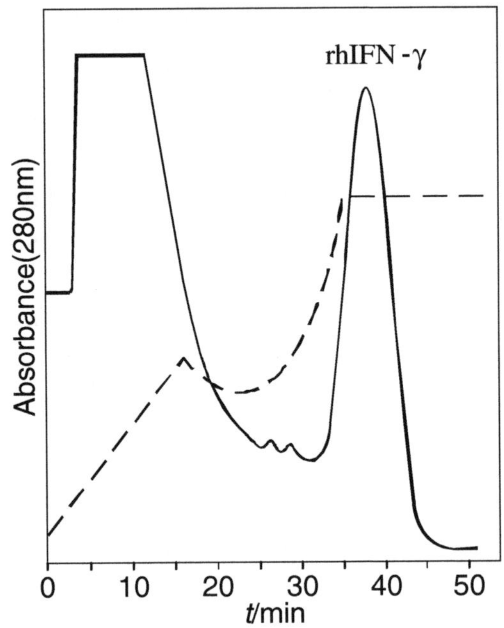

Fig. 5.6. Renaturation with simultaneous purification of rhINF-γ by using semi-scale preparative USRPP (10 mm × 50 mm ID). Chromatographic conditions: Sample size: 500 μL (5 × 100μL injections); mobile phase A, 3.0 mol/L ammonium sulfate + 0.05 mol/L potassium dihydrogen phosphate, pH 7.0; mobile phase B, 0.05 mol/L potassium dihydrogen phosphate, pH 7.0; 35-min non-linear gradient elution of 0–100% B (*dash line*), 0–15 min, 0–50% B; 15 min, curve −2; 15–23 min, 50–40% B; 23 min, curve 2; 23–35 min, 40–100% B; flow rate, 5.0 mL/min; detection wavelength, 280 nm.

measurements of the recoveries of bioactivity and mass of the rhIFN-γ and SDS-PAGE analysis after desalting (*see* **Note 11**).

3.6.4. "Cake" Cleaning (see Note 17)

1. Because of very strong hydrophobicity, the precipitate of rhIFN-γ can be formed when the sample loaded on "cake." It can not only block the column but also lead to the very serious loss of the recombinant proteins.
2. A "cake" cleaning buffer containing denaturing agent, such as 8.0 M urea solution, in the presence of DTT should be used to periodically dissolve the precipitates deposited on the stationary phase to prolong the USRPP life.
3. The "cake" or column was washed with the cleaning buffer, or let the cleaning buffer to stay in the "cake" or column overnight, and then wash it next day and collect the fractions containing the target proteins.
4. With re-injecting the washed out solution containing the rhIFN-γ (of course, it is in unfolded state) into the "cake" to be renatured again, the losses of the mass and bioactivity of the rhIFN-γ would decrease (*see* **Notes 18 and 19**).

3.7. Bradford Protein Assay

1. Preparation of Bradford reagent: dissolve 100 mg Coomassie Blue G-250 (Fluka, MO, USA) in 50 mL 95% ethanol, add 100 mL 85% (w/v) phosphoric acid to this solution, and dilute the mixture to 1 L with water. After filtration with filter paper, store at –4°C (23).
2. Pipette 0, 20, 40, 60, 80, 100, 120 μL of BSA (0.2 mg/mL) into individual 5 mL Eppendorf tube.
3. Add 0.15 M NaCl to all tubes to bring the final volume to 200 μL.
4. Add 2 mL of Bradford reagent into all tubes containing standard or sample.
5. Read absorbance at 595 nm without any prior incubation.

3.8. Determination of Bioactivity of rhIFN-γ

The bioactivity assay for the rhIFN-γ was done by CPE inhibitor with WISH cell and VSV virus (24).

4. Notes

1. The separation principle of a 2D (HIC, IEC) column is briefly explained as that under low salt concentration in mobile phase, protein renaturation, and separation by IEC mode, while under high salt concentration, that by HIC mode. From standpoint of stoichiometric displacement theory for retention of proteins, the changes in salt concentration cause the changes in the concentration of two displacers, salt for IEC mode and water for HIC mode (18, 20, 25).
2. Because proteins have different properties and molecular structures with each other, it is very difficult to employ same experimental conditions for the folding of all protein with PFLC. To any target proteins, the experimental conditions for the renaturation and purification simultaneously with PFLC must be optimized, including the selection of the kinds of LC, stationary phase, mobile phase, flow rate, pH value. The first important thing is to select the kinds of LC. The 2D (HIC, WCX) column should be the first choice for accomplishing PFLC, because IEC is suitable for the oxidation refolding of reduced/denatured proteins, while HIC is valid for the refolding of the proteins with very strong hydrophobicity. Because some aggregates and precipitates in refolding process can block, even damage the employed column, the chromatographic cake is the best way to replace the conventional column.

3. As a powerful tool for the investigation of protein folding, scientist may design any kinds of stationary phase by means of bonding ligands or adsorbing substances on the surface of stationary phase to assisted protein folding, and also selecting some buffers as the employed mobile phase and screening a suitable concentration for the optimization of protein folding.
4. Based on the molecular mechanism of protein refolding in HIC, the refolding process occurs instantaneously (3), while that for the reduced–denatured proteins in IEC, it involves to carry out oxidation–reduction reaction which needs an enough time. Thus the obtained peak shape, peak height, even retention of the refolded proteins in this instance may be a little different from that of the same proteins in N state. Furthermore, it takes a certain time, even over 10 h, to accomplish the whole process of the oxidation–reduction reaction. The contribution of the IEC column is to prevent from the aggregates of the unfolded protein molecules and make the –SH correctly closed together by the recognition of some specific amino acid residues to the stationary phase of the IEC and subsequently wait for accomplishing oxidation. Because of this, compared to HIC, the selection of the kind of IEC is more important. How to accelerate the oxidation of two –SH and correctly bond together is a puzzle in PFLC (5).
5. For a successful oxidative refolding, one of the main points is how to depress the formation of both protein aggregates and intermolecular inappropriate disulfide bonds. If the intermediate states, which have native-like tertiary structure, are separately trapped from each other, the subsequent formation of the disulfide bonds may proceed successfully. Generally, urea decreases Lys and RNase A retentions. The presence of urea in mobile phase, on the one hand, can diminish the aggregate forming, being favorable to protein refolding. On the other hand, urea can increase the eluting strength of mobile phase and destroy the three- or four-dimensional molecular structure of protein, being unfavorable to protein refolding. A suitable range of urea in mobile phase ranges from 2 to 4 mol/L (21).
6. In IEC, sodium chloride is usually used as a common displacer. Ammonium sulfate is a stronger eluting agent. When the same concentration of urea is used, the mass recovery of protein refolding with ammonium sulfate as displacer is more than that of using sodium chloride, especially when using lower urea concentrations in the mobile phase.

7. As expected, the renaturation of Lys by using IEC can be performed at a much higher protein concentration comparing with renaturation in solution under the same conditions.
8. By comparison, dilution method was used at the same time. The denatured Lys and RNase A were more rapidly diluted with the renaturation buffer and renatured by dilution method. The pooled fractions of mobile phase, which ran the same gradient except the sample was not loaded, were used as the renaturation buffer of dilution method, and enzymatic activity of reduced/denatured Lys was measured also after 4- to 5-h incubation for kinetics request.
9. The selection of denaturant for dissolving protein inclusion bodies based on the employed kinds of LC. Because GuaHCl is a kind of salt, it is never employed for IEC mode. Theoretically, urea solution can be valid for dissolving protein inclusion bodies for the renaturation and separation of both HIC and IEC modes, but the denatured extent for a specific protein may be weaker than GuaHCl.
10. In terms of the bioactivity recovery of the renatured rh-IFN-γ by HIC, the denaturant, 7.0 mol/L GuaHCL solution for dissolving its inclusion body, is experimentally proved to be better than 8.0 mol/L urea.
11. The molecules of the rhIFN-γ have very strong hydrophobicity and do not contain disulfide bonding. On the one hand, the refolding of rhIFN-γ should be mainly dominated by hydrophobic interaction forces and has no other problem of misbinding of disulfide bond. On the other hand, the strong hydrophobicity of rhIFN-γ makes it to have strong retention, resulting in a favorable separation from impure proteins with weak or middle hydrophobicity. Thus, HIC would be first chosen.
12. To prevent protein precipitates from either entering the soft media of the cake or from being trapped in the interstitial spaces in the cake it is advisable to use packing material with a rigid matrix. These trapped precipitates can lead to high column pressures and result in protein aggregates that are difficult to dissolve for refolding.
13. An ideal "cake" simultaneously having four functions shown in **Fig. 5.1** should be used. In case of precipitates formed, the "cake" back pressure does not raise up seriously, and thus working well in this instance. They are to completely remove denaturing agent, to renature the

aim protein, to separate the renatured proteins from impure proteins, and to recycle waste denaturing agent easily.

14. The rhIFN-γ originally in the extract of 7 mol/L GuHCl solution can be directed into the HIC "cake," and only with one step and within 40 min, the obtained purity and the specific bioactivity can approach to 95% and 8.7×10^7 IU/mg, respectively.
15. Comparing the total bioactivity in the extract, or before that it was injected into the "cake"-HIC, the bioactivity recovery is 62-fold of the original total bioactivity in the extract of GuHCl solution.
16. A solution containing denaturing agent, such as 8.0 urea solution, in the presence of dithiothreitol (DTT) should be used to periodically dissolve the precipitates deposited on the stationary phase to prolong the "cake" life.
17. In addition, when re-injecting the washed out solution containing the rhIFN-γ (of course, it is in unfolded state) into the "cake" to be renatured again, the losses of the mass and bioactivity of the rhIFN-γ would decrease.
18. In order to increase the mass recovery of protein refolding by PFLC, the un-retained proteins usually are eluted out together with or nearby the solvent. It can be collected and re-injected for the second chromatographic run for recovering the target protein. However, some aggregates and/or precipitates of the target proteins are so tightly adsorbed and/or deposited on the stationary phase that the strong solution of the employed gradient elution cannot elute out them, resulting in mass loss and gradual increase in the column back pressure, even blocking the column. The column can be re-dissolved by a very strong washing solution and to recover it (17, 18).
19. The column regeneration for obtaining a reproducible and clean surface of the employed stationary phase of a column is an indispensable step. Compared to conventional separation of native proteins by LC, it is quite different in this instance that some precipitates from either target proteins or impure proteins may form not only on the filter surface of the column enter side but also on the surface of particles of the packing materials. It is absolutely necessary periodically to clean up the column with very strong washing solutions, such the reduced/denatured buffer for dissolving protein inclusion bodies shown in **Section 2.1**.

References

1. Kane, J. F., Hartley, D. L. (1988) Formation of recombinant protein inclusion bodies in *Escherichia coli. Trends Biotechnol* 6, 95–99.
2. Geng, X. D., Chang, X.-Q. (1992) High-performance hydrophobic interaction chromatography as a tool for protein refolding. *J Chromatogr* 599, 185–194.
3. Geng, X. D., Bai, Q. (2002) Mechanism of simultaneously refolding and purification of proteins by hydrophobic interaction chromatographic unit and applications. *Sci China (Ser. B)* 45, 655–669.
4. Bai, Q., Kong, Y., Geng, X. D. (2002) Studies on the refolding of reduced-denatured insulin with high performance hydrophobic interaction chromatography. *Chem Res Chin Univ* 23, 1483–1488.
5. Geng, X. D., Wang, C. Z. (2007) Protein folding liquid chromatography and its recent developments. *J Chromatogr B* 249, 69–80.
6. Geng, X. D., Bai, Q., Wang C.-Z. (2006) *Protein Folding Liquid Chromatography.* Science Press, Beijing, China, in Chinese.
7. Geng, X. D., Wang, L. L. (2008) Liquid chromatography of recombinant proteins and protein drugs. *J Chromatogr B* 866, 133–153.
8. Guo, L. A., Geng, X. D. (2000) Simultaneous renaturation and purification of proteins by liquid chromatography. *Chin J Biotechnol* 16, 661–666.
9. Li, M., Su, Z. G., Janson, J. C. (2004) In vitro protein refolding by chromatographic procedures. *Protein Expr Purif* 33, 1–10.
10. Jungbauer, A., Kaar, W., Schlegl, R. (2004) Folding and refolding of proteins in chromatographic beds. *Curr Opin Biotechnol* 15, 487–494.
11. Middelberg, A. P. (2002) Preparative protein infolding. *Trends Biotechnol* 20, 437–443.
12. Geng, X. D. (2004) *Stoichiometric Displacement Theory and Its Application.* Science Press, Beijing, China, in Chinese.
13. Geng, X. D., Zhang, Y. J. (2007) Caky chromatographic column and the method for producing it and its applications. *US Patent* 7,208,085 B2.
14. Geng, X. D., Bai, Q., Zhang, Y.-J., Li, X., Wu, D. (2004) Refolding and purification of interferon-gamma in industry by hydrophobic interaction chromatography. *J Biotech* 113, 137–149.
15. Bai, Q., Kong, Y., Geng, X. D. (2003) Studies on renaturation with simultaneous purification of recombinant human proinsulin from *E. coli* with high performance hydrophobic interaction chromatography. *J Liquid Chromatogr Related Tech* 26, 683–695.
16. Wang, C. Z., Liu, J. F., Geng, X. D. (2009) Optimization of refolding with simultaneous purification of recombinant human granulocyte colony-stimulating factor from *Escherichia coli* by immobilized metal ion affinity chromatography. *Biochem Eng J*, 43, 197–202.
17. Wang, L. L., Wang, C. Z., Geng, X. D. (2006) Expression, renaturation and simultaneous purification of recombinant human stem cell factor in *Escherichia coli. Biotechnol Lett* 28, 993–997.
18. Wu, D., Wang, C. Z., Geng, X. D. (2007) An approach for increasing the mass recovery of proteins derived from inclusion bodies in biotechnology. *Biotechnol Prog* 23, 407–412.
19. Ke, C. Y., Geng, X. D. (2008) A new chromatographic method for fast separation of active proteins. *Chin Sci Bull* 53, 1113–1116.
20. Geng, X. D., Ke, C., Chen, G., Liu, P., Wang, F., Zhang, H., Sun, X. (2009) On-line separation of native proteins by two-dimensional liquid chromatography with a single column. *J Chromatogr A*, 1216, 3553–3562.
21. Wang, Y., Gong, B. L., Geng, X. D. (2003) Refolding of denatured/reduced lysozyme using weak-cation exchange chromatography. *Chin Chem Lett* 14, 828–831.
22. Goldberg, M. E., Rudolph, R., Jaenicke, R. (1991) A kinetic study of the competition between renaturation and aggregation during the refolding of denatured-reduced egg white lysozyme. *Biochemistry* 30, 2790–2797.
23. Laskowski, M. (1955) in (Colowick, S. P., Kaplan, N. O. eds.), *Methods in Enzymology.* Chymotrypsinogens and chymotrypsins Academic, New York, NY, pp. 8–26.
24. Lengyel, P. (1981) Enzymology of interferon action – a short survey. *Methods Enzymol* 79, 135–162.
25. Geng, X. D., Guo, L. A., Chang, J. H. (1990) Study of the retention mechanism of proteins in hydrophobic interaction chromatography. *J Chromatogr* 507, 1–23.

Chapter 6

Site-Specific Protein Labeling by Intein-Mediated Protein Ligation

Inca Ghosh, Nancy Considine, Elissa Maunus, Luo Sun, Aihua Zhang, John Buswell, Thomas C. Evans, Jr., and Ming-Qun Xu

Abstract

Intein-mediated protein ligation (IPL) employs an intein to create a protein possessing a C-terminal thioester that can be ligated to a protein or peptide with an amino-terminal cysteine via a native peptide bond. Here we present a procedure to conduct isolation and labeling of recombinant proteins expressed in *E. coli* using synthetic short peptides possessing a fluorescent moiety. This approach can be readily utilized for site-specific conjugation of a fluorophore to the C-terminus of a protein of interest, without the drawback of non-specific chemical labeling. This chapter also gives a general review of the critical parameters of intein-mediated cleavage and ligation reactions.

Key words: Intein-mediated protein ligation, expressed protein ligation, intein, protein labeling.

1. Introduction

E. coli is preferred for heterologous protein production because of its fast growth, simple fermentation, uncomplicated nutritional and sterility requirements, and extensive characterization (1). A protein expression system, IMPACT (intein-mediated purification with an affinity chitin-binding tag), was developed by New England Biolabs by exploiting inteins (2). Inteins were initially engineered as fusion partners with inducible cleavage activity for isolation of recombinant proteins in *E. coli* (3).

There are various methods to site-specifically label a protein, one of which is intein-mediated protein ligation. Intein-mediated protein ligation (IPL) or expressed protein ligation

T.C. Evans, M.-Q. Xu (eds.), *Heterologous Gene Expression in E. coli*, Methods in Molecular Biology 705,
DOI 10.1007/978-1-61737-967-3_6, © Springer Science+Business Media, LLC 2011

(EPL) employs an intein, to create a protein possessing a C-terminal thioester that can be ligated to a protein or peptide possessing an N-terminal cysteine to form a native peptide bond (4, 5). A variety of intein-enabled approaches have broken down the size limitation barrier of traditional chemical methods in protein semisynthesis and has opened new avenues of site-specific labeling of proteins. Unlike some chemical methods of labeling, IPL uses a thiol reagent which can be easily removed. Another advantage is that the ligated peptide or protein needs only to contain an N-terminal cysteine and can be synthesized with various non-standard chemical moieties. Furthermore, a peptide as small as two amino acids can be added to your protein of interest, thus reducing possible perturbations to the native structure of the protein.

1.1. Thiol-Induced Removal of the Intein Tag

In order to conduct IPL, the gene encoding a protein of interest is cloned into an expression vector in which the target protein is fused to the N-terminus of a modified intein and a small chitin-binding domain (CBD, 6 kDa) (*see* **Note 1**). Following expression and affinity purification of the three-part fusion protein from *E. coli*, the intein cleavage is induced with a thiol reagent, such as 2-mercaptoethanesulfonic acid (MESNA), to release the protein of interest. The resulting thioester-tagged protein is mixed with a peptide or protein possessing an amino-terminal cysteine and ligation occurs spontaneously to produce a native peptide bond between the two proteins or the protein and the peptide (*see* **Fig. 6.1**). In general, for protein purification 1,4-dithiothreitol (DTT) is used as the intein cleavage reagent; however, the ligation efficiencies were significantly higher when MESNA or thiophenol was used in place of DTT for inducing cleavage (4–6). We have examined the critical parameters of intein-mediated cleavage and ligation reactions. The choice of amino acid residues at the intein cleavage site and their effect on cleavage and ligation efficiency are presented. We also describe the selection of thiol reagents, 2-mercaptoethanesulfonic acid or 1,4-dithiothreitol, for protein purification and ligation reactions.

Furthermore, in vivo and thiol-induced cleavage at the N-terminus of an intein can be dramatically affected by target protein sequence and, in particular, by the C-terminal amino acid residue of the target sequence (7). It is believed that each intein has evolved in its own host protein context and thus may display different reaction rates in response to non-native amino acid residues adjacent to the scissile peptide bond. A diverse choice of inteins allows cleavage and ligation reactions to be optimized depending on the protein of interest to be isolated.

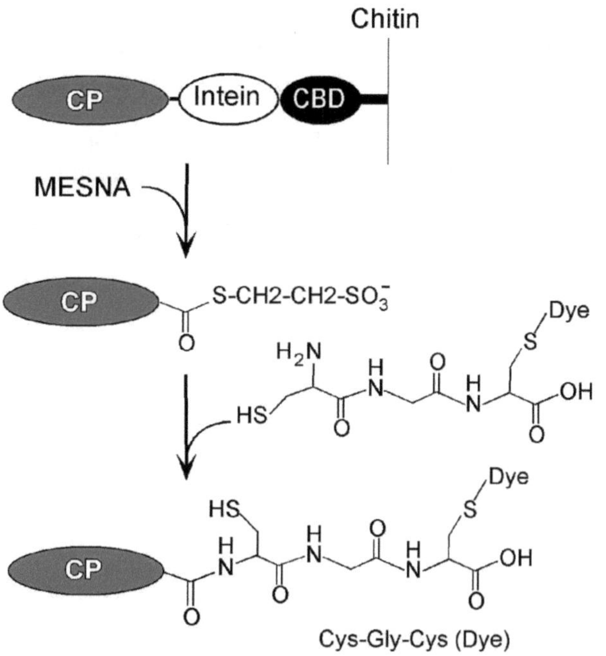

Fig. 6.1. Schematic of intein-mediated protein ligation (IPL). The target carrier protein (CP) is expressed at the N-terminus of an intein–chitin binding domain (CBD) protein fusion. The expressed fusion protein is bound to the chitin resin and cleavage between the CP and the intein is carried out by the thiol reagent 2-mercaptoethanesulfonic acid (MESNA) to release the CP with a C-terminal thioester. The peptide Cys-Gly-Cys (Dye), which contains a free N-terminal cysteine, is ligated to the thioester-tagged protein by a peptide bond to produce a CP protein tagged with a dye.

1.2. Target Protein Effect on Intein-Tag Cleavage

Table 6.1 summarizes the cleavage and ligation efficiencies with regard to different amino acid residues preceding the cleavage site (termed the –1 position) when a modified *Mycobacterium xenopi* GryA intein (198 residues) was used as a fusion partner (7). This corresponds to the C-terminal amino acid of the target protein when it is fused to the Mxe GyrA intein tag. In general, cleavage of a target protein–intein fusion with DTT is more efficient than cleavage with MESNA and therefore is preferred for the purpose of protein isolation in most cases. However, ligation efficiencies using MESNA-tagged paramyosin proteins and a PB1 peptide are generally higher than those with DTT-tagged proteins, probably due to higher stability and reactivity of a MESNA thioester tag. If the target protein possesses an unfavorable residue at its carboxyl terminus, the insertion of a preferred residue at the cleavage site may improve cleavage and ligation efficiency. These data are similar to the data published previously on native chemical ligation (8) with a few differences. The variation may arise

Table 6.1
Amino acid residue effect on thiol-induced cleavage and ligation: Effect of the C-terminal residue of a target protein on DTT- or MESNA-induced cleavage

C-Terminal residue	DTT % cleavage 16 h		DTT % cleavage 40 h		MESNA % cleavage 16 h		MESNA % cleavage 40 h		IPL	
	4°C	23°C	4°C	23°C	4°C	23°C	4°C	23°C	DTT	MESNA
Tyr	++	++	++	+++	+++	+++	+++	+++	++	+++
Leu*	++	+++	+++	+++	+++	+++	+++	+++	–	–
Phe	++	+++	+++	+++	+++	+++	+++	+++	++	+++
Met	++	+++	+++	+++	+++	+++	+++	+++	–	++
Lys	++	+++	+++	+++	+++	+++	+++	+++	++	++
Gln	+++	+++	+++	+++	++	+++	+++	+++	++	+++
Arg	++	+++	++	+++	++	+++	+++	+++	++	++
Trp	++	+++	++	+++	++	++	++	+++	–	+++
His	++	+++	+++	+++	++	+++	+++	+++	++	+++
Val	++	++	++	+++	+	++	+	++	N/D	–
Glu	++	+++	++	+++	+	+	+	++	–	N/D
Thr	++	+++	++	+++	+	+	+	++	–	–
Asn	++	+++	++	+++	–	+	+	++	++	++
Ile	++	++	++	++	–	+	+	++	–	–
Ala	++	++	++	++	–	+	–	++	++	+++
Cys	++	++	++	++	–	–	–	++	++	++
Gly	–	++	–	++	–	–	–	++	++	++
Asp*	–	–	–	–	–	–	–	–	N/D	++
Ser	–	++	–	++	–	–	–	–	–	–
Pro	–	–	–	–	–	–	–	–	N/A	N/A

Note: Paramyosin protein was expressed in pTXB1 vector and mutagenesis was performed to introduce various amino acid residues at the position preceding the cleavage site of the intein. Cleavage of various mutant proteins was induced with DTT or MESNA. Less than 25% (–); 25%–49% (+); 50%–75% (++); 75%–100% (+++). In addition, the presence of Leu at the cleavage junction showed ~50% in vivo cleavage when induced at 15 C; at 37 C in vivo cleavage was less than 5%; Asp showed in vivo cleavage at 15 C and 37 C. N/A, protein concentration no sufficient for ligation. N/D, ligation product not detectable on SDS PAGE gel.

due to the differences in experimental design, thiol reagent, and temperature.

The IPL procedure is very easy to perform. The product contains a native peptide bond between the reactants at the point of ligation and can be used in a variety of biochemical studies. For example, by using IPL for ligation of peptides to carrier proteins a number of limitations of peptides are overcome, resulting in enhanced sensitivity in protein array and ELISA analysis (9). IPL can also be used to generate phosphorylated substrates to study phosphatase activity by western blot analysis (10). We express a carrier protein as an intein fusion using the IMPACT™ system and purify it by inducing cleavage of the fusion protein with MESNA. The purified carrier protein possessing a C-terminal thioester is mixed with a peptide containing an amino-terminal cysteine and a phosphorylated residue at room temperature for 4 h or overnight at 4°C, resulting in a ligated product with a peptide bond between the carrier protein and the phosphorylated peptide for subsequent enzymatic assays (*see* **Fig. 6.1**).

The C-terminal amino acid of the target protein, paramyosin, was mutated immediately upstream of the *Mxe* GyrA intein cleavage site. Cleavage was induced with 40 mM DTT or 40 mM MESNA. Percent cleavage was determined by Coomassie stained SDS-PAGE analysis of chitin beads before and after thiol cleavage. Note that boiling in SDS Sample Buffer containing DTT can cause partial or complete cleavage, resulting in an overestimation of *in vivo* cleavage. If substantial *in vivo* cleavage is observed, the cell extract should be evaluated in a SDS Sample Buffer containing no DTT.

Using IPL, a semisynthesis method has been developed to create and join small or large polypeptides. The IPL approach has been used in various applications including production of toxic proteins; introduction of non-coded amino acids, fatty acids, or chemically modified amino acid residues; segmental protein labeling; generation of positive controls for western blot analysis; production of substrates for enzymatic assays; and epitope mapping (4, 9, 11–19).

In this work we discuss bacterial expression of intein fusion proteins, the effect of the amino acid adjacent to the intein cleav-

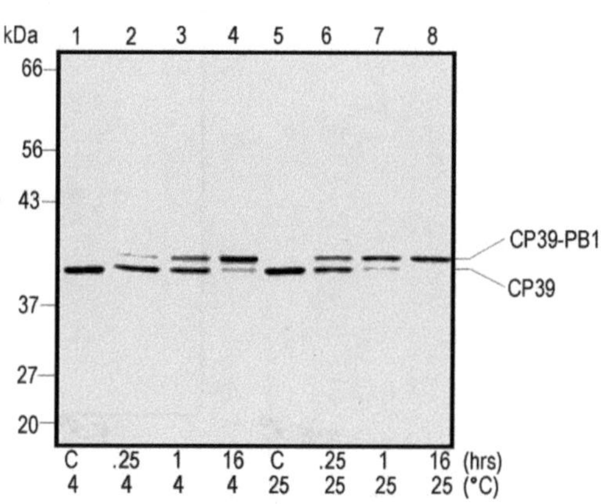

Fig. 6.2. Analysis of intein-mediated protein ligation (IPL) by Coomassie blue-stained SDS-PAGE. PB1 was synthesized with an N-terminal cysteine (CTRSRHSSYPNEYEEDEEMEEEL; MW 2862 Da) and ligated to the C-terminal thioester-tagged carrier protein 39. The reaction times and temperatures are indicated. The samples were electrophoresed on a 12% SDS polyacrylamide gel and examined by Coomassie blue staining. Lanes 1 and 5 are controls (C) that contain only carrier protein 39.

age site, and the effect of using DTT or MESNA on cleavage and ligation (*see* **Table 6.1**). We give an example of labeling a protein with a specific peptide (*see* **Fig. 6.2**) or a peptide containing a fluorescent dye (*see* **Fig. 6.3**) as well as immunoblot analysis to check an anti-peptide antibody of interest (*see* **Fig. 6.4**).

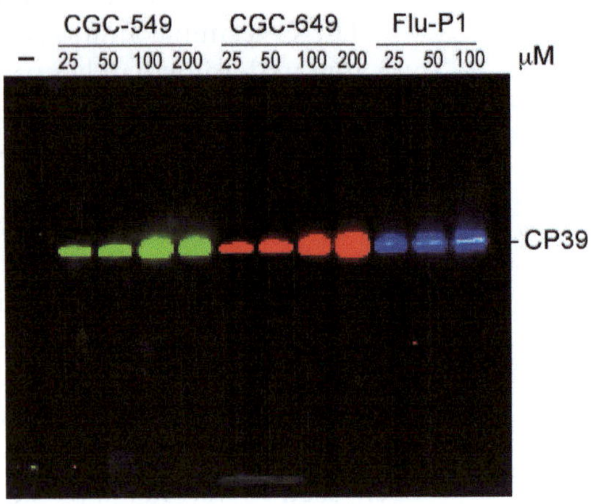

Fig. 6.3. The C-terminal thioester-tagged CP39 was incubated with 0, 25, 50, 100, or 200 μM CGC-549, CGC-649, or Flu-P1 peptide overnight at 4°C. The samples were electrophoresed on a 12% SDS polyacrylamide gel and electroblotted to a nitrocellulose membrane. The membrane was scanned using the Typhoon fluorescent imager with the 532/580 nm excitation/emission filter set for CGC 549 (*green*), 633/670 nm excitation/emission filter set for CGC 649 (*red*), and 488/520 nm for Flu-P1 (*blue*). The merged image for the three channels is shown.

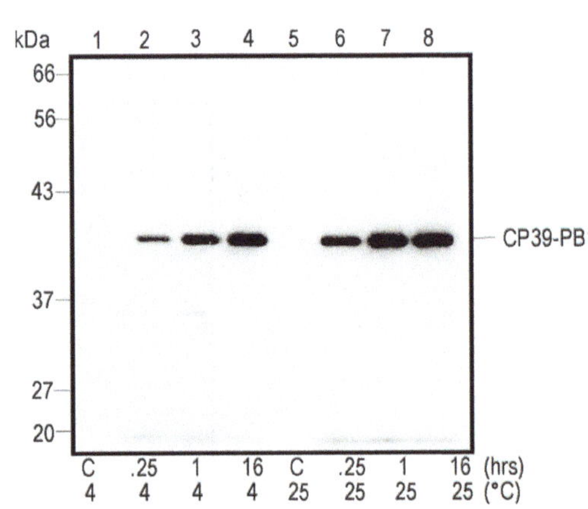

Fig. 6.4. Western blot analysis of the peptide PB1 ligated to the carrier protein (CP39) using anti-PB1 antiserum. PB1 was synthesized with an N-terminal cysteine (CTRSRHSSYPNEYEEDEEMEEEL; MW 2862 Da) and ligated to the C-terminal thioester-tagged carrier protein 39. The reaction times and temperatures are indicated. The samples were electrophoresed on a 12% SDS polyacrylamide gel and transferred to nitrocellulose. Anti-PB1 peptide antibody was used as the primary antibody. Lanes 1 and 5 are controls (C) that contain only carrier protein 39.

2. Materials

2.1. Cloning and Expression

1. IMPACT C-terminal fusion vector, pTXB1, pTXB3, pTYB1, pTYB2, pTYB3, pTYB4, pTWIN1, or pTWIN2 (2). These vectors are commercially available from New England BioLabs, Inc.
2. Restriction enzymes (New England BioLabs, Ipswich, MA).
3. Phusion™ Site-Directed Mutagenesis Kit (New England BioLabs).
4. *E. coli* strain T7 Express (New England BioLabs).
5. Ampicillin: 100 mg/mL stock (American Pharmaceutical Partners, Schaumburg, IL). Filter sterilize and store in 1 mL aliquots at –20°C.
6. Luria–Bertani (LB) media: 10 g tryptone, 5 g yeast extract, and 10 g NaCl are added to H_2O and the pH adjusted to 7.0 with NaOH. After adjusting the pH bring the final volume to 1 L and sterilize by autoclaving.
7. IPTG: Isopropyl β-D-thiogalactopyranoside (American Bioanalytical, Natick, MA). Make up a stock solution of 100 mM, filter sterilize, aliquot, and store at –20°C.

2.2. Purification

1. Chitin resin (New England BioLabs).
2. Column buffer: 20 mM Tris or HEPES, 0.5 M NaCl, pH 8.5 at 25°C.
3. MESNA: 2-mercaptoethanesulfonic acid (Sigma, St. Louis, MO).
4. DTT: 1,4-dithiothreitol (JT Baker, Mallinckrodt Baker, Inc., Phillipsburg, NJ).
5. Protein assay kit (Bio-Rad, Hercules, CA).
6. 10X IPL reaction buffer: 100 mM Tris-HCl, 250 mM NaCl, 10 mM MESNA, pH 8.5 at 25°C.

2.3. SDS Polyacrylamide Gel Electrophoresis (SDS–PAGE)

1. 12 or 10–20% Tris glycine gels (Invitrogen, Carlsbad, CA).
2. ColorPlus Prestained Protein Ladder, broad range (10–230 kDa) (New England BioLabs).
3. Protein ladder (10–250 kDa) (New England BioLabs).
4. Biotinylated protein marker (Cell Signaling Technology, Danvers, MA).
5. Fluorescent marker (Thermo Fisher Scientific Inc. Waltham, MA).
6. Blue loading buffer pack (New England BioLabs): 3X sodium dodecyl sulfate-polyacrylamide gel electrophoresis

reducing sample buffer (3 mM NaCl, 70 mM Tris-HCl, 1 mM EDTA, 2% (w/v) SDS, 0.01 % (w/v) phenol red, 10% glycerol, pH 6.8 at 25°C, 40 mM dithiothreitol).

7. SDS-PAGE Tris glycine buffer: 0.1% sodium dodecyl sulfate (SDS), 0.19 M glycine, 0.025 M Tris, pH 8.3.

2.4. Electroblotting and Western Blotting

1. Transfer buffer: 0.15% ethanolamine, 25 mM glycine, and 20% methanol.
2. Coomassie blue: Brilliant Blue R (Sigma).
3. Anti-biotin horse radish peroxidase (HRP)-linked antibody (Cell Signaling Technology).
4. Anti-peptide antibody.
5. Filter paper, 3 MM (Whatman Inc., Piscataway, NJ).
6. Non-fat dry milk powder.
7. Anti-mouse IgG, horse radish peroxidase (HRP)-linked antibody (Cell Signaling Technology).
8. Anti-rabbit IgG, HRP-linked antibody (Cell Signaling Technology).
9. LumiGLO Reagent and Peroxide (Cell Signaling Technology).
10. Amersham Hyperfilm ECL (GE Healthcare Life Sciences, Chalfont St. Giles, UK).

2.5. Peptide Synthesis

1. PB1: Synthetic peptide with the sequence CTRSRHSSYP-NEYEEDEEMEEEL. This sequence was derived from mouse Bad protein (*see* **Section 3.2.1**).
2. Bio-P1: Commercially available synthetic peptide of sequence CDPEK(Bt)DS, Bt represents biotin (New England BioLabs) (*see* **Section 3.2.1**).
3. Flu-P1: Commercially available synthetic peptide of sequence CDPEK(Fl)DS, Fl represents fluorescein (New England BioLabs) (*see* **Section 3.2.1**).
4. CGC 549: A commercially available synthetic peptide of sequence CGC 549 where C549 represents a fluorophore covalently attached to the C-terminal cysteine (New England BioLabs). The maximum absorption and emission wavelengths of the fluorophore are 553 and 566 nm, respectively, in water (*see* **Section 3.2.2**).
5. CGC 649: A commercially available synthetic peptide of sequence CGC 649 where C649 represents a fluorophore with maximal absorbance at 649 nm covalently attached to the C-terminal cysteine (New England BioLabs). The maximum absorption and emission wavelengths of the flu-

orophore are 655 and 676 nm, respectively, in ethanol (*see* **Section 3.2.2**).

6. Lys(5/6–FAM)–OH (*N*-α-Fmoc-*N*-ε-(5/6-carboxy-fluorescein)-L-lysine) amino acid: Anaspec, San Jose, CA) (*see* **Section 3.2.1**).
7. Lys(Biotin–LC)–OH (*N*-α-Fmoc-*N*--(d-biotin-6-amido-caproate)-L-lysine) amino acid: (Anaspec) (*see* **Section 3.2.1**).
8. Acetonitrile (EMD Chemicals, Gibbstown, NJ).
9. TFA: Trifluoroacetic acid solution (American Bioanalytical).
10. Maleimide-activated Dyomics 549 (DY549) and Dyomics 649 (DY649) (Dyomics GmbH, Jena, Germany).

2.6. Equipment

1. Innova 4230 floor incubator/shaker (New Brunswick Scientific, Edison, NJ).
2. Beckman J2–21 centrifuge (Beckman, Ramsey, MN).
3. Cell disruptor (Active Ultrasonics, La Chaux-de-Fonds, Switzerland).
4. Centriprep YM-10 concentrator (Millipore, Billerica, MA).
5. HPLC system (Waters Corp, Milford, MA).
6. ABI 433A Peptide Synthesizer (Applied Biosystems, Foster City, CA).
7. Vydac semi-preparative RP-C18 column (W.R. Grace and Co., Columbia, MD).
8. Waters MALDI Micro MX (Waters Corp.): The MALDI micro MX™ is a high-performance, matrix-assisted laser desorption/ionization time-of-flight mass spectrometer (MALDI-TOF-MS).
9. 10–20% Novex Tris-Glycine Mini gel (Invitrogen).
10. XCell Surelock Mini-Cell (Invitrogen).
11. TE22 Mini Tank Transfer Unit (Hoefer Scientific Instruments, Harvard Biosciences, Horsham, PA).
12. PowerPac power supply (Bio-Rad, Hercules, CA).
13. Power Supply (VWR, West Chester, PA).
14. Typhoon Imager 9400 (GE Healthcare, Piscataway, NJ).

2.7. Reagent Setup

1. Dissolve the peptide, with the N-terminal cysteine, in water to a concentration of 1–10 mM. If the peptide is insoluble in water, dissolve it in dimethyl sulfoxide (DMSO, Sigma).

2. Column buffer: 20 mM Tris-HCl, pH 8.5 at 25°C, 500 mM NaCl.
3. Cleavage buffer: 50 mM 2-mercaptoethanesulfonic acid or 50 mM DTT in column buffer.
4. 10X IPL reaction buffer: 1 M Tris-HCl, 2.5 M NaCl, 100 mM MESNA, pH 8.5 at 25°C.
5. Coomassie blue staining solution: 0.1% Brilliant Blue R, 10% glacial acetic acid, 50% methanol.
6. Destain: 10% glacial acetic acid, 50% methanol.
7. Transfer buffer: 0.15% ethanolamine, 25 mM glycine, 20% methanol.
8. TBSTT: Tris-buffered saline Triton Tween, 20 mM Tris, pH 7.5 at 25°C, 150 mM NaCl, 0.2% Tween 20, 0.05% Triton X-100.
9. Blocking solution: 5% non-fat dry milk powder in TBSTT.

3. Methods

For IPL, a protein with a C-terminal thioester is generated by expressing the protein as an intein fusion (*see* **Note 1**). To this protein, a peptide or protein with a N-terminal cysteine can be ligated by simply mixing the two components at pH 8.5–9. The reaction can be carried out at room temperature or 4°C (*see* **Fig. 6.1**).

Carrier Protein 39 (CP39) and Carrier Protein 27 (CP27) target proteins were expressed as intein fusions and used as thioester-tagged proteins in the IPL reaction. CP39 is a 39 kDa recombinant protein from *Haemophilus haemolyticus* while CP27 is a 27 kDa recombinant protein from *Dirofilaria immitis*; both have low cross reactivity with mouse and rabbit sera (13).

CP39 and CP27 are expressed as target proteins as C-terminal intein fusions. The C-terminal thioester-tagged protein CP27 is purified after cleavage with a thiol reagent, either DTT or MESNA, and ligated to the PB1 or Bio-P1 peptides. The ligated products are detected by adding 3X SDS sample buffer, subjecting them to electrophoresis on an SDS-PAGE gel and staining with Coomassie blue. The efficiency of cleavage with the DTT and MESNA and the subsequent ligation efficiency is evaluated (*see* **Table 6.1**).

To fluorescently label proteins, CP39 is purified using MESNA as a cleavage reagent and subsequently used for IPL. Fluorescent peptides containing an N-terminal cysteine, such as

Flu-P1, CGC 549 or CGC 649, are ligated to the protein overnight at 4°C and then run on a gel. The fluorescent proteins are detected by fluorescent scanning of the gel or after blotting to a nitrocellulose membrane (*see* **Fig. 6.3**).

3.1. Expression and Purification of a Thioester-Tagged Protein

1. Clone a protein of interest into a suitable C-terminal IMPACT vector (for example, the pTXB or pTYB series) following standard molecular biology procedures and the instruction manual (2). For more information on the vectors please see the IMPACT manual and FAQs at http://www.neb.com/nebecomm/products/productE6901.asp.

   The C-terminal residue of the target protein, adjacent to the intein, plays an important role in cleavage and ligation (*see* **Note 2**). For the study of the effect on cleavage of the amino acid adjacent to the intein the target protein CP27 was used. The mutations were made with the Phusion™ Site-Directed Mutagenesis Kit following the manufacturer's instructions. The plasmid constructs were sequenced to verify the construct. For the labeling of a protein with a PB1 peptide, CP39 was used (*see* **Note 3**).

   The expression of the fusion protein may be affected by a variety of factors such as the *E. coli* host strain, induction temperature, cell density at induction, IPTG concentration, toxicity of the target protein, codon usage and mRNA stability.

2. Transform the plasmid bearing the target gene into *E. coli* T7 Express or another suitable *E. coli* expression strain that expresses T7 RNA polymerase.

3. Inoculate a freshly grown colony into LB medium containing 100 μg/mL ampicillin and grow the cells at 37°C. When the OD_{600} of the culture reaches 0.5–0.7, induce protein expression by adding IPTG to a final concentration of 0.3–0.5 mM. Remove an aliquot of the culture before induction and let it incubate under the same conditions as the induced cells; this will serve as an uninduced control. Incubate the induced cells at 15–37°C (*see* **Note 4**). For an 'uninduced' and 'induced' sample remove 40 μl of the culture and add 20 μl of 3X SDS sample buffer (*see* **Note 5**).

4. Pellet the cells by centrifuging at 5,000×*g* at 4°C for 20 min.

5. At 4°C equilibrate a chitin column (20–40 mL chitin slurry for 1 L culture) with 5 packed column volumes of water. Decant the water and then wash with 5 volumes of column buffer (*see* **Note 6**).

6. Break the induced cells by sonication in column buffer and slowly load the clarified lysate onto the chitin column.
7. Wash the column with at least 20 bed volumes of column buffer to thoroughly remove any unbound proteins.
8. Take a sample of the chitin beads by removing 200 μl of chitin resin and mixing with 100 μl of 3X SDS sample buffer; this is a 'before cleavage' sample. Quickly wash the column with 3 bed volumes of cleavage buffer [column buffer containing 50 mM DTT (for purification) or 50 mM MESNA (for IPL)].
9. Stop the flow and leave the column at 4–23°C for 16–40 h (*see* **Note 7**).
10. Elute the target protein by continuing the column flow with column buffer and collect fractions. The fractions size should be approximately 3–5 mL.
11. For the study of the effect of the –1 amino acid on cleavage and ligation after cleavage at 4°C for 16 h the protein was eluted and IPL was conducted by adding PB1, an N-terminal cysteine containing peptide (*see* **Section 3.3.3**). The samples were incubated at 4°C for 16 h and analyzed by Coomassie blue-stained SDS-PAGE. For samples where inadequate protein was eluted, IPL was conducted on the chitin column by adding peptide directly to the chitin resin.
12. Dialyze the target protein into an appropriate storage buffer for long-term storage (*see* **Note 8**).
13. To examine cleavage efficiency, remove 200 μl of chitin resin and mix with 100 μl of 3X SDS sample buffer. After boiling for 5 min, the supernatant is analyzed on SDS-PAGE to determine the cleavage efficiency (*see* **Note 9**).
14. To regenerate the chitin resin wash the column with 3 bed volumes of 0.3 M NaOH (stripping solution). Allow resin to soak 30 min and wash the resin with an additional 7 bed volumes of 0.3 M NaOH. Wash with 20 bed volumes of water, followed by 5 bed volumes of column buffer.

3.2. Synthesis of Peptides

3.2.1. Synthesis of Peptides PB1, Bio-P1, and Flu-P1

1. Peptides were synthesized using an ABI model 433A peptide synthesizer using FastMoc chemistry.
2. Lysines containing fluorescein (for Flu-P1) or biotin (for Bio-P1) modifications were coupled with an extended activation step on the synthesizer. The peptide Lys(5/6–FAM)–OH was used for Flu-P1 and Lys(Biotin–LC)–OH for Bio-P1.

3. Cleave/deprotect peptides from support using the appropriate cleavage cocktail.
4. Purify the peptides using standard HPLC conditions: a Vydac semi-preparative RP-C18 column using a 60 min linear gradient, 10–100% B at a flow rate of 2 mL/min. Buffer A: 0.1% TFA/H_2O (V/V) and buffer B: 0.1% TFA/60% acetonitrile/40%H_2O (V/V/V). Detection is conducted at 214 nm.

3.2.2. Synthesis of Peptides CGC 549 and CGC 649

Two peptides, CGC 549 and CGC 649, were synthesized using maleimide-activated DY-549 or DY-649. The N-terminal cysteine residue has an unmodified sulfhydryl group whereas the C-terminal cysteine residue possesses a sulfhydryl group linked to DY-549 or DY-649:

1. Peptides were synthesized using an ABI model 433A peptide synthesizer using FastMoc chemistry.
2. Synthesize tripeptide NH_2-cysteine-glycine-cysteine-$CONH_2$ with the sulfhydryl group of the amino-terminal cysteine residue protected with t-butylthio.
3. Cleavage/deprotection is carried out in only TFA/water to leave the N-term Cys(t-buthio) group intact.
4. Confirm mass by mass spectrometry.
5. Conjugate the dye (DY-549 or DY-649) to the sulfhydryl group of the carboxyl-terminal cysteine residue.
6. Purify the labeled peptides using standard HPLC conditions: A Vydac semi-preparative RP-C18 column using a 60 min linear gradient, 10–100% B at a flow rate of 2 mL/min. Buffer A: 0.1% TFA/H_2O (V/V) and buffer B: 0.1% TFA/60% acetonitrile/40%H_2O (V/V/V). Detection is conducted at 214 nm.
7. Confirm mass by mass spectrometry.
8. Lyophilize product.
9. Deprotect the sulfhydryl group of the amino-terminal cysteine residue using tris(hydroxypropyl) phosphine. The deprotection of Cys(tButhio) is carried out according to manufacturer's instructions (Novabiochem) (20, 21).
10. Purify the labeled peptides with free sulfhydryl group using standard HPLC conditions; a Vydac semi-preparative RP-C18 column using a 60 min linear gradient, 10–100% B at a flow rate of 2 mL/min. Buffer A: 0.1% TFA/H_2O (V/V) and buffer B: 0.1% TFA/60% acetonitrile/40%H_2O (V/V/V). Detection is conducted at 214 nm.
11. Confirm mass by mass spectrometry.

12. Determine the amount of peptide using the extinction coefficient and absorption spectra of the specific fluorophore used.
13. Prepare a 10 mM peptide stock and store at –20°C until use.

3.3. Intein-Mediated Protein Ligation (IPL)

3.3.1. Ligation of a Peptide to a Carrier Protein

1. Dissolve the peptide (containing an N-terminal cysteine) in water to a final concentration of 1 mM (*see* **Note 10**).
2. Thaw the carrier protein (CP) and 10X IPL reaction buffer at room temperature and place on ice.
3. Mix the water, buffer, carrier protein and peptide together. Include an unligated control reaction of CP protein (*see* **Note 11**).
4. Place at room temperature for 15 min to 1 h or overnight at 4°C (*see* **Note 12**).
5. Add 3X SDS-PAGE sample buffer and load onto an SDS-PAGE gel (*see* **Note 13**).

3.3.2. For Immunoblotting

1. Thaw at room temperature the carrier protein, peptide, and ligation buffer and place on ice.
2. To a microcentrifuge tube add 1 μL of carrier protein (0.2 mg/mL), 12.5 μL of 2X peptide stock (1–2 mM), 2.5 μL of 10X peptide ligation buffer and sterile water for a total volume of 25 μL. The peptide should be at a final concentration of 0.5–1 mM (*see* **Note 14**).
3. Incubate at room temperature for 15 min to 4 h or at 4°C overnight.
4. Add 12.5 μL of 3X SDS-PAGE sample buffer and store at –20°C.

3.3.3. For Coomassie Blue-Stained SDS–PAGE

1. Thaw at room temperature the carrier protein, peptide, and ligation buffer and place on ice.
2. To a microcentrifuge tube add 10 μL of carrier protein (0.2 mg/mL), 12.5 μL of 2X peptide stock (1–2 mM), 2.5 μL 10X peptide ligation buffer, and sterile water for a total volume of 25 μL.
3. Incubate at room temperature for 15 min to 1 h or 4°C overnight.
4. Add 12.5 μL of 3X SDS-PAGE sample buffer and store at –20°C.

3.3.4. For Detection of Fluorescently Labeled Proteins

1. Mix the thioester-tagged protein and dye-conjugated peptide (CGC 549, CGC 649, or Flu-P1) in the presence of

1X IPL buffer. Use the protein at a final concentration of 10 μM (or 0.4 mg/mL) and peptide at 25–200 μM.

2. Perform a control reaction in the absence of peptide.
3. Perform the reactions overnight at 4°C in the dark.
4. Add 3X SDS-PAGE sample buffer to a final concentration of 1X and store at –20°C.

3.4. SDS–PAGE

3.4.1. For Immunoblotting

1. Heat the samples for 5 min at 95–100°C.
2. Load 5–15 μL of ligation reaction/lane (~25–75 ng) on a 10–20% or 12% tris-glycine gel for immunoblotting (*see* **Note 15**).
3. Electrophorese samples by SDS-PAGE (*see* **Note 16**).
4. Place gel into transfer buffer before immunoblotting.

3.4.2. For Coomassie Blue Staining

1. Load 15 μL of ligation reaction/lane (~0.8 μg carrier protein) on a 10–20 or 12% tris-glycine gel (*see* **Note 17**).
2. Electrophorese samples by SDS-PAGE.
3. Stain with Coomassie blue staining solution for at least 30 min (*see* **Fig. 6.2**).

3.4.3. For Fluorescent Detection

1. If necessary, dilute each ligation sample in 1X SDS sample buffer to 10 ng/μL.
2. Heat the samples for 5 min at 95–100°C.
3. Boil each reaction sample for 5 min.
4. Apply approximately 100 ng of each sample to a SDS-PAGE gel (*see* **Note 18**).
5. Following electrophoretic separation the gel is scanned by a fluorescence imager with appropriate laser and filter settings. Alternatively the proteins can be electroblotted onto a membrane and the membrane scanned by a fluorescent imager (*see* **Fig. 6.3**).

3.5. Immunoblotting

1. Wearing gloves, cut two pieces of filter paper and one piece of nitrocellulose slightly larger than the gel (*see* **Note 19**).
2. Soak the filter paper in a solution of transfer buffer (*see* **Note 20**).
3. Soak the nitrocellulose membrane in water for 1 min and then in transfer buffer. Where appropriate treat the nitrocellulose membrane according to the manufacturer's instructions.
4. Set up the immunoblot transfer with the gel toward the negative electrode so that the proteins are transferred to the nitrocellulose near the positive electrode.

5. On the transfer cassette, which will lie near the positive electrode, place a porous pad or sponge. On it place a sheet of filter paper and then nitrocellulose.
6. On the nitrocellulose membrane place the gel, filter paper, and then the porous pad or sponge (*see* **Note 21**).
7. Transfer in the cold (4°C), stirring the solution at 225 mA for 2 h (*see* **Note 22**).
8. After transfer, open the cassette carefully. Lightly mark the nitrocellulose with a ball-point pen or pencil outlining the dimensions of the gel and the wells.
9. Remove the gel and mark the positions of the prestained molecular weight marker bands.
10. Incubate the membrane in 50–100 mL of blocking solution for 1 h at room temperature (*see* **Note 23**).
11. Wash once with wash buffer.
12. Dilute the antibody in a TBSTT solution containing 2% non-fat dry milk according to the manufacturer's recommended concentration.
13. Incubate the membrane in the primary antibody overnight at 4°C.
14. Wash the membrane three times for 10 min each with TBSTT.
15. Incubate the membrane with HRP-conjugated secondary antibody for 45 min at room temperature. The secondary antibody should be diluted in a 2% non-fat dried milk in TBSTT (*see* **Note 24**).
16. Wash the membrane three times for 10 min each with TBSTT.
17. Use a chemiluminescent substrate to detect the signal.
18. Expose X-ray film to visualize the signal (*see* **Note 25**) (*see* **Fig. 6.4**).

3.6. Detection of Fluorescent Proteins

1. For in-gel detection by a fluorescent imager, subject 100–250 ng of the carrier protein ligation reaction to electrophoretic separation by SDS-PAGE. Alternatively, the protein bands can be electroblotted onto a 0.45 μm nitrocellulose membrane. Store the moist gel or blot in the dark.
2. Scan the gel or blot with a fluorescent scanner, such as the Typhoon Imager 9400, following the manufacturer's instructions. Position the blot with the protein-side down onto the clean glass platen.
3. Select the instrument settings for imaging fluorescent proteins in gels or blots (*see* **Fig. 6.3**). Focal plane: platen; PMT

voltage: 300 V; pixel size: 100 μm. Choose the filter settings for the following fluorescent substrates.

Flu-P1: 488/526 nm, excitation/emission filter set; CGC 549: 532/580 nm excitation/emission filter set; CGC 649: 633/670 nm excitation/emission filter set.

4. Notes

1. For cloning of the gene of interest in to the IMPACT vector please refer to the NEB web site: http://www.neb.com/nebecomm/tech_reference/protein_expression_purification/impact_vectors.asp
2. For IPL the intein tag should be at the C-terminus of the target protein. It is possible to add extra amino acids to the target protein's C-terminus, if necessary, to ensure efficient cleavage. The cleavage is influenced by the amino acid residues at the C-terminus as well as the folding of the entire protein, so it is difficult to predict if a certain fusion protein will spontaneously cleave in vivo; different inteins and induction conditions can be tried to optimize expression and cleavage.
3. We have tested two carrier proteins for IPL, CP27 is a 27 kDa carrier protein, whereas the molecular mass of CP39 is 39 kDa; both have a reactive C-terminal thioester. Ligation of CP27 to a peptide (for example, small peptides of 10 amino acids) can be readily detected by a 12% Coomassie blue-stained SDS-PAGE. The ligated CP39-peptide product can be detected by Western blot using an anti-peptide antibody; however, its detection by Coomassie staining of an SDS-PAGE is difficult when the size of the peptide is smaller than 1.5 kDa. For production of positive controls for Western blot analysis the size of the carrier protein is not important unless you wish to create a different size fusion protein or if you need to determine the ligation efficiency for a small peptide (<10 amino acids). The mobility shift of CP27 after ligation to a small peptide is more easily detected on a 12% Coomassie blue-stained SDS-PAGE, as compared to the mobility shift with CP39. Additionally, it has been observed that in some cases, CP27 is a preferred carrier protein for generating an optimal peptide substrate for kinase assay, arrays, etc. (22). The optimal carrier protein for a particular assay may be determined by testing the different carrier protein–peptide substrates.
4. To optimize expression and solubility for the fusion protein different induction conditions can be tried, such as

15°C overnight, 30°C for 4–6 h or 37°C for 2–4 h, as well as varying IPTG concentrations (from 0.05 to 0.5 mM IPTG). If the protein is folded incorrectly it maybe necessary to try using another intein and/or change from an N- to a C-terminal fusion or vice versa.

5. If the fusion protein is not expressed there are many possible reasons. The DNA should be sequenced to ensure that the reading frame is correct. We also recommend the use of a fresh colony from an overnight transformation. The control plasmid pMYB5 or pMXB10 (New England BioLabs) should be used as an expression control. If the cell culture is overgrown before induction (OD_{600} is over 0.7) it is possible that the protein will not be overexpressed.
6. If some fusion protein is detected in the flow-through fraction it could be that not enough chitin resin was used; the binding capacity of the resin is approximately 2 mg/mL. The cell extract may be very concentrated; we recommend resuspending the pellet from 1 L of culture in at least 100 mL of column buffer and loading slowly.
7. For the study of the effect of the −1 amino acid on cleavage and ligation 40 mM thiol reagent was used and the column was left to incubate at 4 or 23°C for 16 or 40 h.
8. To store the thioester-tagged protein, dialyze into the buffer 5 mM Bis-tris, pH 6, 0.25 M NaCl. Store at –80°C and avoid repeated freeze–thaw cycles. Greater than 50% ligation is observed when the carrier protein (CP) is stored at –80°C in the storage buffer for greater than 12 months.
9. The comparison of the before cleavage chitin bead sample and the after cleavage chitin bead sample allows for an estimation of cleavage efficiency with the thiol reagent by comparing the amounts of the fusion protein and the target protein.
10. Ligation efficiency is optimal when the final peptide concentration is 0.5–1 mM. This peptide concentration will help drive the IPL reaction to completion.
11. The IPL reaction is optimal at pH 8.5–9. The reaction is conducted with the protein concentration of 0.01 mM or higher and the peptide concentration between 0.5 and 1 mM. As a control reaction protein alone, without any peptide, should be included. A peptide without an N-terminal cysteine can be included as a control for IPL.

12. For maximum ligation we have found 4°C overnight or 23°C for 4 h to be optimal.
13. The causes of inefficient ligation could be the following:
    (A) Peptide does not possess an N-terminal cysteine or the sulfhydryl group is oxidized.
    (B) The peptide solution may be very acidic and cause the pH of the reaction to drop significantly. Check the pH of your peptide solution. If the pH is below 6 dissolve the peptide in 1 M Tris (pH 9.0).
    (C) Peptide preparation contains impurities. Purify the peptide. The purity of the peptide effects the extent of ligation. We have found ligation to work with both crude and purified peptides. However, more consistent results are obtained when peptide purity is above 30%. To ensure the greatest amount of ligated product we recommend using a purified peptide.
    (D) Concentration of peptide is incorrect. If the molecular mass of the peptide is 2862 Da then the calculation is as follows:
    1 M = *Mr* g/l
    1 M = 2862 g/l
    1 mM = 2.862 mg/mL
    (E) Peptide or ligation product is insoluble. If a peptide is not soluble, you may first dissolve the peptide in dimethyl sulfoxide (DMSO) and then add water to make a 2–10 mM stock solution. The peptide solution should be stored at –80°C.
    (F) The carrier protein has lost its ligation capability due to repeated freeze–thaw cycles or long-term storage at –20°C. Repurify the target protein and dialyze it into the storage buffer, if it is not going to be ligated immediately.
14. Carrier proteins and IPL can be used in array analysis but should be used at a higher concentration of 1 mg/mL which is necessary for increased sensitivity in array analysis (9). The use of a lower concentration of carrier protein in array analysis resulted in decreased signal.
15. In Western blot analysis, even a small amount of ligated product will give a positive signal; the detection limit primarily depends on the quality of antibody (27 ng of CP39 was loaded per well) (Ghosh, I., unpublished observations).
16. For a Western blot a number of controls can be carried out. Run a prestained protein marker to monitor elec-

trophoresis and transfer on to the nitrocellulose membrane and run a biotinylated marker as a positive control for the immunoblotting and detection.

17. Run a protein marker to monitor electrophoresis and staining.
18. Run a fluorescent marker, if necessary, and a prestained protein marker to monitor electrophoresis and transfer onto the nitrocellulose.
19. Wear gloves, especially when handling nitrocellulose. Use forceps to handle the membranes.
20. All components of the blot should be kept moist at all times. The transfer buffer should be at 4°C.
21. Make sure there are no bubbles between the gel and the membrane. Squeeze out the air bubbles using a pencil or glass pipette as a roller.
22. The transfer buffer should not be used for more than five to ten transfers.
23. All incubations should be done with gentle agitation.
24. Fluorescently labeled secondary antibodies can also be used to detect signal.
25. The chemiluminescent signal can be captured digitally with an instrument such as a FujiFilm LAS 1000-plus.

Acknowledgments

The authors wish to thank New England Biolabs and Donald. G. Comb, Jim Ellard, Richard Roberts, and Christopher Noren for their support and suggestions. We thank the Organic Division for peptide synthesis.

References

1. Rai, M., Padh, H. (2001) Expression systems for production of heterologous proteins. *Curr Sci* 80, 1121–1128.
2. Xu, M. Q., Evans, T. C., Jr. (2003) Purification of recombinant proteins from *E. coli* by engineered inteins. *Methods Mol Biol* 205, 43–68.
3. Chong, S., Mersha, F. B., Comb, D. G., Scott, M. E., Landry, D., Vence, L. M., Perler, F. B., Benner, J., Kucera, R. B., Hirvonen, C. A., Pelletier, J. J., Paulus, H., Xu, M. Q. (1997) Single-column purification of free recombinant proteins using a self-cleavable affinity tag derived from a protein splicing element. *Gene* 192, 271–281.
4. Evans, T. C., Jr., Benner, J., Xu, M. Q. (1998) Semisynthesis of cytotoxic proteins using a modified protein splicing element. *Protein Sci* 7, 2256–2264.
5. Muir, T. W., Sondhi, D., Cole, P. A. (1998) Expressed protein ligation: a general method for protein engineering. *Proc Natl Acad Sci USA* 95, 6705–6710.
6. Xu, M. Q., Evans, T. C., Jr. (2005) Recent advances in protein splicing: manipulating

proteins in vitro and in vivo. *Curr Opin Biotechnol* 16, 440–446.

7. Southworth, M. W., Amaya, K., Evans, T. C., Xu, M. Q., Perler, F. B. (1999) Purification of proteins fused to either the amino or carboxy terminus of the *Mycobacterium xenopi* gyrase A intein. *Biotechniques* 27, 110–114, 116, 118–120.
8. Hackeng, T. M., Griffin, J. H., Dawson, P. E. (1999) Protein synthesis by native chemical ligation: expanded scope by using straightforward methodology. *Proc Natl Acad Sci USA* 96, 10068–10073.
9. Sun, L., Rush, J., Ghosh, I., Maunus, J. R., Xu, M. Q. (2004) Producing peptide arrays for epitope mapping by intein-mediated protein ligation. *Biotechniques* 37, 430–436.
10. Kochinyan, S., Sun, L., Ghosh, I., Barshevsky, T., Xu, J., Xu, M. Q. (2007) Use of intein-mediated phosphoprotein arrays to study substrate specificity of protein phosphatases. *Biotechniques* 42, 63–69.
11. Burbulis, I., Yamaguchi, K., Gordon, A., Carlson, R., Brent, R. (2005) Using protein-DNA chimeras to detect and count small numbers of molecules. *Nat Methods* 2, 31–37.
12. Camarero, J. A., Muir, T. W. (1999) Biosynthesis of a head-to-tail cyclized protein with improved biological activity. *JACS* 121, 5597–5598.
13. Ghosh, I., Sun, L., Evans, T. C., Jr., Xu, M. Q. (2004) An improved method for utilization of peptide substrates for antibody characterization and enzymatic assays. *J Immunol Methods* 293, 85–95.
14. Girish, A., Sun, H., Yeo, D. S., Chen, G. Y., Chua, T. K., Yao, S. Q. (2005) Site-specific immobilization of proteins in a microarray using intein-mediated protein splicing. *Bioorg Med Chem Lett* 15, 2447–2451.
15. Rak, A., Pylypenko, O., Niculae, A., Pyatkov, K., Goody, R. S., Alexandrov, K. (2004) Structure of the Rab7:REP-1 complex: insights into the mechanism of Rab prenylation and choroideremia disease. *Cell* 117, 749–760.
16. Seyedsayamdost, M. R., Yee, C. S., Stubbe, J. (2007) Site-specific incorporation of fluorotyrosines into the R2 subunit of *E. coli* ribonucleotide reductase by expressed protein ligation. *Nat Protocol* 2, 1225–1235.
17. Skrisovska, L., Allain, F. H. (2008) Improved segmental isotope labeling methods for the NMR study of multidomain or large proteins: application to the RRMs of Npl3p and hnRNP L. *J Mol Biol* 375, 151–164.
18. Sun, L., Ghosh, I., Barshevsky, T., Kochinyan, S., Xu, M. Q. (2007) Design, preparation and use of ligated phosphoproteins: a novel approach to study protein phosphatases by dot blot array, ELISA and Western blot assays. *Methods* 42, 220–226.
19. Sun, L., Ghosh, I., Xu, M. Q. (2003) Generation of an affinity column for antibody purification by intein-mediated protein ligation. *J Immunol Methods* 282, 45–52.
20. Beekman, N. J. C. M., Schaaper, W. M. M., Tesser, G. I., Dalsgaard, K., Kamstrup, S., Langeveld, J. P. M., Boshuizen, R. S., Meloen, R.H. (1997) Synthetic peptide vaccines: palmitoylation of peptide antigens by a thioester bond increases immunogenicity. *J Peptide Res* 50, 357–364.
21. Ruegg, U. T., Gattner, H. G. (1975) Reduction of S-sulpho groups by tributylphosphine: an improved method for the recombination of insulin chains. *Hoppe Seylers Z Physiol Chem* 356, 1527–1533.
22. Xu, J., Sun, L., Ghosh, I., Xu, M. Q. (2004) Western blot analysis of Src kinase assays using peptide substrates ligated to a carrier protein. *Biotechniques* 36, 976–978, 980–981.

Chapter 7

Efficient Expression of Human Aromatase (CYP19) in *E. coli*

Norio Kagawa

Abstract

Human aromatase (CYP19, P450arom) is responsible for the conversion of androgens to estrogens. In addition to the estrogen biosynthesis in gonads and adrenals in a classical endocrine manner, this enzyme is widely expressed in various tissues and locally regulates the estrogen level or estrogen/androgen ratio in an intracrine manner. Since estrogen biosynthesis is involved in many essential biological events in humans, aromatase is an attractive target for investigations in basic biomedical and pharmacological sciences. Aromatase is a membrane protein localized on the endoplasmic reticulum, and its instability and hydrophobic nature has hampered the investigation of this important biological system. To investigate the structure–function relationship of human aromatase by obtaining quantities of the purified enzyme, we have developed the expression system of human aromatase in *E. coli*. The human aromatase has two major forms from its genetic polymorphism, arginine (264R) and cysteine (264C) at the amino acid position 264. Although there is only one amino acid difference between the two forms of aromatase, the 264C form was expressed in *E. coli* with the cold stress response induced by chloramphenicol while the 264R form was expressed by the coexpression of heat shock proteins GroES/GroEL. In the case of human aromatase, a clear protocol is important to determine the expression levels by the reduced CO-difference spectrum. Therefore, the expression methods for the two forms of human aromatase as well as a method for the determination of the reduced CO-difference spectrum will be described.

Key words: Aromatase, CYP19, P450, cold stress response, heat shock protein, chloramphenicol, GroES/GroEL, molecular chaperone.

1. Introduction

Aromatase, a member of the cytochrome P450 superfamily, catalyzes the conversion of androgens to estrogens through three consecutive reactions using three molecular oxygens and the equivalent electrons from NADPH via an NADPH-dependent cytochrome P450 reductase. Estrogens play essential roles in

T.C. Evans, M.-Q. Xu (eds.), *Heterologous Gene Expression in E. coli*, Methods in Molecular Biology 705, DOI 10.1007/978-1-61737-967-3_7,

many biological events as well as in the growth of cancer cells. As known in classical endocrinology, estrogens are mostly produced in gonads and adrenals by aromatase and delivered through the blood to organs and tissues in the body where estrogens are required for the development and/or maintenance of normal organ function. In mice, the number of alveoli decreased in the lungs of ovariectomized female mice and it was recovered by the administration of estrogen (1), suggesting that estrogens produced in ovaries may be essential for the maintenance of lungs. On the other hand, aromatase is also expressed in normal lungs, which suggests the capacity of lungs to produce estrogens for the maintenance of lung function in an intracrine manner (2). In addition to the regulation of normal biological events in the body, production of estrogens mediated by aromatase expressed in adipose tissues in the breast is associated with the development of breast cancer. Since over 70% of breast cancers are estrogen dependent, aromatase and estrogen receptors are key therapeutic targets for the treatment of breast cancer. Aromatase is a membrane protein located on the endoplasmic reticulum through an N-terminal membrane anchor. Since human aromatase is abundantly expressed in placenta, the purification of the enzyme from placenta was intensively investigated in the 1980s (3–5). However, instability and the hydrophobic nature of aromatase hampered the purification of the enzyme in quantities sufficient for in-depth analyses.

Because of the increasing interest in aromatase, we have developed a heterologous expression system of the human aromatase in *E. coli* to obtain enough of the purified enzyme for biochemical characterization. The human aromatase is a hemeprotein consisting of 503 amino acids. There are two major forms of human aromatase derived from genetic polymorphism, 264C and 264R forms. Both forms have been expressed by the truncation of the N-terminal membrane anchor and the addition of a histidine tag at the C-terminus (6). Although there is only one amino acid difference in 503 amino acid residues between the two forms, the human aromatase 264C was optimally expressed by inducing the cold stress response using chloramphenicol while the 264R form requires the coexpression of heat shock molecular chaperones GroES/GroEL for the efficient expression in *E. coli*.

2. Materials

2.1. *E. coli* Cell Culture

1. Chemically competent BL21(DE3) cells.
2. Expression vector pET17b (Novagen, Darmstadt, Germany).

3. GroES/GroEL coexpression vectors, pGro12 and pGro7 (Takara Bio Inc., Otsu, Japan)
4. Terrific Broth (1 l): 12 g tryptone (tryptone, pancreatic, without sulfonamide antagonists, powder (VWR, West Chester, PA, USA)), 24 g Bacto™ yeast extract technical (Becton Dickinson, Franklin Lakes, NJ, USA), and 4 ml glycerol are dissolved in 900 ml distilled (or Millipore-filtrated) water and autoclaved at 121°C for 20 min. 10X phosphate buffer (1 l) is made by dissolving 23.1 g KH_2PO_4 (anhydrous) and 125.4 g K_2HPO_4 and is separately autoclaved. One liter Terrific broth (TB) medium can be made by mixing 900 ml of the media and 100 ml 10X phosphate buffer before use (*see* **Note 1**).
5. Fernbach culture flasks (2.8 l): Pyrex® 2800 ml Fernbach-style culture flask (product #4420-2XL) can be purchased from Fisher Scientific, Pittsburgh, PA, USA. When the Pyrex flask is not available, the Fernbach flask from Nalgene (2.8 l polycarbonate, Sigma-Aldrich, St. Louis, MO, USA) is also useful.
6. Culture incubator with refrigeration: Innova 43R or 44R, New Brunswick Scientific, Edison, NJ, USA
7. Ampicillin stock solution (20 mg/ml): 1 g sodium ampicillin (Sigma, St. Louis, MO, USA) is dissolved in 50 ml autoclaved water and stored at –20°C.
8. Kanamycin stock solution (25 mg/ml): 1.25 g kanamycin (Sigma, St. Louis, MO, USA) is dissolved in 50 ml autoclaved water and stored at –20°C.
9. Chloramphenicol stock solution (1 mg/ml in water): 50 mg chloramphenicol (Sigma, St. Louis, MO, USA) is dissolved in 50 ml autoclaved water and stored at –20°C.
10. δ-Aminolevulinic acid: δ-aminolevulinic acid (Fluka Chemie AG, Buchs, Switzerland) is dissolved at 1 M in autoclaved water (10 ml) and stored at –20°C.
11. L-(+)-Arabinose: L-(+)-arabinose (Sigma, St. Louis, MO, USA) is dissolved at 0.2 g/ml in autoclaved water (100 ml), sterilized by filtration (0.2 μm), and stored at 4°C.
12. IPTG: Isopropyl β-D-1-thiogalactopyranoside (Sigma, St. Louis, MO, USA) is dissolved at 100 mM in autoclaved water (50 ml) and stored at –20°C.

2.2. Cell Lysis and Protein Extraction

1. Lysozyme buffer: 50 mM Tris–HCl pH 7.2, 250 mM sucrose, 0.5 mM EDTA.
2. Extraction buffer: 50 mM potassium phosphate, pH 7.4, 500 mM sodium acetate, 0.1 mM EDTA, 0.1 mM DTT,

20 % glycerol, 1.5% sodium cholate, 1.5% Tween 20, and 100 μM PMSF.

2.3. Reduced CO-Difference Spectra

1. Sodium dithionite (sodium hydrosulfite) is a powder and needs to be stored in a desiccator.
2. CO gas: The smallest CO tank should be placed in the ventilation hood, and CO bubbling should be carried out in the hood for safety.
3. 2-Hydroxypropyl-β-cyclodextrin (Sigma, St. Louis, MO, USA) is dissolved at 45% in autoclaved water (1 ml) and stored at −20°C.
4. 19-Norandrostenedione is dissolved at 50 mM in 45% 2-hydroxypropyl-β-cyclodextrin and stored at −20°C.
5. DL-Dithiothreitol is dissolved at 1 M in water and stored at −20°C.
6. Sample dilution buffer: 50 mM potassium phosphate (pH 7.4), 20% glycerol, 0.1 mM EDTA, 500 mM sodium acetate, 1% sodium cholate, 1% Tween 20, 0.1 μM PMSF.

3. Methods

3.1. Plasmid Design for Expression of Human Aromatase

In this section, we describe several key points that are required for the plasmid design to achieve an efficient expression of human aromatase in *E. coli*. As shown in **Fig. 7.1**, the amino-terminus of microsomal P450s consists of three regions, membrane anchor, basic region, and proline-conserved region followed by the core catalytic domain. In most cases, P450s can be expressed in *E. coli* either with or without the membrane anchor although the membrane anchor truncated forms are usually expressed at higher levels than the full-length forms (7). The truncation of membrane anchor was required for the expression of human aromatase.

After the truncation of membrane anchor, several P450s have been expressed in *E. coli* with the replacement of the basic region by the corresponding region from other P450s that are well expressed in *E. coli*. Although either the basic region from the bovine CYP17 or that from the rat CYP2C11 was useful, the basic region from the rat CYP2C11 stably produced the human aromatase in *E. coli* (8). Therefore, the basic region of rat CYP2C11 was used for the expression of human aromatase as shown in **Fig. 7.2**.

In the *E. coli* expression system, the second codon preference has been known (9). Although AAA (Lys) is the most preferable second codon, GCT (Ala) was most frequently used as a sec-

A)

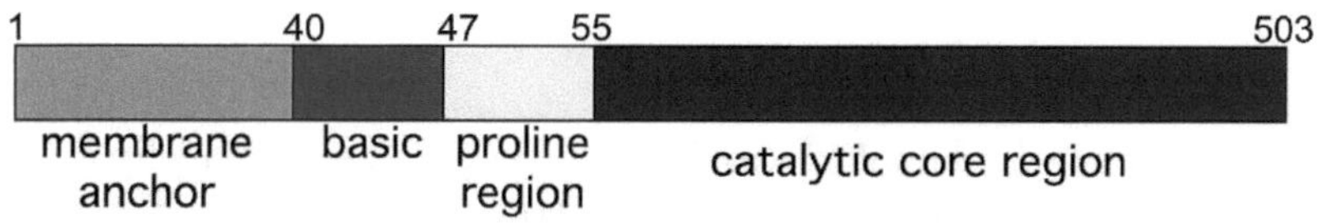

B)

	Membrane Anchor	B-Region	P-Region	
CYP1A1	MLFPISMSATEFLLASVIFCLVFWVI	RASRPQVPKGLKN	PPGPWGWP	LIG
CYP1A2	MALSQSVPFSATELLLASAIFCLVFWVL	KGLRPRVPKGLKS	PPEPWGWP	LLG
CYP1B1	MGTSLSPNDPWPLNPLSIQQTTLLLLLSVLATVHVGQ	RLLRQRRRQLRSA	PPGPFAWP	LIG
CYP2A6	MLASGMLLVALLVCLTVMVLMSVW	QQRKSKGKL	PPGPTPLP	FIG
CYP2B6	MELSVLLFLALLTGLLLLLV	QRHPNTHDRL	PPGPRPLP	LLG
CYP2C8	MEPFVVLVLCLSFMLLFSLW	RQSCRRRKL	PPGPTPLP	IIG
CYP2D6	MGLEALVPLAVIVAIFLLLVDLMH	RRQRWAARY	PPGPLPLP	GLG
CYP2E1	MFALGVTVALLVWAAFLLLVSMW	RQVHSSWNL	PPGPFPLP	IIG
CYP2F1	MDSISTAILLLLLALVCLLLTL	SSRDKGKL	PPGPRPLS	ILG
CYP3A3	MALIPDLAMETWLLLAVSLVLLYLYG	THSHGLFKKLG	IPGPTPLP	FLG
CYP4A11	MSVSVLSPSRLLGDVSGILQAASLLILLLLLIKAVQLYL	HRQWLLKALQQF	PCPPSHWL	FGH
CYP4F2	MSQLSLSWLGLWPVAASPWLLLLLVGASWLLAHVLAWTYAF	YDNCRRLRCF	PQPPRRNW	FWG
CYP5A1	MEALGFLKLEVNGPMVTVALSVALLALLKWYSTSAF	SRLFKLGLRH	PKPSPFIG	NLT
CYP17	MWELVALLLLTLAYLFWP	KRRCPGAKY	PKSLLSLP	LVG
CYP21	MLLLGLLLLPLLAGARLLWNWW	KLRSLHL	PPLAPGFL	HLL
CYP19	MVLEMLNPIHYNITSIVPEAMPAATMSVLLLTGLFLLLW	NYEGTSS	IPGPGYCM	GIG

Fig. 7.1. Basic structure of P450 and aromatase. (**a**) Schematic structure of human aromatase. Human aromatase anchors on the endoplasmic reticulum with the membrane anchor domain consisting of 39 amino acids, followed by the basic region (amino acid at 40–46) and the proline-conserved region (amino acid at 47–54). This characteristic N-terminal sequence is followed by core catalytic domain of the human aromatase. (**b**) Amino terminal sequences and conserved structure of human microsomal P450s. Human microsomal P450s have three characteristic domains at their N-termini. The membrane anchor domain inserted into the endoplasmic reticulum is followed by the basic region (B-region) that is rich in basic amino acids. After B-region, an 8-amino acid domain having conserved prolines (P-region) is present and this domain is followed by the core catalytic sequence.

ond codon for the construction of mammalian P450 expression vectors. Therefore, we used GCT (Ala) as a second codon for the construction of aromatase expression plasmid. Therefore, the N-terminus of the aromatase expression plasmid was designed to start the translational initiation codon ATG (Met), GCT (Ala), followed by the basic region from rat CYP2C11 as shown in **Fig. 7.2**.

To facilitate the purification of aromatase, the sequence encoding the 4×His tag (HisHisHisHis) was inserted right upstream of the translational termination signal at the C-terminus since the C-terminal His tag is frequently used for the expression of P450s. Although 4×His tag was used in the original paper, we now use 6×His tag because 6×His tag is more efficient for

```
                     20                         40     47
P450arom          AMPAATMSVLLLTGLFLLLWNYEGTSSIPGP
2cA                                      MARQSFGRGKLIPGP
CYP2C11        MDPVLVLVLTLSSLLLLSLWRQSFGRGKLPPGP
```

Fig. 7.2. N-terminal modification used for the expression of human aromatase in *E. coli*. For efficient expression, the N-terminal sequence of human aromatase up to the position 46 Ser is replaced with RQSFGRGKL, the basic region from the rat 2C11. The translation initiation codon ATG (Met) and a preferred second codon for the *E. coli* expression, GCT (Ala), were added at the N-terminus, producing 2cA constructs.

purification. In addition to the facilitation of purification, the C-terminal His tag may have a protective effect on the proteins during the expression (10).

3.2. Transformation of BL21 and Small-Scale Expression for the Selection of Colony

1. The *E. coli* BL21(DE3) cells are cotransformed with 2cAr-pET and pGro12 (or the 2cAcpET for the expression of 264C form), and transformants are screened on a LB plate supplemented with 100 μg/ml ampicillin (Amp) and 50 μg/ml kanamycin (Kan) (or a LB plate supplemented with 100 μg/ml ampicillin (Amp) for the 264C form).
2. Several isolated colonies (4–5 colonies) are picked up, inoculated in 1 ml TB medium with 100 μg/ml Amp and 50 μg/ml Kan (or in 1 ml TB medium with 100 μg/ml Amp for the 264C form), and incubated overnight at 37°C.
3. The overnight cultures (0.2 ml each) are mixed with 0.2 ml autoclaved 60% glycerol and stored at –80°C. The cultures are also diluted into 25 ml TB media with 100 μg/ml Amp and 50 μg/ml Kan (or into 25 ml TB media with 100 μg/ml Amp for the 264C form) in 125 ml Erlenmeyer flasks and the expression level of aromatase of each colony is examined in the small-scale experiment. (Since the expression method is similar to the large-scale expression, please refer the large-scale expression.)
4. The frozen stock of the colony showing the highest expression is saved, and other frozen stocks should be discarded. Overnight culture for the large-scale expression is always inoculated from this frozen stock for reproducibility. It works at least several years when it is always frozen in liquid N_2.

3.3. Coexpression of Heat Shock Molecular Chaperone GroES and GroEL for the Expression of Human Aromatase 264R

E. coli cells respond to a higher temperature and produce a series of heat shock proteins. Some of the heat shock proteins are molecular chaperones involved in the protein folding. These molecular chaperones facilitate to fold proteins that are not properly folded at a higher temperature. GroES and GroEL are well-investigated molecular chaperones and the major components of the protein folding machinery. The coexpression of GroES/GroEL is often effective for the *E. coli* expression of foreign proteins (11–14) and also essential to the expression of aromatase 264R form (6, 15).

In the expression experiment using a given flask and shaker incubator, time course of pH in the culture medium is critical for the efficient expression of foreign proteins. The time course of pH is affected by providers of tryptone and yeast extract, induction timing, concentration of inducers, and culture conditions (shaking speed, temperature, shape of flask, and volume of the medium, etc.). As shown in **Fig. 7.3a**, pH of the TB medium goes down to around 6.0 in the first 10 h after induction and

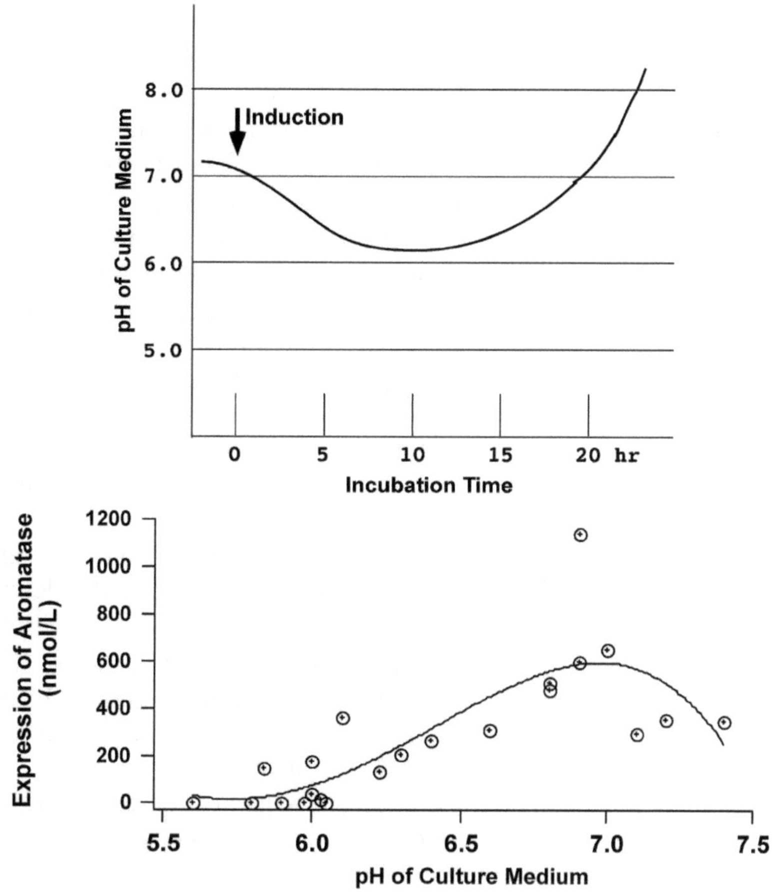

Fig. 7.3. Time course of medium pH and expression of aromatase during the expression culture. (**a**) The time course of pH of TB medium during the expression culture is schematically presented. In preincubation of a large-scale expression culture at 37°C for 3–4 h, pH of TB medium slightly decreases to approximately 7.2. After the addition of inducers, pH gradually goes down to around 6.0 and goes back to 7.0 and quickly increases to a higher pH. (**b**) During the large-scale (250 ml) expression, 10 ml culture media were transferred into test tubes. The expression level of aromatase and pH of each sample were determined. The relationship between the expression of aromatase and pH clearly demonstrates that the culture should be harvested at pH 7.0 for the best expression of aromatase. (The data were obtained from seven independent experiments.)

again goes up above 7.0. Although a detectable amount of aromatase is expressed in the first 10 h after induction, it disappears at lower pH. The expression level of aromatase again increases along with the pH increase, reaches to a peak at pH 7.0, and the accumulated aromatase quickly disappears when the pH goes higher than 7.0 (**Fig. 7.3b**).

1. 1 μl of the frozen stock is inoculated in 5 ml TB supplemented with 100 μg/ml ampicillin and 40 μg/ml kanamycin in a 50 ml conical tube. The culture is

incubated at 37°C overnight. (The culture may be started at 6:00–7:00 PM of day 1.)

2. 4 ml overnight culture is diluted in 400 ml TB (pH 7.2) with 100 μg/ml ampicillin and 40 μg/ml kanamycin in a 2.8 l culture flask (*see* **Note 2**).
3. The culture is incubated at 37°C for 3.5 h with a vigorous shaking (at 240 rpm, Innova 44R) (The incubation may be started at 2:30–3:00 PM of day 2.) (*see* **Notes 2** and **3**).
4. The culture is added with 0.5 mM IPTG for the transcriptional induction of aromatase, 50 μg/ml Amp, 4 mg/ml arabinose for the induction of GroES and GroEL, and 1 mM δ-ALA (a precursor of heme biosynthesis) and further incubated at 27–28°C for 14.5 h with a reduced shaking (190 rpm) (This will be at 6:00–6:30 PM of day 2.) (*see* **Note 4**).
5. A sample (8 ml) of the culture is taken from the flask, spun down, and pH of the media determined (This will be at 8:00–9:00 AM of day 3.) (*see* **Note 5**).
6. When pH reaches 7.0, cells are harvested by a low-speed centrifugation at 4000 rpm × 15 min (tabletop centrifuge Allegra X-15R, Beckman Coulter, Fullerton, CA, USA) at 4°C (*see* **Note 6**).
7. *E. coli* cell pellet is suspended in 40 ml lysozyme buffer with 0.5 mg/ml lysozyme, left on ice for 10 min, spun, and the supernatant is discarded. (After lysozyme treatment, cell pellets can be frozen in liquid N_2 and stored at –80°C.)
8. The cell pellet is carefully sonicated in 30 ml extraction buffer at 4°C without formation of bubbles (*see* **Note 7**).
9. The sonicated solution is fractionated by ultracentrifugation (Beckman TL100; 95,000 rpm × 10 min at 4°C). The supernatant is pooled, frozen in liquid N_2, and stored at –80°C. (The supernatant pool will be approximately 40 ml at this step for a 400 ml culture.)
10. The supernatant (100 μl) is diluted with 5 volumes (500 μl) of the sample dilution buffer for the determination of expression level of human aromatase by the reduced CO-difference spectrum.

3.4. Cold Stress Response Induced by Chloramphenicol for the Expression of Human Aromatase 264C

In response to a lowered temperature, *E. coli* cells repress a series of heat shock proteins and produce cold shock proteins. These proteins help the cell to adapt and survive at low temperatures. CspA is the major cold shock protein of *E. coli*. There is a family of nine CspA homologues in *E. coli* although only some of them are cold shock inducible. CspA and its homologues destabilize secondary structures in both RNA and DNA and are therefore referred to as nucleic acid chaperones (16). One of the detrimental effects of temperature downshift is the stabilization of secondary structures in RNA. This leads to hindering of transcription and translation as solid secondary structures in RNA cause inhibition of transcription elongation by RNA polymerase and movement of ribosomes during translation. CspA and its homologues, CspC and CspE, act as transcription antiterminators by virtue of their RNA chaperone activity (17), and the secondary structure destabilization activity of CspA family proteins is essential for cold acclimation of cells (18).

A subclinical concentration (1 μg/ml) of chloramphenicol and tetracycline, protein synthesis inhibitors, partially inhibits the *E. coli* translation by the interaction with translating ribosomes. This partial translational inhibition by the antibiotics mimics cold shock and induces a series of cold shock proteins including CspA in *E. coli* (19). Intriguingly, the induction of cold stress response by chloramphenicol was required for the expression of aromatase 264C form (6, 8). Based on the DNA sequence of 264 Cys, TGC, compared with that of 264 Arg, CGC, the secondary structure of aromatase 264C mRNA is not more stable than that of the 264R form. Therefore, CspA may not be a key protein at the rate-limiting step for the expression of aromatase 264C form although CspA is induced by chloramphenicol. The molecular mechanism of the expression of aromatase 264C form remains obscure. However, the low concentration (1 μg/ml) of chloramphenicol or tetracycline mimics cold shock, resulting in the cellular response to produce a series of cold shock proteins that enhances the expression of foreign proteins (20). This method is particularly useful for the expression of foreign proteins when the coexpression of GroES and GroEL is not efficacious.

1. 1 μl of the frozen stock is inoculated in 5 ml TB supplemented with 100 μg/ml ampicillin in a 50 ml conical tube. The culture is incubated at 37°C overnight. (The culture may be started at 6:00–7:00 PM of day 1.)
2. 4 ml overnight culture is diluted in 400 ml TB (pH 7.2) supplemented with 100 μg/ml ampicillin in a 2.8 l culture flask (*see* **Note 2**).
3. The culture is incubated at 37°C for 4 h with a vigorous shaking (at 240 rpm, Innova 44R) (The incubation may be started at 10:30–11:00 AM of day 2. Instead of determination of cell density by OD_{600} (approximately 1.0), induction timing may be determined by the incubation time.) (*see* **Note 2**).
4. The culture is added with 0.5 mM IPTG for the expression of aromatase 264C form, 50 μg/ml Amp, 1 μg/ml chloramphenicol for the induction of cold stress response, and 1 mM δ-ALA (a precursor of heme biosynthesis) and further

incubated at 27–28°C for 14.5 h with a reduced shaking (190 rpm) (This will be at 2:30–3:00 PM of day 2.) (*see* **Notes 3 and 4**).

5. A sample (8 ml) of each culture is taken from the flask, spun down, and the pH of the media determined. (This will be at 8:00–9:00 AM of day 3.) When pH of each culture reaches 7.0, cells in each flask are separately harvested by a low-speed centrifugation at 4°C (*see* **Notes 5** and **6**).
6. *E. coli* cell pellet is suspended in 40 ml lysozyme buffer with 0.5 mg/ml lysozyme, left on ice for 10 min, spun, and the supernatant is discarded. (After lysozyme treatment, cell pellets can be frozen in liquid N_2 and stored at –80°C.) After this step, follow steps 7–10 in "**Section 3.3**" (*see* **Note 8**).

3.5. Determination of P450 by the Reduced CO-Difference Spectrum (See Note 9)

1. The extracts are diluted with 5 volumes of the sample dilution buffer with 4 mM DTT in a cuvette. (DTT is freshly added for the stabilization of P450.) Substrates may be added at this step when required (*see* **Note 10**).
2. A small amount of sodium dithionite (20–40 grains of dithionite powder) is added to the cuvette. The cuvette is covered with parafilm and the sample is mixed by repeatedly inverting the cuvette.
3. The sample is quickly scanned (400–500 nm) for baseline memorization.
4. Small bubbles of CO using a Pasteur pipette are introduced in the sample.
5. The sample cuvette is covered with parafilm and repeatedly scanned (400–500 nm) until the peak at 450 nm reaches plateau (*see* **Note 11**).
6. The absorbance value at 450 nm is subtracted from the value at 490 nm. This value $\Delta A_{450\text{–}490}$ of 1 mM P450 solution should be 91 $\text{cm}^{-1}\ \text{mM}^{-1}$.
7. The expression level is calculated as below:

$$\Delta A_{450-490} \times (\text{total volume of the solution in the cuvette/ the volume of extract aadded to the cuvette}) = \Delta A_{450-490} \text{ of the extract before dilution}$$

$$\Delta A_{450-490} \text{ of the extract before dilution}/91 \times 1000 = \text{concentration of P450arom in the extract}(\mu\text{M}) = C_{\text{ext}}\ (\mu\text{M})$$

When the given extract was 40 ml and was extracted from 400 ml culture, the amount of P450 in the 40 ml extract is

$$C_{ext}\,(\mu M) \times 0.04\,(1) = 0.04C_{ext}\,(\mu mol) \times 1000 = 40C_{ext}\,(nmol)$$

The amount of aromatase was expressed in 400 ml culture. Therefore, the expression level (nmol/l culture) of aromatase in this given experiment is

$$40C_{ext}\,(\mu mol) \times 1000/400 = 100C_{ext}\,(nmol/l)$$

4. Notes

1. Qualities of media are quite different from maker to maker. Because the quality of media is very critical for foreign protein expression, the providers and catalog numbers are indicated.
2. The adequate culture volume may be different among incubators, flasks, and the shaking speed. Using an Innova 43R model, we shake the Fernbach flasks at 240 rpm as a maximum shaking speed. At a shaking speed, we determine an adequate culture volume for an individual system by watching the movement of different volumes of water (such as 250, 300, 350, 400, 450, and 500 ml).
3. Instead of determination of cell density by OD_{600}, induction timing may be determined by the incubation time of preculture. Because the measurement of OD_{600} often takes more than 10 min and a 10-min longer incubation results in a faster time course of pH, we determine the incubation timing by the time of preculture.
4. The culture temperature has to be decreased below 30°C to avoid the formation of inclusion bodies. The shaking speed is important to obtain an efficient aeration and minimize the formation of bubbles on the surface of media since the viscosity increases with the growth of bacteria. We use a reduced shaking speed (190 rpm) to avoid or minimize the formation of bubbles.
5. When more than one flask is incubated at a time, media pH of each flask should be monitored. Since *E. coli* growth varies depending on the position of flasks, probably because of temperature distribution within the chamber, the pH time course of each flask shows a considerable difference. Therefore, when cultures are harvested at a time the mean expression level is much lower than the maximum

expression level. Based on the media pH, we separately harvest *E. coli* cells in each flask to obtain the maximum expression of aromatase.

6. After determination of pH, the culture (8 ml) can be treated with lysozyme followed by extraction of proteins. When samples are taken several times, time course of pH and aromatase expression can be determined. The pH value 6.9–7.1 shows the highest expression level of the human aromatase. When pH is higher, P420 increases with a concomitant decrease of P450. When pH is lower, P450 decreases without the formation of P420. The time course of media pH can be adjusted to be slower when the preculture is shortened by 10 min and to be faster when the preculture is elongated by 10 min for the induction.
7. Introduction of air in the buffer at sonication causes serious denaturation of proteins, resulting in the loss of P450 and the formation of P420. Using a tapered microtip (1.6 mm diameter) for sonication, we manually sonicate the cell pellet with always keeping the top of the microtip in an enough depth of the extraction buffer.
8. Compared with the culture for the expression of aromatase 264R form using the GroES and GroEL coexpression, the growth of *E. coli* cells is slow with a subclinical dosage of chloramphenicol. Therefore, the cell pellet for the 264C form is usually less than that for the 264R form.
9. The reduced CO-difference spectrum can be determined using either single beam or double beam spectrophotometers. In this section, we describe a method of the reduced CO-difference spectrum using a single beam spectrophotometer.
10. The human aromatase is unstable and quickly converted to P420 in the presence of sodium dithionite. DTT stabilizes the human aromatase under the reducing condition, perhaps because DTT could protect P450 from hydrogen peroxide or oxygen radical that is produced by P450 using the reducing reagent and oxygen molecule.
11. Human aromatase shows an atypical reduced CO-difference spectrum in the presence of androstenedione (or a substrate having the 19-methyl group). As shown in **Fig. 7.4**, human aromatase shows a large 420 peak with a considerably reduced 450 peak in the presence of androstenedione, which is a characteristic of human aromatase. This atypical spectral property of human aromatase may be due to the space of its heme pocket. The

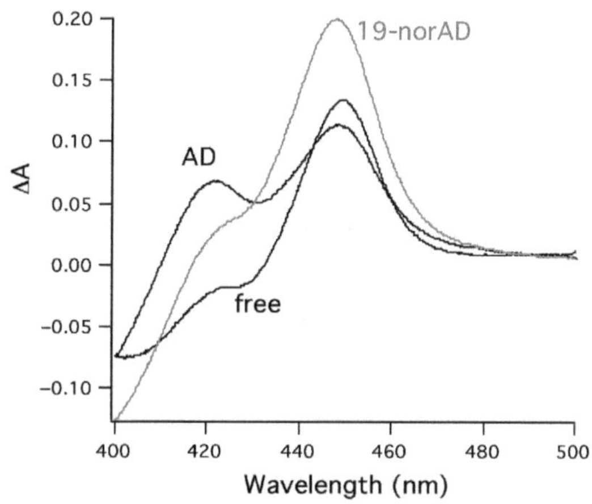

Fig. 7.4. Representative reduced CO-difference spectra of human aromatase using the *E. coli* extract. Proteins were extracted from *E. coli* (400 ml culture) with 40 ml extraction buffer. The extract (100 μl) was diluted in the sample dilution buffer (500 μl) in the absence of substrates (free), in the presence of 250 μM 19-norandrostenedione (19-norAD), or 250 μM androstenedione (AD). The reduced CO-difference spectra were determined as described in **Section 3**.

heme pocket in human aromatase may not have enough space for the CO binding to the heme iron under the steric hindrance of 19-methyl group of substrates. Therefore, unnatural substrates having no 19-methyl group do not cause steric hindrance and can stabilize the CO-aromatase complex, resulting in production of the highest 450 peak during the determination of the reduced CO-difference spectrum. In general, substrates of P450s are added during the purification and determination of reduced CO-difference spectrum, in particular, for stabilization of unstable P450s. However, the addition of androstenedione or testosterone to human aromatase considerably reduced the production of CO-P450 complexes although it may stabilize this enzyme. Because of instability of human aromatase, many investigators tried to purify aromatase from human placenta in the presence of androstenedione or testosterone and had difficulty in obtaining a typical spectrum of CO-P450 complex. Human aromatase is also unstable for the reduced CO-difference spectral analysis, resulting in a quick production of P420 with a concomitant decrease of P450. Therefore, the highest 450 peak can be obtained when the spectrum is determined in the presence of 0.25 mM 19-norandrostenedione. (19-norandrostenedione is dissolved in 45% 2-hydroxypropyl-β-cyclodextrin at 50 mM and stored at 4°C).

References

1. Massaro, D., Massaro, G. D. (2004) Estrogen regulates pulmonary alveolar formation, loss, and regeneration in mice. *Am J Physiol Lung Cell Mol Physiol* 287, L1154–L1159.
2. Martel, C., Melner, M. H., Gagne, D., Simard, J., Labrie, F. (1994) Widespread tissue distribution of steroid sulfatase, 3 beta-hydroxysteroid dehydrogenase/delta 5-delta 4 isomerase (3 beta-HSD), 17 beta-HSD 5 alpha-reductase and aromatase activities in the rhesus monkey. *Mol Cell Endocrinol* 104, 103–111.
3. Nakajin, S., Shinoda, M., Hall, P. F. (1986) Purification to homogeneity of aromatase from human placenta. *Biochem Biophys Res Commun* 134, 704–710.
4. Hagerman, D. D. (1987) Human placenta estrogen synthetase (aromatase) purified by affinity chromatography. *J Biol Chem* 262, 2398–2400.
5. Kellis, J. T., Jr., Vickery, L. E. (1987) Purification and characterization of human placental aromatase cytochrome P-450. *J Biol Chem* 262, 4413–4420.
6. Kagawa, N., Hori, H., Waterman, M. R., Yoshioka, S. (2004) Characterization of stable human aromatase expressed in *E. coli. Steroids* 69, 235–243.
7. Kagawa, N., Kusano, K., Nonaka, Y. (2000) Heterologous expression of mammalian cytochrome P450 as active forms in *Escherichia coli. Seibutsu-kogaku* 78, 82–93.
8. Kagawa, N., Cao, Q., Kusano, K. (2003) Expression of human aromatase (CYP19) in *Escherichia coli* by N-terminal replacement and induction of cold stress response. *Steroids* 68, 205–209.
9. Looman, A. C., Bodlaender, J., Comstock, L. J., Eaton, D., Jhurani, P., de Boer, H. A., van Knippenberg, P. H. (1987) Influence of the codon following the AUG initiation codon on the expression of a modified lacZ gene in *Escherichia coli. EMBO J* 6, 2489–2492.
10. Nonaka, Y., Fujii, T., Kagawa, N., Waterman, M. R., Takemori, H., Okamoto, M. (1998) Structure/function relationship of CYP11B1 associated with Dahl's salt-resistant rats – expression of rat CYP11B1 and CYP11B2 in *Escherichia coli. Eur J Biochem* 258, 869–878.
11. Kagawa, N., Katagiri, M. (2003) Expression of foreign proteins in *E. coli*: through the expression of P450s. *Seikagaku* 75, 279–284.
12. Uchida, E., Kagawa, N., Sakaki, T., Urushino, N., Sawada, N., Kamakura, M., Ohta, M., Kato, S., Inouye, K. (2004) Purification and characterization of mouse CYP27B1 overproduced by an *Escherichia coli* system coexpressing molecular chaperonins GroEL/ES. *Biochem Biophys Res Commun* 323, 505–511.
13. Arase, M., Waterman, M. R., Kagawa, N. (2006) Purification and characterization of bovine steroid 21-hydroxylase (P450c21) efficiently expressed in *Escherichia coli. Biochem Biophys Res Commun* 344, 400–405.
14. Zollner, A., Kagawa, N., Waterman, M. R., Nonaka, Y., Takio, K., Shiro, Y., Hannemann, F., Bernhardt, R. (2008) Purification and functional characterization of human 11beta hydroxylase expressed in *Escherichia coli. FEBS J* 275, 799–810.
15. Oyama, T., Kagawa, N., Sugio, K., Uramoto, H., Hatano, O., Harada, N., Kaneko, K., Kawamoto, T., Yasumoto, K. (2009) Expression of aromatase CYP19 and its relationship with parameters in NSCLC. *Front Biosci* 14, 2285–2292.
16. Jiang, W., Hou, Y., Inouye, M. (1997) CspA, the major cold-shock protein of *Escherichia coli*, is an RNA chaperone. *J Biol Chem* 272, 196–202.
17. Bae, W., Xia, B., Inouye, M., Severinov, K. (2000) *Escherichia coli* CspA-family RNA chaperones are transcription antiterminators. *Proc Natl Acad Sci USA* 97, 7784–7789.
18. Phadtare, S., Tyagi, S., Inouye, M., Severinov, K. (2002) Three amino acids in *Escherichia coli* CspE surface-exposed aromatic patch are critical for nucleic acid melting activity leading to transcription antitermination and cold acclimation of cells. *J Biol Chem* 277, 46706–46711.
19. Jiang, W., Jones, P., Inouye, M. (1993) Chloramphenicol induces the transcription of the major cold shock gene of *Escherichia coli*, cspA. *J Bacteriol* 175, 5824–5828.
20. Kusano, K., Waterman, M. R., Sakaguchi, M., Omura, T., Kagawa, N. (1999) Protein synthesis inhibitors and ethanol selectively enhance heterologous expression of P450s and related proteins in *Escherichia coli. Arch Biochem Biophys* 367, 129–136.

Chapter 8

Expression of Recombinant Cytochromes *c* in *E. coli*

Yuri Y. Londer

Abstract

Answering questions about proteins' structures and functions in the new era of systems biology and genomics requires the development of new methods for heterologous production of numerous proteins from newly sequenced genomes. Cytochromes *c* – electron transfer proteins carrying one or more hemes covalently bound to the polypeptide chain – are one of the most recalcitrant classes of proteins with respect to heterologous expression because post-translational incorporation of hemes is required for proper folding and stability. However, significant advances in expression of recombinant cytochromes *c* have been made during the last decade. It has been shown that a single gene cluster, *ccmA–H*, is responsible for cytochrome *c* maturation in *Escherichia coli* under anaerobic conditions and that constitutive co-expression of this cluster under aerobic conditions is sufficient to provide heme incorporation in many different types of cytochromes *c*, regardless of their origin, as long as the nascent polypeptide is translocated to the periplasm. Using conditions that result in sub-maximal protein induction can dramatically increase the yield of mature protein. The intrinsic peroxidase activity of hemes can be used as a highly selective and sensitive detection method of mature cytochromes in samples resolved by gel electrophoresis.

Key words: Cytochrome *c*, *Escherichia coli*, heterologous expression, periplasmic expression, peroxidase reaction, UV–Vis spectroscopy.

1. Introduction

Cytochromes are electron transfer proteins carrying one or more hemes as prosthetic groups. They function in aerobic and anaerobic respiratory chains as well as in photosynthetic electron transport. In cytochromes *c*, unlike other cytochromes, hemes are covalently bound to the polypeptide chain. Cytochromes *c* have a characteristic sequence pattern CXXCH, where the cysteines form two thioether bonds to the heme and the histidine serves as one

T.C. Evans, M.-Q. Xu (eds.), *Heterologous Gene Expression in E. coli*, Methods in Molecular Biology 705,
DOI 10.1007/978-1-61737-967-3_8, © Springer Science+Business Media, LLC 2011

of the two axial (extraplanar) ligands to the heme iron. The other ligand is usually a methionine or another histidine located farther away in the amino acid sequence (1, 2). A single polypeptide chain can have multiple hemes covalently attached, in some cases as many as 35–40.

A major challenge of post-genomic biology is the development of efficient methods for heterologous production of hypothetical proteins from newly sequenced genomes – an important enabling feature for structural and functional studies of proteins. Cytochromes *c* are one of the most recalcitrant classes of proteins with respect to heterologous expression due to the necessity of post-translational modification (covalent heme attachment) for proper folding and stability. Although overproduction of recombinant proteins in *Escherichia coli* has proven successful for numerous proteins from eukaryotic and prokaryotic organisms, attempts of heterologous expression of cytochromes *c* often failed, mainly due to the low efficiency of post-translational covalent modification (e.g., 3, 4). Indeed, an organism producing cytochromes *c* faces a number of challenges during the biogenesis of these proteins. First, all known cytochromes *c* are extracytoplasmic proteins. They are either soluble periplasmic proteins (or soluble proteins residing in the intermembrane space of mitochondria) or membrane-anchored (in the inner or outer membrane) via an N-terminal transmembrane helix or a lipid anchor. There are also examples of cytochrome *c*-like domains that are anchored by two transmembrane helices, at the N- and C-termini of the domain[1] (e.g., 5, 6). Regardless of a cytochrome's final topology, both the apoprotein and free heme must be translocated through the inner membrane. After the translocation of the apoprotein, the cysteines destined to covalently bind heme(s) must be kept reduced in the oxidative periplasmic environment. Finally, the apoprotein and heme(s) are brought together and thioether bonds are formed in an enzyme-catalyzed step (see refs. 2, 7–9 for detailed review).

All organisms studied to date use one of three pathways, or systems, to make *c*-type cytochromes (8, 10). *E. coli* (as well as many other Gram-negative bacteria) employs the so-called system I, which includes an ABC transporter, a thioredoxin-like

---

[1] By definition, cytochromes *c* are ***electron transfer proteins*** in which heme is covalently bound to the polypeptide. This definition excludes proteins that contain covalently bound heme(s) but have functions other than electron transfer, such as nitrite reductase or a recently described family of sensor proteins (5, 6). However, the *E. coli* expression system described here has been shown to provide covalent heme attachment in many different types of proteins, regardless of their function. For the purpose of brevity, the term "cytochrome *c*" will be used throughout this chapter but all approaches described here should be applicable to all proteins containing covalently bound hemes.

redox system, heme chaperone, and heme lyase. Normally, *E. coli* has eight proteins, consecutively named CcmA through CcmH, involved in maturation of cytochromes *c* (2, 7–9). These proteins are located in the inner membrane and on the periplasmic side of the inner membrane. They are encoded by gene cluster *ccmABCDEFGH* that is a part of the *aeg46.5* operon and is active only under anaerobic conditions (11). Cloning of this cluster into pACYC184 (a plasmid compatible with many commercially available expression vectors) allowed its constitutive co-expression under control of the tetracycline promoter (12) and made possible the production of properly matured *c*-type cytochromes in *E. coli* under normal aerobic conditions (13, 14, and references therein). The only other major requirement for successful synthesis of recombinant cytochromes (besides co-expression of the *ccm* genes) is the presence of a leader peptide to target the nascent polypeptide to the periplasm (15). For expression in rich medium, such as 2×YT, it is not necessary to supply extra iron or heme precursors.

2. Materials

2.1. Bacterial Strains and Plasmids

1. Typical cloning strains, such as DH5α (Invitrogen), XL1-Blue (Stratagene), NEB 5-alpha (New England Biolabs), or the like.
2. Expression strains, such as BL21(DE3), JCB7123, HM125, and SF110. Other strains that are not chloramphenicol-resistant and do not harbor pACYC or other plasmids with p15A replicon can be used as well.
3. Plasmid pEC86, a pACYC derivative that carries the *ccm* gene cluster. The plasmid carries the gene for chloramphenicol transferase providing resistance to chloramphenicol.
4. The cloning vector(s) of your choice. Several vectors specifically designed for periplasmic targeting include pET-22b(+)[2] (Novagen), pLBM, pLSM2, and pFCM21 (ligation-independent vectors designed by the author; 14, 16). pLSM2 and pFCM21 carry constitutively expressed genes for periplasmic chaperones Skp (17–19) and FkpA (20, 21), respectively. Dedicated vectors for periplasmic targeting of proteins fused to affinity handles are available (*see* **Note 1**).

[2] pET-22b(+) has the T7 promoter (see **Section 3.1** below).

2.2. Expression

1. 2×YT medium: 16 g/L tryptone, 10 g/L yeast extract, 5 g/L NaCl. Pre-mixed powder is also available from several vendors, e.g., Fisher Scientific (*see* **Note 2**).
2. Ampicillin (or carbenicillin): 100 or 200 mg/mL in water (*see* **Note 3**). Store at –20°C.
3. Chloramphenicol: 34 or 68 mg/mL in ethanol. Store at –20°C.
4. IPTG[3]: 100 m*M* and 1 *M* in water (*see* **Notes 3** and **4**). Both stock solutions should be stored at –20°C.

2.3. Cell Lysis

1. TES buffer: 100 m*M* Tris–HCl, pH 8.0, 20% sucrose, 0.5 m*M* EDTA. Store at 4°C.
2. Chicken egg lysozyme: 50 mg/mL in water. Lyophilized protein can be purchased from several suppliers, e.g., Sigma. The stock solution should be stored frozen at –20°C as single-use aliquots (100–600 μL).
3. Rubber policemen attached to a glass rod or a glass/plastic pipet (*see* **Fig. 8.1**).
4. Protease inhibitors of your choice (*see* **Note 5**).
5. Benzonase Nuclease: Available from Novagen. This reagent is optional.
6. Sodium chloride: 1 or 5 *M* solution (optional).
7. Detergent (*see* **Section 3.5**).

2.4. Gel Separation, Transfer, and Heme Staining

1. Polyacrylamide gels of appropriate percentage, depending on the size of the protein of interest. Pre-cast gels are fine; 4–20% gradient gels are almost perfect one-size-fits-all solution.

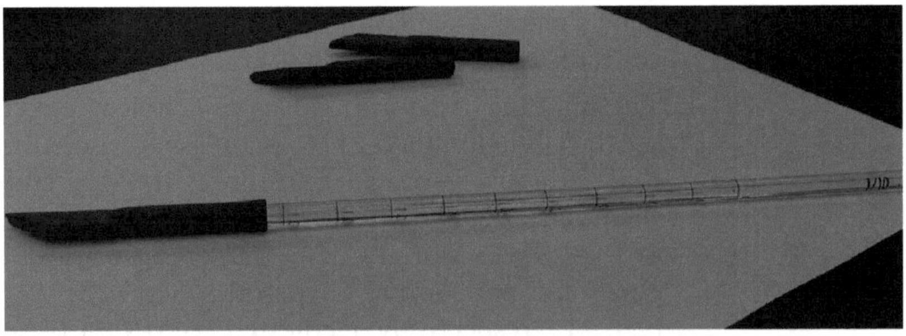

Fig. 8.1. Rubber policemen (*top*) and a rubber policeman attached to a 1 mL glass pipet (*bottom*).

[3] *Abbreviations:* CAPS, 3-(cyclohexylamino)-1-propanesulfonic acid; CMC, critical micelle concentration; IPTG, isopropyl-β-D-thiogalactoside; MBP, maltose-binding protein; TEV, tobacco etch virus; TM, transmembrane.

2. Running buffer: 25 m*M* Tris, 192 m*M* glycine, 0.1% (w/v) SDS. Running buffer can be prepared as a 5X or 10X stock solution and stored at room temperature. Alternatively, a pre-mixed stock solution can be purchased from several manufacturers, e.g., Bio-Rad (*see* **Note 6**).
3. SDS-PAGE sample buffer: 50 m*M* Tris–HCl, pH 6.8, 2% SDS, 0.1% bromophenol blue, 10% glycerol. *No reducing agents should be used.* It is usually made as a 2X stock solution and stored at room temperature.
4. Transfer buffer: 10 m*M* 3-(cyclohexylamino)-1-propanesulfonic acid (CAPS) in 10% methanol, pH 11.0 (*see* **Notes 7** and **8**). Store at 4°C.
5. Hybond-C Extra nitrocellulose membranes (GE Healthcare, formerly Amersham Biosciences) (*see* **Note 9**).
6. Filter paper (cut to size slightly larger than the size of the gel).
7. Transfer apparatus.
8. PBS buffer: 1 m*M* KH_2PO_4, 10 m*M* Na_2HPO_4, 137 m*M* NaCl, 2.7 m*M* KCl. Large packs (e.g., 4 L) of a pre-mixed 10X stock solution are available from several manufacturers, e.g., Roche Applied Science.
9. SuperSignal West Femto Maximum Sensitivity Substrate (Pierce Biotechnology). It is actually a kit containing two reagents: stabilized hydrogen peroxide solution and luminol. Store at 4°C.
10. Ziploc bags of size slightly larger than the size of the gels you plan to use for SDS-PAGE analysis of your samples (*see* **Note 10**).
11. X-ray film cassette.
12. X-ray films (CL-XPosure films from Pierce produce consistently good results but other brands can be used as well).
13. X-ray film processor (optional but highly desirable; *see* **Notes 11** and **12**).

2.5. Spectral Measurements

1. UV–Vis spectrophotometer.
2. UV transparent quartz cuvettes, 0.5–1.0 mL, with a path length of 10 mm (*see* **Note 13**).
3. Sodium dithionite. Also known as sodium hydrosulfite, NaO_2SSO_2Na, it can be purchased from several manufacturers, e.g., Sigma. This is a strong reducing agent used to ensure the complete reduction of the hemes, regardless of their redox potential (*see* **Note 14**).

3. Methods

3.1. Choice of Vector and Cloning

When considering heterologous expression of a cytochrome *c*, one should address the following issues:

(a) *Choice of expression strain.* BL21(DE3) works very well with many cytochromes *c*, both in lac promoter- and T7 promoter-based systems. However, for polyheme cytochromes *E. coli* strain JCB7123 (22) may be a better option. This strain has a *narL*$^-$ genotype, which results in a derepression of the *aeg46.5* operon and increased expression of the *ccm* genes from the chromosome (22). If proteolytic degradation in the periplasm is an issue, strain SF110 or HM125 deficient in different periplasmic proteases may be used (23). These last three strains should not be used with T7 promoter-based vectors as they do not have the gene for T7 RNA polymerase. Other *E. coli* strains may be employed as long as they do not have intrinsic chloramphenicol resistance (plasmid pEC86 harboring the *ccm* genes carries chloramphenicol transferase as a selective marker).

(b) *Purification approach.* Many cytochromes *c* have a high isoelectric point, which makes it possible to isolate them from the periplasmic fraction in a single cation-exchange step (the overwhelming majority of periplasmic proteins are acidic and they do not even bind to a cation exchanger at circumneutral pH). If this is not the case, affinity tags can be used, but with caution. His-tag may present a problem as it has been reported that histidine clusters seem to interfere with proper maturation of cytochromes *c*, especially those with multiple hemes (24, 25), apparently because the histidines compete with the native heme ligands for binding the iron atom(s). However, a hybrid 6×His/MBP tag has been successfully used for expression of recombinant cytochromes *c* (5, 16). MBP acts as a spacer that prevents the His-tag from interacting with the cytochrome component of the fusion. Moreover, MBP is a known solubility enhancer and can facilitate the expression of some targets in the soluble form. MBP alone (without a His-tag) can also be used as an affinity handle as it binds to amylose. Tags containing cysteine residues (e.g., GST-tag) are not recommended as it has been shown that extra cysteines in the polypeptide interfere with cytochrome maturation (Y.Y.L., unpublished data), apparently, due to formation of incorrect disulfide bonds between the extra cysteines and the cysteines supposed to bind the hemes. If this is the case, adding a spacer domain, such as MBP, may alleviate the problem. However, this has not been experimentally tested.

(c) *Presence of a signal peptide, TM helix (helices), or lipid anchor.* If a cytochrome of interest is (or predicted to be) membrane-bound, for the majority of physicochemical studies it is more convenient to express and purify the soluble domain only, which will require predicting the exact location of the membrane anchor in the sequence (especially its C-terminus). The predicted soluble domain then can be cloned into a vector containing a signal peptide.

All the above considerations will dictate the choice of expression vector. It is likely that more than one vector has to be tested before optimal expression and purification protocols can be developed. In the author's opinion, the anticipated purification approach is the most important criterion determining the choice of vector. As mentioned above, the majority of cytochromes discovered and studied so far are relatively basic, with p*I* > 6.5, and can be purified by simple cation exchange as a large fraction of other cellular proteins have p*I* in the range 4–6. In addition, the cytochromes are purified from the periplasmic fraction rather than the whole cell lysate, which automatically eliminates a number of basic DNA- and RNA-binding cytoplasmic proteins. Therefore, if a cytochrome of interest is predicted[4] (or experimentally shown) to have a relatively high p*I*, any vector for periplasmic targeting would be suitable (though, as mentioned above, vectors with T7 promoters would have a limited range of host strains). Alternatively, if the cytochrome has its own cleavable signal sequence (rather than a membrane anchor), it can be cloned into any expression vector as the mechanisms of Sec-dependent translocation are well conserved in bacteria. If the above strategy does not work or if the cytochrome is acidic, then affinity tags should be seriously considered.

Cloning of the target should be performed according to standard procedures. PCR amplification of the target gene from genomic DNA is usually the most straightforward approach. When designing the primers, it is imperative to make sure that the leader sequence targeting the recombinant protein to the periplasm is present in the cloned product, either from the vector or from the original gene (but not both!). If the genomic DNA is not easily available or amplification repeatedly fails, the gene can be assembled de novo from oligonucleotides (which also

---

[4] It should be noted that theoretically predicted p*I* is less accurate for cytochromes than for other proteins as heme group contains iron that can be in oxidation state +2 or +3, depending on the redox potential of the given heme (which is determined by many factors, including the heme environment, its ligands, and the degree of its exposure to solvent) and two carboxylic groups, whose p*I* can vary significantly, depending on their environment and degree of exposure to solvent.

allows codon-optimizing it for *E. coli*). There are free Web-based tools to facilitate the design of oligonucleotides for gene assembly, e.g., DNAWorks (http://helixweb.nih.gov/dnaworks/, 26). Gene synthesis can also be outsourced, though it may be the most expensive alternative (~ $1 per bp).

3.2. Transformation

Since expression of cytochromes requires the presence of two plasmids in the cell, pEC86 that carries the *ccm* gene cluster and the plasmid with the gene of interest, two rounds of transformation are necessary. It is convenient, especially when considering expression of several cytochrome targets, to prepare a stock of competent cells of the expression strain already harboring pEC86 and then transform them with the expression constructs (*see* **Notes 15** and **16**). Competent cells, both with and without pEC86, can be prepared by any method (a typical calcium chloride protocol is adequate). All growth procedures for strains harboring pEC86 should be performed in medium containing chloramphenicol 34 μg/mL (except during the post-heat shock recovery step).

1. Transform pEC86 into an expression strain of your choice following standard protocols for plasmid transformation and plate onto LB or 2×YT agar containing chloramphenicol. Incubate the plate overnight at 37°C (*see* **Notes 17** and **18**).
2. Prepare competent cells of the expression strain harboring pEC86. In parallel, prepare frozen stocks, at least 2–3 separate vials (aliquots of the culture intended for preparation of competent cells can be used for the frozen stocks). Competent cells can be frozen as 200 μL aliquots and stored at –80°C (*see* **Note 19**).
3. Transform the expression construct into the expression strain harboring pEC86 and plate onto LB or 2×YT agar containing both ampicillin (100 μg/mL) and chloramphenicol (*see* **Note 20**). Incubate the plate overnight at 37°C (*see* **Notes 17** and **18**).

3.3. Expression and Analysis

Even with co-expression of the *ccm* genes, functional holocytochromes are produced in *E. coli* at relatively low levels, which makes it difficult or impossible to analyze expression clones using Coomassie-stained polyacrylamide gels. One indication of successful expression is coloration of the cell pellet – depending on a particular target and expression levels, the color may vary from pink to brown/rust-red. However, the presence of color only confirms heme incorporation but is not informative as to whether the protein is soluble and whether it has the correct molecular weight. The color of the periplasmic fraction (as well as its

UV–visible spectrum) can indicate the presence of a soluble heme protein or, strictly speaking, the presence of heme in the soluble form. One particular unfortunate case when this distinction is relevant is when the protein of interest is degraded by proteases to fragments, one or more of which still have covalently bound hemes. Much more accurate results can be obtained by separating cell lysates in a denaturing polyacrylamide gel and applying a shortcut version of Western blotting. Heme possesses peroxidase activity that is not affected by the protein denaturation and can be used for detection of cytochromes *c* in the presence of chemiluminescent peroxidase substrates, e.g., those used for ELISA or Western blotting (27). Detection of chemiluminescence as a result of heme peroxidase activity can be performed using X-ray films or a charge-coupled device.

Another advantage of this approach with respect to expression of recombinant cytochromes *c* in *E. coli* is that *E. coli* does not make its own cytochromes *c* under aerobic conditions. Therefore, all bands observed after gel separation, transfer, and development are due to expression of foreign *c*-type cytochromes. A limitation of this method is that heme must be in the ferric (oxidized) state and, therefore, reducing agents, such as DTT or mercaptoethanol, CANNOT be used in the sample buffer (or anywhere in the procedure). A consequence of this is that immunoblotting with peroxidase-conjugated antibodies cannot be carried out on the same gel (or membrane) as used for heme "staining" (if the samples are treated with a reducing agent, then there will be no signal from the hemes; if there is no reducing agent, the signal from the hemes would interfere with the signal from peroxidase). Using alkaline phosphatase conjugated antibodies is an option if both an immunoblot and heme staining of a given gel are desired.

The procedure also requires special molecular weight markers, such as several (or at least one) purified cytochromes *c*. Monoheme mitochondrial cytochromes *c* (MW $\sim$ 13 kDa) can be purchased from different manufacturers, e.g., Sigma. Several cytochrome constructs of different molecular weights are available from the author:

- *Geobacter sulfurreducens* cytochrome c_7 (9.5 kDa; 24)
- *G. sulfurreducens* dodecaheme cytochromes GSU0592 and GSU1996 (40.5 and 42 kDa, respectively; 14)
- the N-terminal half of the latter (tandem AB, 21.5 kDa; 13)
- MBP fusion with heme-containing sensor domain of *G. sulfurreducens* sensor protein GSU0303 (63 kDa; 5)

In principle, heme detection can be performed directly in gels, although it will require very careful handling of the gel to preserve its integrity and significantly greater amounts of the reagents. The advantage of direct in-gel staining is that prestained protein markers can be used. The protocol described below does assume that the proteins will be transferred onto a nitrocellulose membrane and the chemiluminescence will be monitored using X-ray films. If digital image capture is preferred, the method described by Feissner and coworkers can be used (27).

In the case of multiheme cytochromes *c* and affinity-purified monoheme cytochromes, the completeness of heme incorporation should be eventually determined. Although the presence of heme(s) is usually indispensable for proper folding and stability, occasionally isoforms containing fewer hemes are co-purified, especially if affinity chromatography has been used as the only purification step (24). A convenient way to assess heme content is mass-spectrometry. Each covalently bound heme adds 616 Da to the molecular weight of the polypeptide and, therefore, the mass numbers obtained from mass-spectrometry experiments allow precise determination of the number of hemes per polypeptide. If immature species are present in the sample, they will show up as separate peaks with masses less than that of the fully mature protein by 616 Da (or 2×616, 3×616, etc.).

This chapter will present protocols for large-scale (1 L cultures) and small-scale (25 mL or less) expression and analysis of cytochromes. The choice between large- and small-scale test formats is predominantly determined by the need to optimize conditions for a given expression construct in a given strain versus a simultaneous analysis of multiple samples, at least 5–10 or more (e.g., several different cytochromes or expression in different vectors and/or strains). It is strongly recommended not to conduct large-scale experiments until a small-scale experiment confirms the expression of a cytochrome with the desired molecular weight. The expression levels of many cytochromes are very sensitive to the IPTG concentration used for induction, with the best concentration often in the low micromolar range (13, 24). However, results of small scale tests do not always scale up to larger volumes, that is, if small-scale experiments show that the highest expression level is achieved at, for instance, 25 μ*M* IPTG, it is not necessarily the case after switching to large-scale cultures. One reason for that may be differences in oxygenation levels between the large-scale and small-scale cultures. Therefore, it is highly desirable to set up optimization experiments under exactly the same conditions as will be used for the eventual large-scale growth and purification – the same flasks, shakers, batches of IPTG stock solutions, etc.

3.3.1. Small-Scale Expression and Analysis

This procedure was developed for 25 mL cultures grown in 50 mL Erlenmeyer flasks but can be scaled down, if necessary, to as little as 2–3 mL in 10 mL culture tubes (especially if spectral measurements are not a high priority or if small cuvettes, 100–200 μL, are available).

1. Transform the expression construct into the chosen expression strain harboring pEC86 as described above (*see* **Note 21**).
2. Use a single colony to inoculate a 2 mL culture in 2×YT containing both ampicillin (100 μg/mL) and chloramphenicol (34 μg/mL). Incubate overnight at 30°C and 250 rpm.
3. Next morning, use 250 μL of the overnight culture to inoculate 25 mL of 2×YT containing ampicillin and chloramphenicol. Incubate at 30°C and 250 rpm for 8–10 h. Before leaving for the day, add IPTG to a final concentration of 30 μ*M* (e.g., 7.5 μL of a 100 m*M* stock solution), reduce the shaking speed to 200 rpm, and allow to grow overnight at the same temperature (*see* **Note 22**).
4. Next morning take out a 100 μL aliquot from each culture, centrifuge 5 min in a benchtop centrifuge, discard the supernatant, and freeze the pellet (it will be used to prepare gel samples of total cell lysate).
5. Harvest the bulk of the cells by centrifugation for 15 min at 4000 rpm at 4°C and then remove the supernatant (*see* **Notes 23** and **24**).
6. Resuspend the pellets (by gentle pipetting) in ice-cold TES buffer containing the amount of protease inhibitors recommended by the manufacturer (*see* **Note 25**). Use 2 mL of the buffer to resuspend the pellet from 25 mL of culture.
7. Add lysozyme to a final concentration of 0.5 mg/mL (20 μL of a 50 mg/mL stock for 2 mL suspensions) and incubate 15 min at room temperature (*see* **Note 26**).
8. Add an equal volume (2 mL) of ice-cold water, place the tubes horizontally in a bucket of ice, and shake gently (~100 rpm) for 15 min.
9. Centrifuge at 12,000×*g* for 20 min at 4°C (*see* **Notes 27** and **28**).
10. Transfer the supernatant to a fresh tube (*see* **Note 28**). It is the soluble periplasmic fraction that can be used for gel analysis and spectral measurements.
11. Mix 10 μL aliquots of each periplasmic fraction with 10 μL of 2× sample buffer and boil (*see* **Note 29**).
12. Resuspend the pellets from the 100 μL aliquots collected in step 4 in 20 μL of 1× sample buffer by pipetting and/or vortexing. Make sure that there are no visible clumps. Boil for 5 min and centrifuge in a benchtop centrifuge for 1 min.
13. Apply 10 μL of each sample to the gel, together with the molecular weight markers, and run under the

manufacturer-recommended conditions for this type of gel (*see* **Notes 30** and **31**).

14. While the gel is running, spectral measurements of the periplasmic fraction(s) can be taken (*see* **Section 3.4**).
15. Disassemble the gel cassette and cut a small piece off one corner of the gel (e.g., the lower right corner) to track the gel orientation at all times.
16. Assemble the transfer unit according to the manufacturer's instructions (*see* **Note 32**).
17. Run a transfer procedure under conditions specified by the manufacturer.
18. Disassemble the transfer unit and wash the membrane in PBS for 5–10 min with gentle shaking (100–150 rpm).
19. Blot the membrane with filter paper.
20. Place the membrane in a Ziploc bag. Add 2 mL of hydrogen peroxide solution and 2 mL of luminol solution (both from SuperSignal West kit). Close the bag, making sure that the membrane is completely covered with the solution and there are no air bubbles. Place the bag flat on a shaking platform and incubate for 5–10 min with gentle shaking (100–150 rpm) at room temperature.
21. Open the bag and take out the membrane (*see* **Note 33**). Blot the membrane with filter paper. Place the membrane between two transparent plastic sheets and put the sheets with the membrane into an X-ray film cassette. Make sure that the side of the membrane that has been in contact with the gel is facing upward.
22. *The following steps must be done in a dark room with a safe light*. Take an X-ray film and cut off a corner. Place the film on top of the plastic sheet covering the membrane, making sure that the cut-off corner of the film corresponds to that of the membrane. Lock the cassette.
23. Expose for 30 s, then take the film out, and develop it in the X-ray film processor. The relative intensities of heme signals for the total cell lysate and for the corresponding periplasmic fraction can give an estimate of the percentage of the heme-containing protein that is soluble: if the bands are of comparable intensity, the protein is mostly soluble; if the band in the periplasmic fraction is significantly weaker than the band in the total cell lysate, that means either the protein forms inclusion bodies or it strongly associates with the cell membranes. If the latter is the case, the protein may be solubilized in the presence of detergents and/or NaCl (*see* **Section 3.5**).

3.3.1.1. Troubleshooting

It is likely that more exposures will be needed, either to resolve overexposed bands or to observe weak bands. If the lanes are clearly overloaded (**Fig. 8.2a**), then several more exposures can be taken every 15–30 min, until the bands are resolved. The chemiluminescent signal weakens over time (compare panels A–C in **Fig. 8.2**) and disappears completely after 2–4 h, depending on the amount and the nature of the protein in the band (therefore, do NOT postpone further exposures until next morning). It is possible that different bands will become resolved after different exposures.

If there are weak bands or no bands at all on the developed film, it may mean that either the transfer procedure or peroxidase

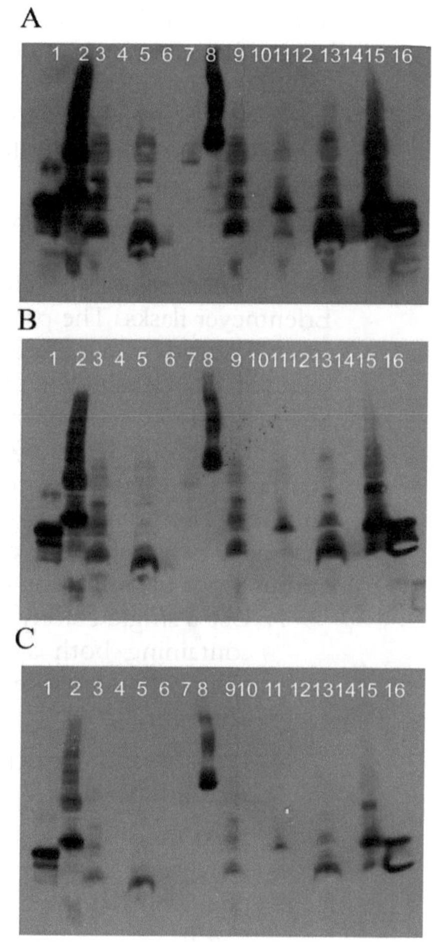

Fig. 8.2. Time dependence of the chemiluminescent signal. Total cell lysates and periplasmic fractions of *E. coli* cells expressing several cytochromes *c* from *Shewanella oneidensis* were separated in a 4–20% gradient gel, transferred onto a Hybond-C membrane, and treated with the SuperSignal Femto Kit reagents as described in the text. Thirty-second exposures were taken in 3 min (**a**), 30 min (**b**), and 1 h 20 min (**c**) after treatment with the SuperSignal reagents.

reaction or film development failed. Lanes with molecular weight markers can serve as positive controls for transfer/peroxidase reaction/development steps: if the markers are hardly visible, obviously something went wrong with the procedure. The success of the transfer procedure can be monitored by the appearance of "rusty" stains on the membrane in marker lanes (*see* **Note 34**). The SuperSignal kit reagents are stable for 6 months at 4°C. If there is a reason to believe that the kit is significantly older or that it has been left at room temperature for a prolonged period of time, a new kit should be ordered and tested. Finally, there may be a problem with the X-ray processing machine. If other people (who are doing different experiments) have problems with it as well, a service technician should be called.

If the marker bands are strong but lanes with actual expression samples have no bands at all, that indicates low expression levels of heme-containing species or no expression at all (*see* **Note 35**). To distinguish between the two possibilities, another exposure should be taken as soon as possible, for 3–5 min. If after a 5-min exposure there is still no band of the appropriate molecular weight, it is safe to assume that no stable heme-containing protein (or no soluble protein) has been produced (*see* **Note 36**).

3.3.2. Large-Scale Expression and Lysis

This procedure was developed for 1 L cultures grown in 2 L Erlenmeyer flasks. The protocol described here involves four cultures induced with different IPTG concentrations. The recommended cell density for induction is $A_{600}=1.0$ (*see* **Note 37**). Suggested IPTG concentrations for the first experiment are 10, 20, 30, and 50 μM (*see* **Note 22**). When the optimal IPTG concentration is established, use the same amount of IPTG for all cultures. The periplasmic fractions after a large-scale experiment that have relatively high cytochrome content can be combined and used for optimization of purification conditions.

1. Use a single colony to inoculate a 50 mL culture in 2×YT containing both ampicillin (100 μg/mL) and chloramphenicol (34 μg/mL). Incubate overnight at 30°C and 250 rpm.
2. Prepare and autoclave four flasks with 1 L of 2×YT medium per flask.
3. Next morning use 10 mL of the overnight culture to inoculate each 1 L culture containing ampicillin and chloramphenicol. Let the cultures grow at 30°C and 250 rpm for ~ 4 h.
4. Measure A_{600} for each culture.
5. Continue monitoring A_{600} every 30–60 min (*see* **Note 38**).
6. When A_{600} for a given culture reaches the range of 0.9–1.0, induce the culture with the appropriate amount of IPTG.

7. After all cultures have been induced, reduce the shaking speed to 200 rpm and continue incubation overnight at the same temperature.
8. Next morning take out a 100 μL aliquot from each culture, centrifuge 5 min in a benchtop centrifuge, discard the supernatant, and freeze the pellet.
9. Harvest the bulk of the cells by centrifugation for 15 min at 4000 rpm at 4°C and then remove supernatant (*see* **Note 23**).
10. Resuspend the cells with a rubber policeman (*see* **Note 39**) in ice-cold TES buffer containing the amount of protease inhibitors recommended by the manufacturer (use 30 mL of TES buffer per pellet from 1 L of culture).
11. Transfer the suspension to 250 mL centrifuge bottles (*see* **Note 40**). Add lysozyme to a final concentration of 0.5 mg/mL (300 μL of a 50 mg/mL stock solution per bottle) and incubate for 15 min at room temperature (*see* **Notes 26** and **41**).
12. Add an equal volume (30 mL/bottle) of ice-cold water. Make sure that the lids are closed tightly and put the bottles horizontally in a bucket of ice. Shake gently (~ 100 rpm) for 15 min.
13. Centrifuge at 12,000×*g* for 20 min at 4°C. Transfer the supernatants to fresh tubes or beakers (*see* **Note 42**).
14. Determine the spectra and A_{280} for each sample as described below (*see* **Section 3.4**). Calculate the ratio of the Soret band absorbance over A_{280} (*see* **Note 43**).

3.4. Spectral Measurements

Spectral measurements can be taken while the gel and/or transfer is run. The following instructions are for a two-beam spectrophotometer with 1 cm/1 mL cuvettes but can be easily modified for other instruments. Before measurements, make sure that all samples are clear (do not have visible particles suspended in the solution). Centrifuge 1.5 mL aliquots 10 min in a benchtop centrifuge at 4°C, if necessary.

The spectral measurements described in this chapter include two components: (1) recording of the UV–visible spectra of each periplasmic fraction in oxidized and reduced forms and (2) measuring absorbance of each fraction at 280 nm. The latter is a measure of the number of cells in a given culture and the efficiency of lysis of the corresponding pellet (for A_{280} measurements periplasmic fractions should be diluted at least fivefold – *see* step 11 of this section). Sometimes cultures grown in the same conditions and induced at the same cell density yield periplasmic lysates with A_{280} differing by a factor of 2 or more. Therefore, even though the Soret band absorbance is the primary measure of the cytochrome expression level for a given construct/culture, it should be interpreted together with A_{280} for the periplasmic fraction. In other words, A_{280} is a "common denominator" used to compare lysates prepared from different cultures. As expression levels for cytochromes *c* are low, they do not contribute significantly to the total A_{280} of the periplasmic fraction.

The steps below describe the measurements for a single periplasmic sample. Usually, there are several samples to be analyzed and in this case it is convenient to measure both oxidized and reduced spectra for all of them (steps 1–8), then re-blank the spectrophotometer at 280 nm, and measure A_{280} for all samples (steps 9–15).

1. Fill two cuvettes with the same buffer as was used to prepare the periplasmic fraction(s) under investigation (*see* **Note 44**).
2. Blank the spectrophotometer over the range of wavelengths from 200 to 800 nm (*see* **Notes 45** and **46**).
3. Remove the buffer from the cuvette in the sample position (it is convenient to use a Pasteur pipet with a rubber bulb or a mechanical pipetting device). It is not necessary to take out the cuvette but make sure that all the buffer is removed.
4. Fill the cuvette with 1 mL of a given periplasmic fraction, make sure that the cuvette compartment lid is closed, and record a spectrum.
5. The major band (Soret band) is expected to be between ~ 400 and 420 nm. If the maximum absorbance in this range exceeds 1–1.5, it is recommended that the sample be diluted, usually somewhere between twofold and fivefold, to bring the Soret band absorbance to within the linear range of the instrument (*see* **Note 47**).
6. Transfer a small amount of sodium dithionite to the cuvette (*see* **Note 48**).
7. Place the cuvette back in the cuvette compartment, make sure that the compartment lid is closed, and record a reduced spectrum (*see* **Note 49**). Normally reduction results in a red shift for the Soret band (~ 10–20 nm) and a replacement of a broad peak around ~ 530 nm by two sharper peaks at ~ 550 and 520 nm (known as α- and β-bands, respectively). Typical spectra for reduced and oxidized forms of a cytochrome *c* are shown in **Fig. 8.3** (*see* **Note 50**).
8. If there is little or no change in the spectrum, several more aliquots of dithionite may be added (*see* **Note 51**).

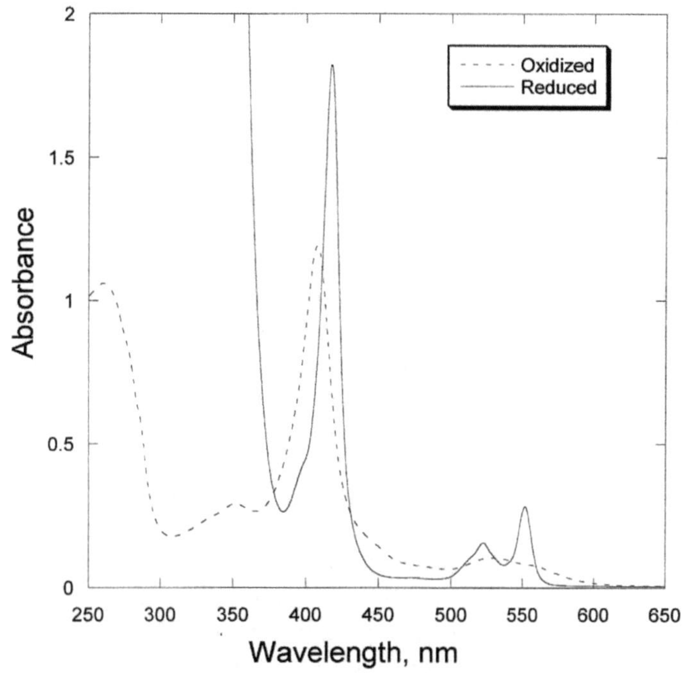

Fig. 8.3. Reduced (*solid line*) and oxidized (*dashed line*) spectra of the periplasmic fraction of *E. coli* cells expressing tri-heme cytochrome PpcA from *Geobacter sulfurreducens*. Note a large peak below ~ 360 nm in the reduced spectrum (products of dithionite oxidation).

9. Take the sample out of the cuvette and fill the cuvette with 1–1.5 mL of deionized water to rinse it (this step is important for removing traces of sodium dithionite). Remove the water with a Pasteur pipet.
10. Remove the blanking buffer from the cuvette in the "blank" position. Rinse it with water as described above.
11. Take a 200 μL aliquot of the rest of the periplasmic fraction (that has not been treated with sodium dithionite) and transfer to a fresh microtube. Add 800 μL of 20 m*M* Tris–HCl with circumneutral pH (*see* **Note 52**). This constitutes a fivefold dilution of the given sample.
12. Prepare 2 mL of the buffer for blanking at 280 nm: mix 400 μL of the buffer used in step 1 and 1600 μL of the buffer used in step 11 for diluting, briefly vortex, and centrifuge for 5–10 s.
13. Add 1 mL of the blanking buffer to both cuvettes and follow the instructions for the spectrophotometer to blank it at a single wavelength, 280 nm.

14. Remove the buffer from the cuvette in the sample position. Fill the cuvette with the diluted periplasmic sample, close the compartment lid, and take absorbance readings.
15. If the absorbance at 280 nm exceeds 1–1.5, the sample should be further diluted (*see* **Note 47**), and new readings should be taken.

3.5. Solubilization Experiments

If small-scale tests show that a large fraction of the target cytochrome is associated with the spheroplast pellet (whether or not the protein has a membrane anchor), solubilization with different agents may be attempted. While extraction of membrane-anchored proteins is traditionally done with various detergents, there is also a possibility that the protein of interest is strongly associated with membranes indirectly, due to its interactions with other membrane-anchored proteins, and these interactions may be predominantly electrostatic in nature. Therefore, solubilization may be accomplished by simply increasing the salt concentration in the lysis buffer to 0.5 *M* (14). This solubilization method is also much more selective than detergent-driven extraction (14). However, it is not possible to predict a priori which approach will work best and, therefore, both should be attempted. Moreover, the solubilization efficiency of different detergents varies depending on the particular protein to be solubilized and the selection of the best detergent is usually done by a trial-and-error method. Several most popular detergents include Triton X-100, lauryldimethylamine-*N*-oxide (LDAO), dodecylmaltoside (*n*-dodecyl-β-D-maltoside), and octylglucoside (*n*-octyl-β-D-glucopyranoside). These chemicals can be purchased from a number of manufacturers, e.g., Sigma. Anatrace (www.anatrace.com) specializes in surfactants and offers the largest selection of ionic, non-ionic, and zwitter-ionic detergents. The recommended working concentration for a detergent is 10X the critical micelle concentration (CMC). CMC numbers can be found in the manufacturer's catalog.

The instructions below assume the expression has been carried out in 25 mL cultures but the protocol can be easily scaled up or down (*see* **Note 53**).

1. Take out a 100 μL aliquot from each culture, centrifuge 5 min in a benchtop centrifuge, discard the supernatant, and freeze the pellet (it will be used to prepare gel samples of total cell lysate).
2. Harvest the bulk of the cells by centrifugation for 15 min at 4000 rpm at 4°C and then remove the supernatant (*see* **Note 23**). Pictures of the pellets can be taken at this point.
3. Gently resuspend the pellets in ice-cold TES buffer containing the amount of protease inhibitors recommended by the

manufacturer (*see* **Note 25**). Use 2 mL of the buffer per pellet from 25 mL of culture.

4. Add lysozyme to the final concentration of 0.5 mg/mL (20 μL of a 50 mg/mL stock for 2 mL suspensions) and incubate 15 min at room temperature (*see* **Note 26**).
5. Add 2 mL of ice-cold 1 *M* NaCl *or* a detergent solution at 20X CMC (*see* **Note 54**). Place the tubes horizontally in a bucket of ice and shake gently (~ 100 rpm) for 15 min.
6. Centrifuge at 12,000×*g* for 20 min at 4°C (*see* **Notes 27** and **28**). Transfer the supernatant to a fresh tube. Continue with gel/Western blot analysis and spectral measurements as described above.

4. Notes

1. The pMAL system (New England Biolabs) features MBP as a carrier protein. In the latest generation of pMAL vectors a genetically engineered version of MBP is used that was specially designed for tighter binding to amylose resin. Vectors pMAL-p5X, pMAL-p5E, and pMAL-p5G include recognition sites for site-specific proteases Factor Xa, Enterokinase, and Genenase I, respectively, which allows the protein of interest to be liberated from the affinity tag after purification. Another vector, pMKL1 (developed by the author), has a hybrid 6×His/MBP tag, a cleavage site for highly specific TEV protease, and a ligation-independent cloning site (16).
2. Other media, such as LB, TB, or CircleGrow, can be used. LB usually produces lower yields than 2×YT. In case isotopic or selenomethionine labeling is required, expression in a minimal medium is feasible if the medium is supplemented with 1 m*M* δ-aminolevulinic acid, a heme precursor (28).
3. The protocol assumes that the expression vector used provides resistance to ampicillin and contains an IPTG-inducible promoter, such as T7, lac, or tac. If the vector carries the resistance to another antibiotic and/or promoter induced with other compounds, e.g., anhydrotetracycline, then the protocol should be adjusted accordingly. It should be noted that the arabinose promoter (P_{BAD}) is not suitable for a tunable induction due to its "all-or-nothing" (autocatalytic) behavior (29).

4. As discussed in **Section 3.3**, the highest expression levels are often obtained with IPTG concentrations in a low micromolar range. This is why a more dilute stock solution (100 m*M*) may be useful.
5. Do not use protease inhibitor cocktails that contain EDTA if a nickel (or cobalt) affinity resin is to be used for purification of the target protein.
6. If you have pre-cast gels for which the manufacturer suggests a different buffer, then follow the manufacturer's instructions.
7. It is convenient to prepare a methanol-free 100 m*M* stock, adjust pH to 11.0 with NaOH, and then make dilutions as necessary. To prepare 1 L of CAPS working solution, mix 100 mL of the 100 m*M* stock and 100 mL of methanol and adjust the volume to 1 L with deionized water.
8. Methanol is poisonous. It may cause blindness and even death if ingested. Avoid skin contact and inhaling vapors. If possible, use a fume hood when preparing the working solution. Always wear gloves when working with methanol or methanol-containing solutions. Check your institution's safety policies on how to dispose of methanol-containing waste.
9. Hybond-C is sold either as rolls or 20 cm × 20 cm sheets. Pre-cut the membrane into pieces corresponding in size to the size of the gel you plan to use for separation (i.e., if you use 9 cm × 10 cm gels, prepare several 9 cm × 10 cm pieces of membrane).
10. The nitrocellulose membrane after the transfer will be placed in the bag and incubated with the SuperSignal solutions. This minimizes the consumption of solutions.
11. Also known as "X-ray film developer" or "X-omat." The latter name is a trademark of X-ray film processors made by Kodak.
12. In the absence of an X-ray film processor, X-ray films used to detect chemiluminescence may be developed manually.
13. Disposable plastic cuvettes are not transparent below 300–350 nm but may be used for measurements in the visible range. Quartz cuvettes are necessary for measuring absorbance at 280 nm.
14. Sodium dithionite is extremely air-sensitive and hygroscopic and must be stored in a desiccator box over a layer of desiccant (e.g., Drierite). It is sold in relatively large containers, 50 or 100 g, but a single instance of reduction only requires a few small crystals. To make the batch last longer, it is recommended to transfer a small amount into a

microtube (no more than one-third full or half full) and use it as necessary. If you use a spatula for transferring, make sure it is absolutely dry (e.g., blot it with a paper towel or a Kleenex sheet). When the tube becomes empty or if there is a reason to believe that the dithionite in the tube is no longer active (*see* **Section 3.4** and **Note 51**), refill it with a fresh aliquot from the batch container. This approach minimizes the number of instances of opening the container and exposing its contents to air.

15. For unknown reasons, cultures grown from fresh transformants often produce higher yields of cytochromes *c* than cultures originated from frozen stocks (even if a frozen stock has been streaked out on a plate and a liquid culture has been started from a single colony). Therefore, at least during the preliminary experiments, it is recommended to carry out a new transformation every time for each series of experiments with a given vector/strain combination.
16. If you only have a strain harboring pEC86 rather than the plasmid itself and desire to transform another strain, the plasmid can be isolated following standard protocols for plasmid purification. Keep in mind that it is a low-copy number plasmid and its yield will be low, as compared to typical pET or pUC vectors, but still more than adequate to transform another strain.
17. *E. coli* strain HM125 CANNOT grow at 37°C. Incubate it at 30°C for at least 36 h.
18. The plates can also be incubated at room temperature, ~ 25°C, for 2–3 days (e.g., the transformation can be done on Friday and left at room temperature until Monday).
19. Usually as little as 50 μL is more than enough for a single transformation. For a number of strains, e.g., BL21(DE3), aliquots of competent cells can be thawed on ice and re-frozen 3–4 times without a significant decrease in competence. It is easier to prepare and store, for example, 50 aliquots of 200 μL than 200 aliquots of 50 μL. When a single transformation reaction is needed, one can thaw a 200 μL aliquot on ice, take out 50 μL, and use it for transformation, then re-freeze the remaining 150 μL until next time.
20. If you are starting to work with a new expression construct, it may be helpful to include a negative control in the experiment: the cells harboring pEC86 but transformed with the "empty" vector.
21. Because of the periplasmic localization of cytochromes *c*, the cells after expression should be immediately treated for isolation of the periplasmic fraction and should NOT be

frozen! (Otherwise, the inner membrane will break when the cells are thawed and the periplasmic fraction will essentially become total cell lysate). Therefore, keep in mind that the entire transformation/expression/analysis procedure will take 4 days. In other words, do not start up a transformation on Wednesday or Thursday unless you plan to work during the weekend. However, a transformation can be done on Friday and the plate(s) can be left at room temperature over the weekend.

22. The suggested IPTG concentration(s) is indicated here as a starting point for a new target that has not been expressed before. It can be adjusted if there is previous data on similar proteins, e.g., if the samples being tested are point mutants of a cytochrome for which the optimal IPTG concentration is 0.1 m*M*, then it is sensible to use 0.1 m*M* for the mutants as well.
23. If the cytochrome is expected to be an outer membrane protein, then the supernatant should be saved and analyzed together with the other fractions as it may contain a significant fraction of the cytochrome (it may be necessary to concentrate the supernatant 10-fold or even more to obtain a detectable signal).
24. A rough estimate of expression levels can be done at this point based on the color of each pellet (and pictures can be taken to document that!). Cells that produce relatively high levels of a cytochrome would have a strong pink color. However, cells may also have reddish, brown, or even greenish color. The lack of distinct color does not necessarily mean that the expression completely failed. Therefore, the rest of the protocol should be carried out in any case.
25. Pipet with 1 mL tips to break down clumps of cells. Vortexing can be used but it usually fails to break down clumps completely. Ideally, the suspension should look homogeneous, without any visible chunks of pellet.
26. Avoid higher lysozyme concentrations and much longer incubation times as they may result in breakage of the inner membrane and contaminating the periplasmic fraction with the cytoplasmic content, including the chromosomal DNA.
27. As 2 mL of periplasmic fractions is usually more than enough for analytical purposes, 2 mL aliquots of each sample can be transferred to 2 mL microtubes and centrifuged in a benchtop centrifuge at 4°C at maximal speed for 15 min.
28. If a sample is viscous and difficult or impossible to transfer, add 1 μL Benzonase Nuclease and continue with centrifugation. Benzonase is active enough at 4°C to digest

the chromosomal DNA to the point where it no longer increases viscosity.

29. The peroxidase activity of denatured cytochromes gradually decays with time and in 2–3 days often becomes undetectable. Therefore, make every effort to run the gel and Western blotting as soon as possible, at least within 24 h. If you do not plan a gel analysis within a day, do NOT make denatured gel samples but freeze aliquots of each periplasmic fraction and keep them frozen until you are ready to start the gel analysis. Purified marker proteins should be stored at 4°C and small aliquots of gel samples (200–500 ng per load) should be prepared together with the rest of the samples.
30. If possible, load the markers in both outside lanes AND at least in one lane in the middle. The marker lanes will provide reference points and facilitate assigning of signals on the film in cases when only a few lanes manifest any signal at all.
31. If there is a reason to believe that the cytochrome(s) under study may be extensively degraded, it is sensible to run the gel for a shorter time, so that the dye front advances only two-thirds or three-fourths of the distance. This will allow visualization of short fragments (2–5 kDa) resulting from degradation. Note, however, that total cell lysates very often have a strong band near the dye front, probably due to non-covalently bound hemes released from cytochromes *b* and *d* during denaturation.
32. The exact assembling procedure varies depending on the type of transfer apparatus used but the following recommendations should always be heeded. First, always wear gloves when handling membranes or gels, both to avoid contamination of membranes and to protect the hands from methanol and acrylamide. Second, when assembling the transfer unit, make sure that the membrane is on the side of the gel facing the anode. Third, cut a small piece off the corner of the membrane that corresponds to the cut-off corner of the gel and note the relative orientation of the membrane with respect to the gel (e.g., when the side of the membrane contacting the gel is facing the worker, the cut-off corner is lower right).
33. The bag can be cut with scissors or a razor, which makes it easier to extract the membrane.
34. If you are just learning the transfer procedure, it may be helpful to deliberately overload one or two lanes with heme-containing markers (1–2 μg per lane). The intense "rusty" bands on the membrane after the transfer

procedure will indicate successful transfer, while the absence of such bands will signal a failed transfer (and will prevent you from wasting time and supplies to carry out the rest of the protocol). Alternatively, pre-stained molecular weight markers can be used.

35. Accordingly, the absence of signal for periplasmic fraction indicates that little or no protein is present in the soluble form.
36. The author has encountered at least one case of very low transfer efficiency for the protein of interest. If there is a reason to suspect that, e.g., the cells have a distinct coloration but no signal is detected for the total cell lysate, peroxidase detection can be done directly in gel, although it will require larger volumes of both reagents, ~ 5–10 mL, depending on the size of the gel.
37. The optimization procedure may also include induction at different cell densities. However, it is possible that IPTG titrations produce different results at different cell densities, i.e., induction at A_{600}=0.75 works best with, for example, 50 μ*M* IPTG and induction at A_{600}=2.0 works best with 25 μ*M* IPTG.
38. It usually takes between 5 and 8 h for a culture to reach A_{600}=1.0 under these conditions. If the reading for a given flask is below 0.2, take another reading in 1 h. If the reading is greater than 0.2, take readings every 30 min.
39. Use the flattened tip of the rubber policeman to rub the cell pellet against the bottom (or walls) of the centrifuge bottle until there are no visible clumps of cells. The efficiency of lysis depends on the efficiency of resuspending the pellet to near homogeneity.
40. Many centrifuge rotors cannot safely spin large (1 L) bottles at speeds greater than 5000–6000 rpm. Therefore, if steps 9 and 10 have been carried out with such large bottles, the cell suspensions should be transferred to 250 mL centrifuge bottles compatible with high-speed rotors.
41. It is recommended that the suspensions be transferred to 250 mL bottles BEFORE (or immediately after) adding lysozyme; otherwise, if lysozyme concentration was too high and/or incubation continued too long, the cells will break and release chromosomal DNA, making the suspension viscous and difficult to handle.
42. If a supernatant is viscous and difficult to transfer, add 5 μL of Benzonase Nuclease, incubate 15–30 min at room temperature, and repeat the centrifugation step.

43. If the goal is to optimize expression conditions for a given cytochrome, then it is strongly recommended that IPTG titration of large-scale cultures be repeated at least twice or even thrice, especially if there is a large scattering of results (which is not unusual for cytochromes). However, in a number of cases the amount of cytochrome produced in the very first large-scale experiments (4–6 L of culture induced with different IPTG concentrations) may be sufficient (especially if the purification protocol is simple, such as affinity chromatography) for such purposes as crystallization or antibody production. In this case there is obviously no need to further optimize the protocol.
44. Normally, the buffer would be TES (containing 0.5 mg/mL lysozyme and the protease inhibitors) diluted 1:1 with water. However, if the lysis procedure involved NaCl, detergents, or other additives (*see* **Section 3.5**), they should be included as well.
45. The exact range will depend on the spectrophotometer (as some spectrophotometers do not detect below 250 nm) and how much information one desires to obtain (some cytochromes have small peaks between 600 and 700 nm but for initial experiments the α-, β-, and Soret bands are usually enough).
46. For many spectrophotometers the procedure to blank over a range of wavelengths is slightly different from the blanking procedure at a single wavelength. Follow the manufacturer's instructions.
47. Many older models of spectrophotometers are incapable of accurately measuring samples that have relatively high absorbencies ($> \sim 1$). A simple test of whether a given absorbance value is still within "accuracy limits" for a given machine is to dilute the sample 1:1 and measure again. The new absorbance value should be approximately half the absorbance of the undiluted sample ± 5%. If it is significantly higher, then the new (diluted) sample should be diluted 1:1 again and a new measurement should be taken. The same buffer as was used for blanking should be used for making dilutions. After some practice one should be able to estimate the Soret band absorbance and the required dilution factor simply by the color intensity of the periplasmic fraction.
48. A convenient way to add dithionite is to "poke" the solid substance in the microtube several times with a 200 μL pipet tip attached to a pipet. As a result, a small amount of the salt will remain in the tip and can be transferred to the cuvette. (This manipulation requires that the tube have

dithionite at least 0.5–1 mm deep in the bottom. In other words, if there is only a thin layer of crystals in the bottom, then one may not be able to pick up a sufficient amount of dithionite. If this is the case, refill the tube.) Pipetting movements can then be used to dissolve the crystals and mix the solution to homogeneity. If you take the cuvette out of the cuvette compartment and look at it from the side (not from the top down), while pipetting the solution, very often a very distinct change of color of the solution can be observed as dithionite dissolves and reduces the heme(s).

49. The products of dithionite oxidation have a very strong absorbance below 350 nm. This will appear as a large peak, usually much bigger than the Soret band. The occurrence of this peak is normal and no reason for concern.

50. Depending on the nature of a particular protein, other peaks may be present.

51. If after adding 3–4 portions of solid dithionite there is still no change in the spectrum, it indicates that either the cytochrome under study was already in the reduced (ferrous) state or the dithionite in the tube has oxidized and is no longer usable. Refill the tube (or fill another tube) from the batch container as described above (*see* **Note 14**) and add an aliquot of fresh dithionite to the sample. The absence of changes in the spectrum after that is still not fully conclusive as it is possible that the entire batch of dithionite has oxidized (usually as a result of improper storage). It is helpful to have a known sample of a cytochrome *c* to test the dithionite. Commercially available mitochondrial cytochromes (e.g., from horse heart) are suitable substrates.

52. Instead of Tris–HCl, other buffers can be used, e.g., HEPES or phosphate, and concentrations as low as 10 m*M* are acceptable. TES buffer itself can also be used.

53. Many detergents are relatively expensive. Therefore, it is recommended to use small-scale tests for evaluation of different solubilizing agents. When the best solubilization approach is selected and you switch to large-scale experiments, be guided by the volumes described in **Section 3.3.2** (30 mL of TES per pellet from 1 L, etc.) but make sure that the final concentration of the solubilizing agent is the same as established in small-scale trials.

54. The goal here is to have NaCl at a final concentration of 0.5 *M* or the detergent at 10X CMC. It can also be accomplished by adding separately ice-cold water and a concentrated, ice-cold solution of a solubilizing agent, e.g., 1.6 mL of water and 0.4 mL of 5 *M* NaCl.

Acknowledgments

The author would like to thank Dr. L. Thöny-Meyer (ETH, Zürich, Switzerland) for plasmid pEC86 carrying the *ccm* genes; Drs. P.R. Pokkuluri and M. Schiffer (Argonne National Laboratory) and C.A. Salgueiro (Universidade Nova de Lisboa, Lisbon, Portugal) for helpful discussions and encouragement; Dr. T. Khare (Argonne National Laboratory) for help in mastering heme staining procedures; and Drs. T. Evans and P. Weigele (New England Biolabs) for critical reading of the manuscript.

References

1. Mathews, F. S. (1985) The structure, function and evolution of cytochromes. *Prog Biophys Mol Biol* 45, 1–56.
2. Thöny-Meyer, L. (1997) Biogenesis of respiratory cytochromes in bacteria. *Microbiol Mol Biol Rev* 61, 337–376.
3. Pollock, W. B., Chemerika, P. J., Forrest, M. E., Beatty, J. T., Voordouw, G. (1989) Expression of the gene encoding cytochrome c_3 from *Desulfovibrio vulgaris* (Hildenborough) in *Escherichia coli*: export and processing of the apoprotein. *J Gen Microbiol* 135, 2319–2328.
4. Grisshammer, R., Oeckl, C., Michel, H. (1991) Expression in *Escherichia coli* of *c*-type cytochrome genes from *Rhodopseudomonas viridis*. *Biochim Biophys Acta* 1088, 183–190.
5. Londer, Y. Y., Dementieva, I. S., D'Ausilio, C. A., Pokkuluri, P. R., Schiffer, M. (2006) Characterization of a *c*-type heme-containing PAS sensor domain from *Geobacter sulfurreducens* representing a novel family of periplasmic sensors in *Geobacteraceae* and other bacteria. *FEMS Microbiol Lett* 258, 173–181.
6. Pokkuluri, P. R., Pessanha, M., Londer, Y. Y., Wood, S. J., Duke, N. E. C., Wilton, R., Catarino, T., Salgueiro, C. A., Schiffer, M. (2008) Structures and solution properties of two novel periplasmic sensor domains with *c*-type heme from chemotaxis proteins of *Geobacter sulfurreducens*: implications for signal transduction. *J Mol Biol* 377, 1498–1517.
7. Thöny-Meyer, L., Fischer, F., Kunzler, P., Ritz, D., Hennecke, H. (1995) *Escherichia coli* genes required for cytochrome *c* maturation, *J Bacteriol* 177, 4321–4326.
8. Page, M. D., Sambongi, Y., Ferguson, S. J. (1998) Contrasting routes of *c*-type cytochrome assembly in mitochondria, chloroplasts and bacteria. *Trends Biochem Sci* 23, 103–108.
9. Thöny-Meyer, L. (2000) Haem-polypeptide interactions during cytochrome *c* maturation, *Biochim Biophys Acta* 1459, 316–324.
10. Kranz, R., Lill, R., Goldman, B., Bonnard, G., Merchant, S. (1998) Molecular mechanisms of cytochrome *c* biogenesis: three distinct systems. *Mol Microbiol* 29, 383–396.
11. Grove, J., Tanapongpipat, S., Thomas, G., Griffiths, L., Crooke, H., Cole, J. (1996) *Escherichia coli* K-12 genes essential for the synthesis of *c*-type cytochromes and a third nitrate reductase located in the periplasm. *Mol Microbiol* 19, 467–481.
12. Arslan, E., Schulz, H., Zufferey, R., Kunzler, P., Thöny-Meyer, L. (1998) Overproduction of the *Bradyrhizobium japonicum c*-type cytochrome subunits of the *cbb*$_3$ oxidase in *Escherichia coli*. *Biochem Biophys Res Commun* 251, 744–747.
13. Londer, Y. Y., Pokkuluri, P. R., Erickson, J., Orshonsky, V., Schiffer, M. (2005) Heterologous expression of hexaheme fragments of a multidomain cytochrome from *Geobacter sulfurreducens* representing a novel class of cytochromes *c*. *Protein Expr Purif* 39, 254–260.
14. Londer, Y. Y., Pokkuluri, P. R., Orshonsky, V., Orshonsky, L., Schiffer, M. (2006) Heterologous expression of dodecaheme "nanowire" cytochromes *c* from *Geobacter sulfurreducens*. *Protein Expr Purif* 47, 241–248.
15. Sanders, C., Lill, H. (2000) Expression of prokaryotic and eukaryotic cytochromes *c* in *Escherichia coli*. *Biochim Biophys Acta* 1459, 131–138.
16. Londer, Y. Y., Giuliani, S. E., Peppler, T., Collart, F. R. (2008) Addressing *Shewanella oneidensis* "cytochromome": the first

step towards high-throughput expression of cytochromes *c*. *Protein Expr Purif* 62, 128–137.
17. Bothmann, H., Pluckthun, A. (1998) Selection for a periplasmic factor improving phage display and functional periplasmic expression. *Nat Biotechnol* 16, 376–380.
18. Schäfer, U., Beck, K., and Müller, M. (1999) Skp, a molecular chaperone of gram-negative bacteria, is required for the formation of soluble periplasmic intermediates of outer membrane proteins. *J Biol Chem* 274, 24567–24574.
19. Mavrangelos, C., Thiel, M., Adamson, P. J., Millard, D. J., Nobbs, S., Zola, H., Nicholson, I. C. (2001) Increased yield and activity of soluble single-chain antibody fragments by combining high-level expression and the Skp periplasmic chaperonin. *Protein Expr Purif* 23, 289–295.
20. Bothmann, H., Pluckthun, A. (2000) The periplasmic *Escherichia coli* peptidylprolyl *cis,trans*-isomerase FkpA. I. Increased functional expression of antibody fragments with and without *cis*-prolines. *J Biol Chem* 275, 17100–17105.
21. Arie, J. P., Sassoon, N., Betton, J. M. (2001) Chaperone function of FkpA, a heat shock prolyl isomerase, in the periplasm of *Escherichia coli*. *Mol Microbiol* 39, 199–210.
22. Gordon, E. H., Steensma, E., Ferguson, S. J. (2001) The cytochrome *c* domain of dimeric cytochrome cd_1 of *Paracoccus pantotrophus* can be produced at high levels as a monomeric holoprotein using an improved *c*-type cytochrome expression system in *Escherichia coli*. *Biochem Biophys Res Commun* 281, 788–794.
23. Meerman, H. J., Georgiou, G. (1994) Construction and characterization of a set of *E. coli* strains deficient in all known loci affecting the proteolytic stability of secreted recombinant proteins. *Biotechnology (New York)* 12, 1107–1110.
24. Londer, Y. Y., Pokkuluri, P. R., Tiede, D. M., Schiffer, M. (2002) Production and preliminary characterization of a recombinant triheme cytochrome c_7 from *Geobacter sulfurreducens* in *Escherichia coli*. *Biochim Biophys Acta* 1554, 202–211.
25. Brigé, A., Cole, J. A., Hagen, W. R., Guisez, Y., Van Beeumen, J. J. (2001) Overproduction, purification and novel redox properties of the dihaem cytochrome *c* NapB, from *Haemophilus influenzae*. *Biochem J* 356, 851–858.
26. Hoover, D. M., Lubkowski, J. (2002) DNAWorks: an automated method for designing oligonucleotides for PCR-based gene synthesis. *Nucleic Acids Res* 30, e43.
27. Feissner R., Xiang, Y., Kranz, R. G. (2003) Chemiluminescent-based methods to detect subpicomole levels of *c*-type cytochromes. *Anal Biochem* 315, 90–94.
28. Fernandes, A. P., Couto, I., Morgado, L., Londer, Y. Y., Salgueiro, C. A. (2008) Isotopic labeling of *c*-type multiheme cytochromes overexpressed in *E. coli*. *Protein Expr Purif* 59, 182–188.
29. Siegele, D. A., Hu, J. C. (1997) Gene expression from plasmids containing the *araBAD* promoter at subsaturating inducer concentrations represents mixed populations. *Proc Natl Acad Sci USA* 94, 8168–8172.

Chapter 9

Semi-synthesis of Glycoproteins from *E. coli* Through Native Chemical Ligation

Jonathan P. Richardson and Derek Macmillan

Abstract

Sufficient quantities of homogeneous samples of post-translationally modified proteins are often not readily available from biological sources to facilitate structure–function investigations. Native chemical ligation (NCL) is a convenient method for the production of biologically active proteins from smaller fragments. Such an approach allows protein modifications to be introduced in a controlled fashion into smaller peptide fragments which are amenable to total chemical synthesis. These fragments of defined sequence and structure can be elaborated to full-length proteins through NCL reactions with suitable components derived from bacterial origin. This report describes methods for the bacterial production of components for NCL and their use in typical reactions.

Key words: Glycoprotein, intein, thioester, glycopeptide, native chemical ligation.

1. Introduction

Glycoproteins are characterized by a covalent linkage between carbohydrates, usually *N*-acetyl glucosamine, *N*-acetyl galactosamine (but also mannose, xylose and fucose), and the amino acids asparagine and serine or threonine, respectively (1, 2). Examples of N-linked and O-linked oligosaccharides are shown in **Fig. 9.1a**. A handful of groundbreaking studies have shown that the total chemical synthesis of certain glycopeptide components of glycoproteins is possible, though still far from routine (3–10). Indeed, the difficulty associated with glycopeptide synthesis has precluded much biological research using entirely chemical approaches. To address this, numerous innovative

T.C. Evans, M.-Q. Xu (eds.), *Heterologous Gene Expression in E. coli*, Methods in Molecular Biology 705,
DOI 10.1007/978-1-61737-967-3_9,

strategies have been developed. For decades researchers have attempted to site-specifically modify bacterially derived proteins with synthetic saccharides (11–15). While an attractive route, since *E. coli* can produce vast quantities of protein, bacteria often struggle to fold proteins efficiently and considerable losses of material can result during refolding steps (16). More importantly the resulting molecules derived, for example, by comprehensive modification of lysine residues, or selective modification of introduced cysteine residues are unnatural and mere analogues of native glycoproteins. An early breakthrough was the attempted use of insect cells (17, 18) to produce simply glycosylated proteins for further engineering using available enzymes, but this has largely been superseded by the use of engineered yeast. Yeast-based production of mammalian glycoproteins is relatively inexpensive when compared to tissue culture, produces folded and glycosylated proteins in quantity and with some further engineering (transfection with exogenous enzymes such as GlcNAcT I and α(2,3)SiaT), a human-like glycosylation profile can be achieved. An inherent disadvantage of yeast-based systems is that, like glycoproteins of human origin, they are heterogeneous with respect to the site occupancy and identity of individual carbohydrate appendages. Additional strategies including the transfer of genes encoding entire glycosylation pathways into bacteria (19, 20) and unnatural amino acid mutagenesis (21) may also provide access to simply glycosylated proteins in the future.

Native chemical ligation (NCL) (22) is an extremely useful method for the production of synthetic and semi-synthetic proteins (*see* **Fig. 9.1b**). A powerful feature of NCL is that a synthetic component can be site-specifically phosphorylated or glycosylated, retaining native sugar–protein linkages and allowing the assembly of homogeneously phosphorylated/glycosylated proteins that are not readily available from biological sources. The use of NCL to control the identity and the position of glycosylation allows control and versatility that is only limited by the imagination and capability of the synthetic chemist, rendering NCL the most widely used non-cell-based approach to glycoprotein synthesis (3, 23–26). Since it is widely accepted that approximately 50 amino acid residues represent the limit for efficient automated solid-phase peptide synthesis (SPPS), the pursuit of recombinant methods for the production of the required thioester and cysteine-containing components is particularly attractive (27, 28). Microorganism-derived thioesters are most frequently generated by a commercially available intein-fusion expression system, though the use of such vectors for protein semi-synthesis dictates that target proteins must be specifically cloned such that no vector-encoded amino acids intervene between ligated fragments (29). Furthermore, most proteins are not readily expressed with a free N-terminal cysteine residue and consequently a

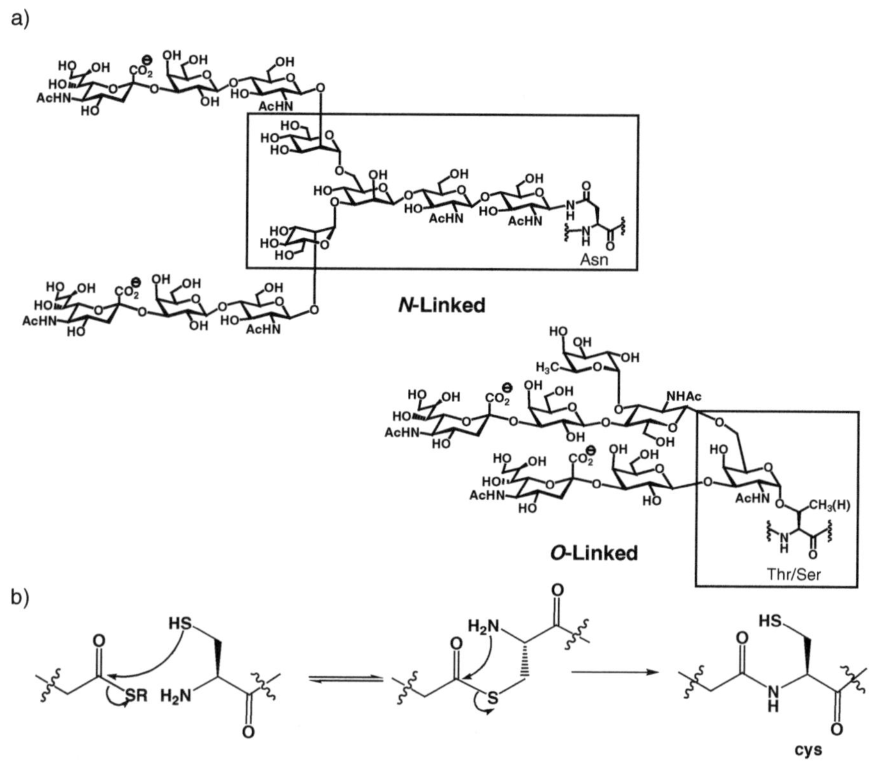

Fig. 9.1. (**a**) Typical *N*- and *O*-linked oligosaccharides of glycoproteins. (**b**) The mechanism of native chemical ligation.

pre-cysteine sequence, that can be later proteolytically processed to the free N-terminal cysteine, is required. In several instances protein fragments, which are the tools for NCL, are highly insoluble and incompatible with many proteases. Thankfully we can resort to chemical methods for selective protein cleavage in such cases. Below, we describe methods that are commonly used in our laboratory, and others, which can be used reliably to access bacterially produced precursor components for native chemical ligation.

2. Materials

2.1. Biological Materials

1. Plasmid DNA containing the gene of interest prepared by Qiagen Mini prep kit.
2. Plasmid for recombinant thioester production (pTYB-1, New England Biolabs) and N-terminal cysteine containing protein fragments (e.g. pTYB-1 or pET-16b (Novagen) for insoluble protein fragments).

3. High-fidelity DNA polymerase, e.g. Deep Vent™ DNA Polymerase (New England Biolabs) or *Pfu* Turbo DNA polymerase (Stratagene) with appropriate reaction buffers.
4. Deoxynucleoside triphosphates: dATP, dTTP, dCTP, dGTP (Roche) prepared in dH_2O as an equimolar stock solution containing 10 mM of each dNTP.
5. Oligonucleotide primers prepared as 10 μM stocks in dH_2O.
6. *Nde*I, *Bam*HI, *Dpn*I, *Sap*I restriction endonucleases (New England Biolabs).
7. 0.2 mL thin-walled PCR tubes (Eppendorf).
8. Mutagenic oligonucleotide primers (Sigma Genosys) each prepared in dH_2O as a 25 ng/μL stock solution.
9. Competent *E. coli* cells (cloning host XL-1 Blue or expression hosts B834 DE3/ER2566).
10. Sterile LB medium: Mix (per litre) 5 g yeast extract, 10 g tryptone and 10 g NaCl. Autoclave at 121°C for 20 min and let it cool down to room temperature.
11. Sterile LB-agar medium: Mix (per litre) 5 g yeast extract, 10 g tryptone, 10 g NaCl and 15 g agar. Autoclave at 121°C for 20 min and let cool down to 50°C before addition of antibiotic. Store plates at 4°C.
12. Plastibrand® disposable 1.5 mL semi-micro cuvettes.
13. Ampicillin: 100 mg/mL stock solution. Filter sterilize and use at a final concentration of 100 μg/mL.
14. IPTG: Isopropyl-thio-β-D-galactopyranoside prepared as a 1 M solution in dH_2O.
15. PMSF: Phenylmethanesulfonyl fluoride (Sigma) prepared as a 0.1 M stock solution in isopropanol. Store at –20°C, bring to room temperature before use.
16. Econo-Pac® chromatography column with a 20 mL bed volume and 30 μM polyethylene bed support (BioRad).
17. Ni-NTA His-Bind Resin (Novagen).
18. Urea stock solution: 99.5%, analytical reagent grade (BDH Chemicals) prepared as an 8 M stock in dH_2O.
19. DTT: Dithiothreitol 99% (Fisher Scientific) prepared as a 1 M stock solution in dH_2O.
20. SPECTRA/Por® dialysis membrane, MWCO = 6–8000 (Spectrum Laboratories, Inc.).
21. Chitin Beads (New England Biolabs).
22. T4 DNA ligase (New England Biolabs).
23. Factor Xa protease (Novagen).

24. Transformation buffer: 50 mM $CaCl_2$, 10 mM Tris–HCl; pH 8.0.
25. SDS-PAGE sample buffer: 10% w/v SDS, 50% v/v glycerol, 400 mM Tris–HCl; pH 6.8, 5 mM EDTA, 0.1% w/v 4-bromophenol blue. Add fresh DTT to an aliquot of this stock to give a concentration of 250 mM.
26. Cell lysis buffer (*see* **Section 3.5.2**): 100 mM sodium phosphate; pH 8.0, 100 mM NaCl, 0.1% v/v Tween 20.
27. Wash buffer: 100 mM sodium phosphate; pH 8.0, 100 mM NaCl.
28. Ligation buffer (*see* **Section 3.9.1**): 100 mM sodium phosphate; pH 8.0, 100 mM NaCl, 2% w/v MESNA, 10 mM TCEP.
29. Cell lysis buffer II (*see* **Section 3.7.2**: 20 mM Tris–HCl pH 8.0, 500 mM NaCl, 5 mM imidazole prepared in 6 M GnHCl.
30. Wash buffer II: 20 mM Tris–HCl pH 8.0, 500 mM NaCl, 50 mM imidazole, prepared in 6 M GnHCl.
31. Elution buffer: 20 mM Tris–HCl pH 8.0, 500 mM NaCl, 500 mM imidazole, prepared in 6 M GnHCl.

2.2. Chemical Synthesis

1. Pre-loaded 4-sulfamylbutyryl-AM resin and NovaSyn®TGT resins for the chemical synthesis of peptide C-terminal thioesters and acids, respectively (*see* **Note 1**).
2. Suitably protected (Fmoc) amino acids and glycoamino acids (Merck Biosciences) (*see* **Note 2**).
3. Coupling reagents for peptide synthesis, e.g. HBTU/HOBt or DCC/HOBt (*see* **Note 3**).
4. Solvents for solid phase peptide synthesis (dichloromethane, NMP, DMF).
5. Piperidine (Sigma).
6. Iodoacetonitrile (Alfa Aesar).
7. *N*, *N*-Diisopropylethylamine (Aldrich).
8. Benzylmercaptan (Alfa Aesar).
9. Benzenethiol, sodium salt.
10. Trifluoroacetic acid (Alfa Aesar).
11. 1,2-Ethanedithiol (Sigma).
12. Diethyl ether (VWR).
13. Guanidine hydrochloride (GnHCl) 98% purum (Fluka).
14. Formic acid 98–100%.

15. Cyanogen bromide 97% (Sigma). Store at −20°C (SAFETY! *see* **Note 4**).
16. Nitrogen gas, reagent grade (BOC).
17. Hydrochloric acid (3 M stock in dH_2O).
18. Sodium hydroxide (10 M stock in dH_2O).
19. 2-Mercaptoethanol, 98% (Lancaster Synthesis).
20. 2-Mercaptoethanesulfonic acid sodium salt (MESNA) (Sigma).
21. 4-Mercaptophenylacetic acid (MPAA) (Aldrich).
22. Tris-carboxyethylphosphine (TCEP) hydrochloride (Aldrich).

2.3. Equipment

1. Applied Biosystems model 433A automated peptide synthesizer (*see* **Note 2**).
2. UPLC-MS system or equivalent for synthetic peptide and glycopeptide analysis. We use a Waters SQD-LC-MS system with C_{18} LC-MS column (2.1×50 mm) and a gradient of 5–95% acetonitrile containing 0.1% TFA over 8 min (flow rate of 0.6 mL/min).
3. HPLC system for Semi-preparative HPLC fitted with a Phenomenex Proteo column (10×210 mm) and a gradient of 5–60% acetonitrile containing 0.1% TFA over 50 min, flow rate = 4.0 mL/min.

3. Methods

We describe procedures for the preparation of protein fragments for NCL and their subsequent use in typical NCL reactions (*see* **Fig. 9.2**). The most important feature when cloning protein fragments for NCL is that there are no additional intervening vector-derived amino acids at the ligation junction. When using expression vector pTYB-1 it is essential that the *Sap*I restriction site is used at the 3′-end of any amplified fragment. It is also important to choose the ligation site with care, utilizing native cysteine residues in the protein sequence for the ligation site where possible. Additionally, the pTYB-1 vector should not be used to access recombinant thioesters of Arg, Asp, Asn, Cys or Pro (corresponding to R-C, D-C, N-C, C-C or P-C ligation junctions, respectively) due to extensive in vivo cleavage in the case of the first two and poor in vitro cleavage in the case of the latter three.

Oligonucleotide primers were obtained fully desalted and deprotected from Sigma-Genosys. De-ionized water (dH_2O) was

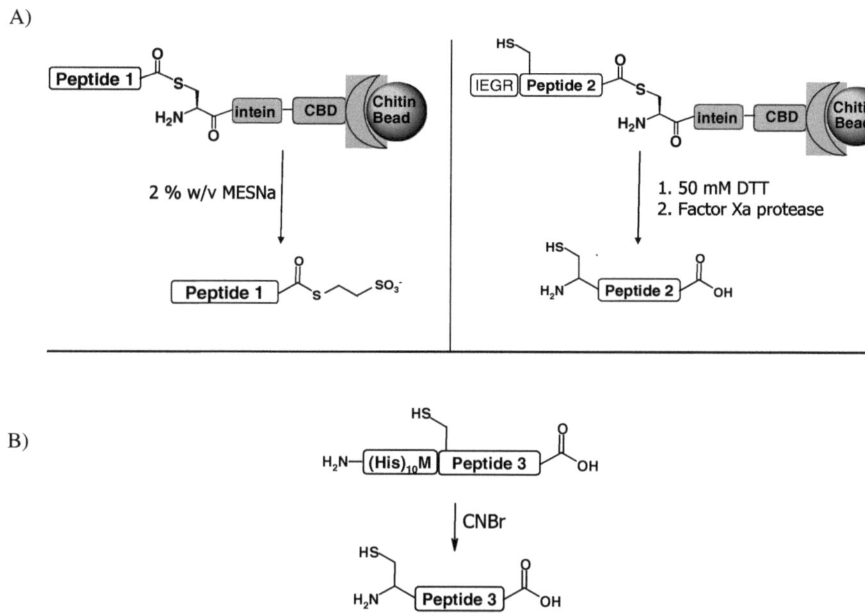

Fig. 9.2. (**a**) Generation of the required C-terminal thioester and N-terminal cysteine containing precursors for NCL from soluble intein-fused polypeptide precursors. The C-terminal thioester can be released upon exposure to 2% w/v MESNA (32) and the N-terminal cysteine residue can be obtained following hydrolysis of the intein-fused precursor and exposure to Factor Xa protease which cleaves an introduced IEGR tetrapeptide sequence. (**b**) Where peptide fragments are highly insoluble the N-terminal cys component can be conveniently generated from a poly-histidine tagged precursor using cyanogen bromide.

obtained using an ELGA OPTION-E 10BP apparatus (Vivendi Water Systems Ltd). Materials listed as sterile were rendered so by autoclaving at 121°C for 15 min. DNA was prepared from bacterial culture using the QIAprep® Spin Miniprep Kit and isolated following electrophoresis using the QIAquick® Gel Extraction Kit (Qiagen) in accordance with the manufacturer's instructions. *Escherichia coli* XL-1 Blue, ER2566 and B834(DE3) cells were obtained from Stratagene, New England Biolabs and Novagen, respectively.

3.1. Vector Construction

Primers used to clone protein fragments into pTYB-1 for the production of soluble peptide thioesters and *soluble* protein fragments containing an N-terminal cysteine are depicted in **Table 9.1** (*see* **Fig. 9.2a**).

Typical primers for construction of pET-16b-derived *insoluble* protein fragments containing an N-terminal cysteine are depicted in **Table 9.2**. As this chemical method for N-terminal cysteine production cleaves proteins at Met residues then internal Met residues need to be mutated (to Leu residues in this case) by site-directed mutagenesis (SDM). The N-terminal poly-histidine tag, employed for purification, is also removed during the protein cleavage reaction (*see* **Fig. 9.2b**).

Table 9.1
Primers required for gene insertion into vector pTYB-1[a,b]

Nde*I forward primer*
5′-GGT GGT CAT ATG (NNN)$_5$-3′.
Nde*I forward primer for N-terminal cysteine*[c]
5′-GGT GGT CAT ATG **ATC GAA GGT CGT** TGT (NNN)$_5$-3′.
Sap*I reverse primer*[d]
5′-GTT GTT TGC TCT TCC GCA (NNN)$_5$-3′.

[a](NNN)$_x$ corresponds to flanking codons from the target gene
[b]The endonuclease recognition site is underlined in each case
[c]The codons in bold correspond to the factor Xa protein recognition sequence Ile-Glu-Gly-Arg placed adjacent to the N-terminal cys (encoded by TGT)
[d]The reverse primer should *not* contain a stop codon

Table 9.2
Primers required for gene insertion into pET-16b[a,b]

Nde*I forward primer*[c]
5′-GGT GGT CAT ATG **TGC** (NNN)$_5$-3′.
Bam*HI reverse primer*[d]
5′-GGT GGT GGA TCC (NNN)$_5$-3′.
SDM removal of internal Met residues (M→L mutation)
Forward primer
5′-(NNN)$_5$ GAA (NNN)$_5$-3′.
Reverse primer
5′-(NNN)$_5$ TTC (NNN)$_5$-3′.

[a](NNN)$_x$ corresponds to flanking codons from the target gene
[b]The endonuclease recognition site is underlined in each case
[c]The codon encoding the N-terminal cys is shown in bold
[d] This reverse primer should contain a stop codon

1. In a PCR tube combine 1 μL of plasmid template DNA (containing the gene of interest), 5 μL forward primer, 5 μL reverse primer (to a final primer concentration of 1 μM), 5 μL dNTP stock solution, 5 μL of 10X Thermopol buffer, 28 μL of sterile dH_2O and 1 μL (1 unit) Deep Vent DNA polymerase.
2. Perform PCR in a thermal cycler using the following parameters: 1 cycle of 96°C, 2 min followed by 30 cycles of 96°C for 1 min, 54°C for 30 s, 60°C for 30 s and a final cycle of 60°C for 5 min followed by a 4°C hold. PCR parameters will vary with the characteristics of the template and primers.
3. Purify the PCR product by agarose gel electrophoresis and extract DNA using the QIAquick® Gel Extraction Kit according to the manufacturer's instructions.

4. Digest isolated PCR product with *Nde*I and *Sap*I for PCR products to be inserted into pTYB-1 or *Nde*I and *Bam*HI for PCR products to be inserted into pET-16b.
5. Purify digested product as described in step 3.
6. Ligate the purified *Nde*I *Bam*HI PCR fragment into similarly digested pET-16b and proceed to **Section 3.2** (*see* **Note 5**).

3.2. Mutagenesis of Internal Methionine Residues

1. In a PCR tube combine 1 μL of a 1:10 dilution of plasmid DNA (pET-16b containing sequenced gene of interest, approximately 12 ng), 5 μL of 10X cloned *Pfu* reaction buffer (*see* **Note 6**), 2 μL of dNTP stock, 5 μL of forward primer, 5 μL of reverse primer (to a final primer concentration of 1 μM), 1 μL of $MgCl_2$ stock, 30 μL of sterile dH_2O and 1 μL (2.5 units) of *Pfu* Turbo DNA polymerase.
2. Perform SDM in a thermal cycler using the following parameters: 1 cycle of 95°C for 2 min, 60°C for 30 s, 69°C for 14 min followed by 13 cycles of 95°C for 30 s, 60°C for 30 s, 69°C for 14 min and a final cycle of 72°C, 10 min followed by a 4°C hold.
3. Once amplification is complete, transfer reactions to a 1.5 mL Eppendorf tube.
4. Add 1 μL (20 units) of *Dpn*I and incubate at 37°C for 1 h (*see* **Note 7**).
5. Transform 2 μL of PCR reaction into *E. coli* XL-1 blue cells and plate out on LB agar plates containing ampicillin.
6. Extract plasmid DNA from ampicillin-resistant colonies by Qiagen mini-prep of a 10 mL overnight culture in LB ampicillin and confirm sequence of mutation (*see* **Note 5**).

3.3. Preparation of Transformation-Competent Bacterial Cells

1. Aseptically transfer 10 mL of sterile LB liquid to a sterile 50 mL tube and inoculate with *E. coli* cells.
2. Grow culture to saturation in an orbital incubator at 37°C overnight with shaking (200 rpm).
3. Next day, inoculate 100 μL of the saturated bacterial culture into 10 mL of fresh sterile LB liquid containing 200 μL of sterile 1 M $MgCl_2$.
4. Grow culture at 37°C with shaking (200 rpm) until $OD_{600} = 0.2$.
5. Harvest cells by centrifugation (3000 rpm, 10 min, 4°C).
6. Discard supernatant and gently re-suspend cells in 4 mL of ice-cold transformation buffer.
7. Place on ice for 30 min and mix occasionally by gentle agitation.

8. Harvest cells by centrifugation as described above.
9. Discard supernatant and gently re-suspend cells in 400 μL of transformation buffer.
10. Place on ice. The cells are now transformation competent and should be used immediately.

3.4. Transformation of Competent Cells

1. Transfer 50 μL of *E. coli*-competent cells to a pre-chilled Eppendorf tube.
2. Add 1 μL of plasmid DNA (or a ligation reaction) and mix gently. Do not vortex.
3. Place on ice for 30 min.
4. Heat shock cells at 37°C for 5 min.
5. Place cells back on ice for 2 min.
6. Add 700 μL of sterile LB liquid and incubate at 37°C for 1 h.
7. Spread 150 μL of solution onto each LB ampicillin agar plate.
8. Incubate plates inverted in a static 37°C incubator for 16 h (*see* **Note 8**).

3.5. Thioester-Tagged Protein Expression and Purification

3.5.1. Protein Expression Using Vector pTYB-1

1. From the plate containing fresh colonies transformed with the target protein-intein tag expression construct use a sterile loop to inoculate each of 4 × 10 mL of sterile LB (in 4 × 50 mL tubes). Inoculate each tube with a single ampicillin-resistant ER2566 colony and add 10 μL of ampicillin stock to each tube.
2. Grow cultures overnight at 37°C with 200 rpm shaking.
3. Add 4 × 500 μL of ampicillin stock to 4 × 500 mL sterile LB prepared in 4 × 2 L flasks, then transfer 5 mL of each overnight culture into each flask.
4. Incubate flasks at 37°C with shaking (125 rpm) until an OD_{600} = 0.6.
5. Once OD_{600} = 0.6, remove 1 mL of uninduced cell culture, harvest cells by centrifugation and discard supernatant. Add 100 μL of SDS-PAGE sample buffer to pelleted cells and boil for 5 min. Store sample at –20°C for SDS-PAGE analysis.
6. Add IPTG stock to the remaining cultures at a final concentration of 1 mM (*see* **Note 9**).

7. Continue to incubate flasks at 30°C with shaking for a further 5 h (*see* **Note 10**).
8. Remove 1 mL of induced cell culture, harvest the cells by centrifugation and discard supernatant. Add 100 μL of SDS-PAGE sample buffer to pelleted cells and boil for 5 min. Store sample at –20°C for SDS-PAGE analysis.
9. Harvest cells by centrifugation at 8000 rpm for 15 min at 4°C in a JLA 16.250 rotor (Beckman) and discard supernatant (*see* **Note 11**).

3.5.2. Isolation of CBD-Tagged Protein

1. Re-suspend the harvested cells in approximately 50 mL of cold cell lysis buffer (100 mM sodium phosphate; pH 8.0, 100 mM NaCl, 0.1% v/v Tween 20).
2. Divide the re-suspended cell solution equally into two polycarbonate centrifuge tubes.
3. Add 1 mL of PMSF stock solution to each tube and mix by inversion.
4. Sonicate tubes on ice for 15 × 1 min (*see* **Note 12**).
5. Pellet the cellular debris by centrifugation at 13,000 rpm in a JA 25.50 rotor for 15 min at 4°C.
6. Carefully decant and retain the clarified cell extract. This is the column input fraction containing the chitin-binding domain (CBD)-fused target protein. Retain a sample of this fraction for SDS-PAGE analysis and store at –20°C.

3.5.3. Purification of Thioester-Tagged Protein

1. Fill chromatography column with 2 mL (settled bed volume) of chitin beads and allow to drain by gravity flow.
2. Wash resin with 5 column volumes (10 mL) of cell lysis buffer and allow to drain.
3. Gently apply the cleared extract to the column and allow to drain by gravity flow.
4. Collect a sample of the flow-through for SDS-PAGE analysis and store at –20°C.
5. Wash resin with 50 mL wash buffer. Retain a sample for SDS-PAGE analysis and store at –20°C.
6. After washing, apply 5 mL of wash buffer containing 2% w/v 2-mercaptoethanesulfonic acid (MESNA) to the column and allow to drain.
7. Plug the end of the column, then apply 2 mL wash buffer containing 2% w/v MESNA (*see* **Note 13**).
8. Cap the column and then allow to stand at room temperature for 16–48 h.
9. Remove cap and allow eluted thioester fractions to drain into 1.5 mL Eppendorf tubes (*see* **Fig. 9.3**).

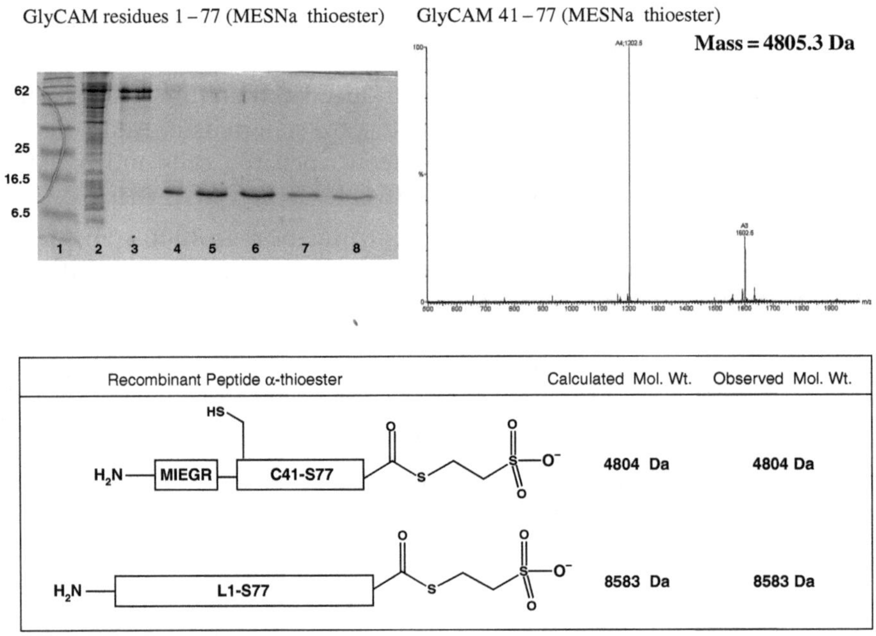

Fig 9.3. (**a**) Purification of recombinant glycosylation-dependant cell adhesion molecule-1 (GlyCAM-1) thioesters for NCL reactions. SDS-PAGE gel (*top left*) showing representative stages of the purification. *Lane 1*, molecular weight markers. *Lane 2*, column input. *Lane 3*, chitin beads after wash. *Lanes 4–8*, GlyCAM-1 (residues 1–77) thioester elution fractions. Mass spectrum (*top right*) of IEGR-labelled GlyCAM-1 thioester residues 41–77. Calculated mass = 4804 Da. Observed mass = 4805 Da.

10. Wash column by applying a further 3 mL wash buffer containing 2% w/v MESNA, collecting eluted protein thioester fractions.
11. Analyze fractions by SDS-PAGE and LC-MS to confirm the presence of the eluted thioester fractions and to assess the degree of thioester hydrolysis (*see* **Note 14**). If thioester hydrolysis is not evident, then these fractions can be used directly in NCL reactions (*see* **Section 3.9.1**).

3.6. Generation of a Protein with an N-Terminal Cysteine

3.6.1. Expression of Soluble N-Terminal Cysteine Component

The appropriate strain of *E. coli* is transformed with the expression plasmid that contains the target protein-intein tag fusion for generation of an N-terminal cysteine (*see* **Fig. 9.2a**). The heterologous protein expression and preparation of the crude cell extract is the same as described in **Sections 3.5.1** and **3.5.2**.

3.6.2. Purification of Latent N-Terminal Cysteine Component

1. Purify the clarified cell extract as described by steps 1–5 in **Section 3.5.3**.
2. After washing, apply 5 mL of wash buffer containing 50 mM dithiothreitol (DTT) to the column and allow to drain.

3. Plug the end of the column and then apply 2 mL wash buffer containing 50 mM DTT.
4. Cap the column and then allow to stand at 4°C for 16–48 h.
5. Remove cap and allow eluted thioester fractions to drain into 1.5 mL Eppendorf tubes.
6. Wash column by applying a further 3 mL wash buffer and collect eluted protein fractions.
7. Analyse fractions by SDS-PAGE and LC-MS to confirm the identity of the eluted product.
8. Purify by dialysis against wash buffer followed by concentration in a centrifuge concentrator. Alternatively purify by HPLC and lyophilize.

3.7. pET-16b for Heterologous Protein Expression

3.7.1. Heterologous Protein Expression Using Vector pET-16b

The appropriate strain of *E. coli* is transformed with the pET-16b expression plasmid. The heterologous protein expression is the same as described in **Section 3.5.1** *except the induction of protein expression with 1 mM IPTG is conducted at 37 °C for 3 h.*

3.7.2. Extract Preparation for Cells Transformed with pET-16b

1. Re-suspend the harvested cells in 50 mL of cold cell lysis buffer II.
2. Divide the re-suspended cell solution into two polycarbonate centrifuge tubes.
3. Add 1 mL of PMSF stock solution to each tube and mix by inversion.
4. Sonicate tubes on ice for 20 × 1 min (*see* **Note 12**).
5. Pellet the cellular debris by centrifugation at 13,000 rpm in a JA 25.50 rotor for 15 min at 4°C.
6. *Discard* the supernatant and add 15 mL of cell lysis buffer II, 1 mL of PMSF stock and a magnetic stir bar to each of the debris pellets (*see* **Note 15**).
7. Gently stir to ensure that all of the pelleted material is re-solubilized.
8. Remove stir bar and centrifuge extract at 19,000 rpm in a JA 25.50 rotor for 15 min at 4°C.
9. Carefully remove the cleared extract. This is the column input fraction. Retain a sample of the input fraction for SDS-PAGE analysis and store at –20°C.

3.7.3. Purification of His-Tagged Proteins

1. Fill chromatography column with 4 mL of Ni-NTA His-Bind resin and allow to drain by gravity flow.

2. Wash resin with 10 column volumes (40 mL) of dH_2O.
3. Equilibrate resin with 3 column volumes (12 mL) of cell lysis buffer II.
4. Gently apply the cleared extract to the column and allow to drain by gravity flow.
5. Collect a sample of the flow-through for SDS-PAGE analysis and store at –20°C.
6. Wash the resin with 100 mL of wash buffer II. Collect a sample of the wash for SDS-PAGE analysis and store at –20°C.
7. After washing, elute bound protein from the resin with 10 mL of elution buffer II.
8. Collect eluted proteins in 10 × 1 mL fractions and analyse by SDS-PAGE and LC-MS (*see* **Fig. 9.4.** and **Note 16**).

3.7.4. CNBr Cleavage to Liberate an N-Terminal Cysteine

1. Fill dialysis tubing with each of the eluted protein fractions.
2. Seal tubing, leaving approximately 1 volume of free space above the liquid.

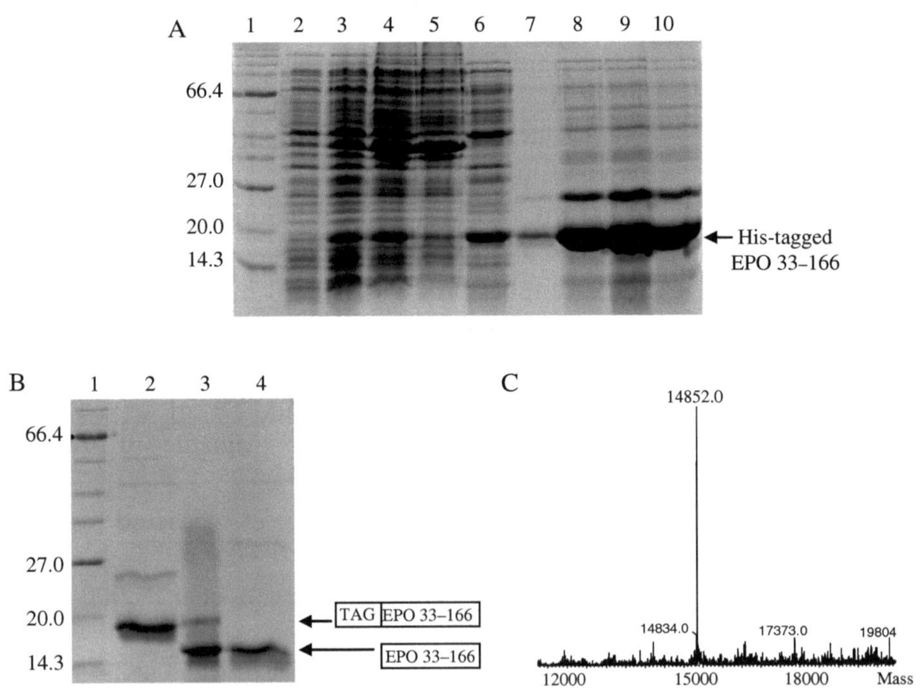

Fig. 9.4. Purification and LC-MS analysis of His_{10}-tagged erythropoietin (residues 33–166). (**a**) SDS-PAGE gel showing representative stages of the purification. *Lane 1*, molecular weight markers. *Lane 2*, uninduced bacterial cells. *Lane 3*, induced bacterial cells. *Lane 4*, column input. *Lane 5*, column flow-through. *Lane 6*, column wash. *Lanes 7–10*, elution fractions 2–5. (**b**) SDS-PAGE gel showing representative stages of CNBr cleavage and recovery of cleaved protein. *Lane 1*, molecular weight markers. *Lane 2*, His-tagged EPO 33-166. *Lane 3*, EPO 33–166 after CNBr treatment. *Lane 4*, recovery of cleaved protein. (**c**) De-convoluted LC-MS spectrum of cleaved EPO residues 33–166. Calculated mass = 14,852.2 Da. Observed mass = 14,852.0 Da.

3. Dialyse against dH_2O (2 L) with gentle stirring for 1 h at 4°C.
4. Discard dH_2O, refill beaker with fresh dH_2O and continue dialysis overnight.
5. Transfer the dialysed solution to a 15 mL centrifuge tube and pellet precipitated protein by centrifugation at 3000 rpm for 15 min.
6. Carefully decant and discard dH_2O.
7. Re-solubilize the pelleted protein in 1 mL of freshly prepared 80% v/v formic acid (*see* **Note 17**).
8. Weigh out cyanogen bromide (5 mg) into a glass vial (SAFETY! *see* **Note 4**).
9. Dissolve CNBr in the formic acid/protein solution (*see* **Note 17**).
10. Transfer to a clean 10 mL round-bottomed flask containing a magnetic stir bar.
11. Flush the flask with nitrogen gas, wrap the flask with tin foil and stir at room temperature for 24 h.
12. Remove formic acid from the flask under reduced atmospheric pressure using a rotary evaporator (*see* **Note 18**).
13. To hydrolyse any formyl adducts that may have formed during the reaction, which may interfere with subsequent NCL reactions, re-solubilize protein in 1 mL of 6 M GnHCl and transfer solution to an Eppendorf tube.
14. Add HCl from a 3 M stock solution to a final concentration of 100 mM.
15. Incubate the acidified protein solution at 37°C for 24 h with shaking.
16. After incubation, adjust pH of solution to 8.0 with concentrated (10 M) NaOH solution (*see* **Note 19**). Retain a sample of this protein solution for SDS-PAGE analysis.
17. Transfer protein solution to a 15 mL centrifuge tube and dilute to a volume of 5 mL with cell lysis buffer II.
18. Add 2-mercaptoethanol, prepared in cell lysis buffer II, to a final concentration of 5 mM (*see* **Note 20**).
19. Incubate solution at 37°C for 1 h with shaking.
20. Equilibrate Ni-NTA His-Bind resin (2 mL) with 10 column volumes (20 mL) of cell lysis buffer II containing 5 mM 2-mercaptoethanol.
21. Apply the protein solution to the His-Bind resin and allow to drain by gravity flow, collecting the flow-through in a 15 mL centrifuge tube.

22. Wash the resin with 2 × 5 mL of cell lysis buffer II containing 5 mM 2-mercaptoethanol and collect all of the flow-through in the same tube.
23. Dialyse this solution against dH_2O as described in steps 1–4.
24. Collect precipitated protein by centrifugation as described (steps 5 and 6) and analyse by SDS-PAGE (*see* **Fig. 9.4**).
25. A small amount of the freeze-dried protein can be re-solubilized in 6 M GnHCl and reduced with DTT prior to LC-MS analysis. A mass yield can be determined from the mass of the peptide recovered after lyophilization of the pelleted protein.

3.8. Solid-Phase Glycopeptide Synthesis

Automated solid-phase peptide synthesis was carried out on a 0.1 mmol scale using 1.0 mmol of each Fmoc amino acid (*see* **Note 3**). Manual solid-phase peptide synthesis (for incorporation of glycoamino acids) was conducted using 0.5 mmol of Fmoc-Ser($(AcO)_3$GalNAc)-OH or Fmoc-Thr($(AcO)_3$GalNAc)-OH) and HBTU/HOBt as coupling reagents. The average coupling time for manual steps was 1 h and the reaction progress was monitored using LC-MS and the Kaiser ninhydrin test.

3.8.1. Solid-Phase Synthesis of Peptides Containing an N-Terminal Cysteine

1. Dispense 0.1 mmol of the appropriate pre-loaded resin into reaction vessel and wash resin to the bottom with approximately 3 mL dimethylformamide (DMF).
2. Cap the vessel and allow to stand for 0.5 h.
3. Transfer the reaction vessel to automated peptide synthesizer and complete coupling and double coupling cycles as necessary.
4. Remove the dry resin from the reaction vessel and dispense into a 25 mL round-bottomed flask or vial.
5. Add 5 mL of a mixture containing 2.5% v/v H_2O, 2.5% 1,2-ethanedithiol and 95% v/v trifluoroacetic acid (TFA) and stir at room temperature for 4 h.
6. Filter the resin with suction, collecting the filtrate and wash the dry resin with further neat TFA (2 mL).
7. Pour the filtrate into 60 mL ice-cold diethyl ether and divide the resulting suspension between two 50 mL disposable centrifuge tubes and store on ice for 10 min.
8. Collect the precipitated peptide by centrifugation at 3000 rpm for 15 min.
9. Decant the supernatant by pouring into a dilute bleach solution (in a fume hood).

10. Add a further 60 mL cold ether and gently vortex the precipitated peptide.
11. Collect the crude peptide by repeating steps 8 and 9 (above) and dry under a stream of nitrogen.
12. Dissolve the crude peptide in the minimum volume of water or water/acetonitrile and purify by semi-preparative reversed-phase HPLC.
13. Lyophilize fractions containing the purified protein as a fluffy white solid that can be used directly in ligations.

3.8.2. Solid-Phase Synthesis of Peptides Containing a C-Terminal Thioester

1. Dispense 0.1 mmol of the appropriate pre-loaded 4-sulfamylbutyryl resin into reaction vessel and wash resin to the bottom with approximately 3 mL DMF.
2. Cap the vessel and allow to stand for 0.5 h.
3. Follow steps 3 and 4 as described in **Section 3.8.1**.
4. Transfer the dry resin to a 5 mL vial or round-bottomed flask and re-suspend in anhydrous DMF (4 mL).
5. Add 0.2 mL *N*,*N*-diisopropylethylamine (DIPEA) and then 0.2 mL iodoacetonitrile (ICH_2CN) (*see* **Note 21**) and stir the reaction mixture at room temperature with the exclusion of light for 24 h.
6. Filter off the resin and wash resin exhaustively with DMF and then dichloromethane (DCM).
7. Resuspend the alkylated resin in anhydrous 4 mL DMF and add 0.3 mL benzylmercaptan and approximately 6 mg sodium thiophenoxide (NaSPh).
8. Stir the reaction at room temperature for 36 h.
9. Filter off the resin and wash with 5 mL DMF, then 5 mL DCM, combining the eluents and evaporating to dryness under vacuum.
10. Now process the residue according to step 5 and then steps 7–13 of **Section 3.8.1** (step 6 is not necessary here).

3.9. Native Chemical Ligation Reactions

3.9.1. Ligation of a Recombinant Thioester and a Synthetic (Glyco)peptide

1. Determine the concentration or weight of dry recombinant peptide thioester obtained from **Section 3.5.3** (*see* **Note 13**).
2. Dissolve the dry peptide thioester in the minimum volume of ligation buffer to a final concentration of 5–10 mg/mL (*see* **Note 22**).
3. Add TCEP to a final concentration of 20 mM (*see* **Note 23**).

4. Weigh lyophilized synthetic (glyco)peptide containing an N-terminal cysteine (2–10 M equivalents) into a clean Eppendorf tube.
5. Transfer the protein thioester solution to the Eppendorf tube containing the dry synthetic (glyco) peptide.
6. Shake the NCL reaction (700 rpm) at 25°C in an Eppendorf thermomixer for 16 h, periodically removing aliquots (1–2 μL) to monitor reaction progress by SDS-PAGE or LC-MS.
7. Purify the ligation product by loading the reaction mixture directly onto a semi-preparative reversed-phase HPLC column.
8. Fractions containing the ligation product (identified by SDS-PAGE or LC-MS) are lyophilized to store the purified glyco(peptide) product.

3.9.2. Ligation of a Synthetic (Glyco)peptide Thioester to a Soluble Peptide

1. Determine the concentration or weight of dry recombinant peptide containing the N-terminal factor Xa cleavage motif (Ile-Glu-Gly-Arg) obtained as described in **Section 3.6.2**.
2. Dissolve the dry peptide thioester in the minimum volume of ligation buffer to a final concentration of 5–10 mg/mL (*see* **Note 22**).
3. Add TCEP to a final concentration of 20 mM (*see* **Note 23**).
4. Weigh lyophilized synthetic (glyco)peptide thioester (2–10 molar equivalents) into a clean Eppendorf tube.
5. Transfer the recombinant protein solution to the Eppendorf tube containing the dry synthetic (glyco) peptide thioester.
6. Add $CaCl_2$ to a final concentration of 2.5 mM.
7. Add 4.5 U factor Xa protease (*see* **Note 24**).
8. Shake the NCL reaction (700 rpm) at 25°C in an Eppendorf thermomixer for 16 h, periodically removing aliquots (1–2 μL) to monitor reaction progress by SDS-PAGE or LC-MS.
9. Purify the ligation product by loading the reaction mixture directly onto a semi-preparative reversed-phase HPLC column.
10. Fractions containing the ligation product (identified by SDS-PAGE or LC-MS) are lyophilized to afford the purified glyco(peptide) product.

3.9.3. Ligation of a Synthetic (Glyco)peptide Thioester to an Insoluble Peptide

1. Dissolve approximately 3 mg of the dry protein sample from **Section 3.7.4** in 0.5 mL of 6 M guanidine hydrochloride containing 300 mM Na phosphate buffer; pH 7.5.
2. Add TCEP to a final concentration of 20 mM.
3. Add MESNA to a final concentration of 50 mM (*see* **Note 25**).
4. Weigh solid (glyco)peptide thioester in a 1.5 mL Eppendorf tube.
5. Add the protein solution to the solid (glyco)peptide thioester in a single portion and shake at room temperature for 4–16 h periodically removing aliquots (1–2 μL) to monitor reaction progress by SDS-PAGE or LC-MS.
6. Purify the ligation product by loading the reaction mixture directly onto a semi-preparative reversed-phase HPLC column (*see* **Note 26**).

3.10. Protein Refolding

If proteins have been subjected to NCL reactions under denaturing conditions then the assembled full-length protein will require refolding. This is often accomplished by dialysis into an appropriate buffer in the presence of additives to prevent aggregation or to correctly form disulfide bonds. However, refolding conditions may vary for different proteins and optimization will be required each time.

4. Notes

1. It saves a significant amount of time if pre-loaded resins are used. Furthermore, loading of the 4-sulfamylbutyryl resin (safety-catch) resin is not always trivial and if a pre-loaded resin is available then its use is recommended.
2. We use standard Fmoc amino acids, an Fmoc-based solid-phase peptide synthesis protocol and an automated peptide synthesizer. However, there are alternative strategies and several reliable manual solid-phase peptide synthesis protocols such that peptide synthesis need not be conducted in automated fashion at all. Additionally, and depending on the complexity of the synthetic peptide fragment, the synthetic peptide thioester or peptide containing an N-terminal cysteine can be purchased from a commercial custom peptide synthesis source. Several simple Fmoc-protected glycoamino acids are also commercially available and these can potentially be elaborated enzymatically at a later stage (30, 31).

3. If a peptide is synthesized (rather than purchased) the coupling reagents used and synthetic protocol adopted depend on the complexity of the target and the instrumentation/method practiced in a particular laboratory. We use the FastMoc protocol (0.1 mmol scale) which employs HBTU/HOBt as coupling reagents in the presence of DIPEA activator base and in NMP as solvent. The coupling time is approximately 0.5 h.
4. SAFETY! Cyanogen bromide is an extremely poisonous pliable crystalline solid that is best dispensed after it has reached room temperature. All manipulations involving CNBr must be performed in a fume hood.
5. The nucleotide sequence of inserts cloned into pET-16b can be determined using the T7 promoter primer (forward): 5′-TAATACGACTCACTATAGGG-3′) and T7 terminator primer (reverse): 5′-GCTAGTT ATTGCTCAGCGG-3′). These primers can be custom synthesized or purchased as a product from a commercial source.
6. The use of 10X *Pfu* reaction buffer results in a final $MgSO_4$ concentration of 2 mM. Supplementing the PCR reaction with additional Mg^{2+} can sometimes enhance amplification.
7. Amplification of plasmid DNA can be checked by analysing 20 μL of the *Dpn*I-treated reaction by agarose electrophoresis.
8. LB ampicillin plates supporting transformed *E. coli* B834(DE3) cells should be stored inverted at 4°C.
9. Prepare IPTG solution immediately before use. Add 0.238 g of IPTG to 1 mL of dH_2O and dissolve by vortexing. The total volume of this solution is 1130 μL. For induction of bacterial cultures, add half of the IPTG solution (565 μL) to each flask giving a final induction concentration of 1 mM.
10. Depending on the stability of the target protein–intein fusion, it may be essential to conduct this step at lower temperatures (e.g. 25°C for 16 h). Consult the IMPACT-CN™ users manual for further details.
11. Harvested cell pellets can be conveniently stored at –20°C.
12. SAFETY! The sonicator should be equipped with a shielded audio cabinet to protect ears from damage.
13. This and proceeding steps (steps 8–11) can be conducted in a number of ways depending on protein thioester

stability. In our case, the first 4.5 mL eluted from the chitin column were loaded directly onto a semi-preparative reversed-phase HPLC column. Fractions containing the peptide thioester were lyophilized to afford the purified peptide (3.0 mg) as a white solid that was used in ligations. Where thioester hydrolysis is severe during release of the thioester from the intein it is necessary to add the peptide component containing an N-terminal cysteine (dissolved in the minimum volume of wash buffer containing 2% w/v MESNA) directly to the chitin column to maximize ligation and minimize hydrolysis. When applying the N-terminal cysteine-containing component directly to the column, mixing (if required) is often achieved by transferring the entire column contents to a 15 mL Falcon tube and conducting the ligation on a blood rotator or horizontal roller.

14. Depending on thioester stability the eluted thioester can be used directly (or after concentration in a centrifuge concentrator) in NCL reactions or after purification by HPLC.
15. Prepare pellet re-solubilization, column wash and protein elution buffers using solid NaCl and imidazole rather than from pre-prepared liquid stocks to minimize dilution of the 6 M GnHCl.
16. While GnHCl is used to re-solubilize insoluble proteins, the use of GnHCl in SDS-PAGE is not recommended. Insoluble protein fragments can be separated from GnHCl by the addition of 10 volumes of 1:1 acetone/MeOH or 10 volumes of water. After placing samples on ice for 30 min the pelleted protein can be collected by centrifugation at maximum speed in a refrigerated (4°C) bench top centrifuge (repeating if required). The pellet can be re-solubilized in 20 μL of 8 M urea and analyzed by SDS-PAGE (*see* **Fig. 9.4**). To confirm mass by LC-MS, take 30 μL of the most concentrated elution fraction (visualized by SDS-PAGE) and reduce with 50 mM DTT for 1 h at 37°C prior to analysis (**Fig. 9.4**).
17. This is best achieved by transferring the formic acid/protein solution to the glass vial containing CNBr and gently pipetting up and down until the CNBr is dissolved.
18. After removal of formic acid from the flask the protein may be observed as a pale yellow-brown film. However, the protein may not be clearly visible at this stage.
19. After pH adjustment, the protein can be reduced with DTT and analysed by LC-MS.

20. When recovering cleaved protein after CNBr treatment, the use of DTT as a reducing agent is not advisable due to its incompatibility with the His-binding resin. We observe no adverse effects on binding when using 2-mercaptoethanol at a concentration of 5 mM.
21. Iodoacetonitrile should be filtered through a short plug of activated basic alumina prior to use.
22. It is important to obtain as high a protein concentration as possible and it may be essential to prepare this buffer in 6 M guanidine HCl if either the protein thioester or N-terminal cysteine containing component is not soluble at this concentration.
23. It is important that the addition of TCEP does not lower the pH of the ligation reaction such that the reaction does not proceed. When TCEP–HCl is dissolved directly in water the solution is strongly acidic. We prepare a stock solution in water and neutralize with 10 M NaOH before making up to the appropriate volume for a 1 M stock solution. The solution is stored at –20°C.
24. We conduct reactions in 2% MESNA with no evidence of non-specific proteolysis at room temperature (23°C). However, the promiscuity of factor Xa for cleavage sites similar to IEGR is well known and the "one-pot" procedure is likely to be protein dependant.
25. (Glyco)peptide thioesters dissolved in the presence of excess MESNA may not be sufficiently reactive to undergo NCL reactions efficiently. In these cases it is recommended that 25–50 mM MPAA is used in place of 2% w/v MESNA, conducting the ligation reaction between pH 6.5 and 7.5, rather than at pH 8.0.
26. If the protein is highly insoluble an alternative isolation method is to collect the protein as a precipitate after pouring into cold water (10 volumes) allowing to stand at 4°C for 16 h and centrifugation at 3000 rpm for 15 min. This process often also serves to remove excess unreacted or hydrolyzed thioester (if water soluble). The crude product can then be taken up in 6 M GnHCl (2.0 mL) and treated with 50 mM DTT at 37°C for 1 h prior to refolding.

Acknowledgments

The authors acknowledge financial support from The Royal Society and The Wellcome Trust.

References

1. Spiro, R. G. (2002) Protein glycosylation: nature, distribution, enzymatic formation, and disease implications of glycopeptide bonds. *Glycobiology* 12, 43R–56R.
2. Macmillan, D., Daines, A. M. (2003) Recent developments in the synthesis and discovery of oligosaccharides and glycoconjugates for the treatment of disease. *Curr Med Chem* 10, 2733–2773.
3. Wu, B., Chen, J., Warren, J. D., Chen, G., Hua, Z., Danishefsky, S. J. (2006) Building complex glycopeptides: development of a cysteine-free native chemical ligation protocol. *Angew Chem Int Ed* 45(25), 4116–4125.
4. Chen, J., Chen, G., Wu, B., Wan, Q., Tan, Z., Hua, Z., Danishefsky, S. J. (2006) Mature homogeneous erythropoietin-level building blocks by chemical synthesis: the EPO 114-166 glycopeptide domain, presenting the O-linked glycophorin. *Tetrahedron Lett* 47, 8013.
5. Wu, B., Tan, Z., Chen, G., Chen, J., Hua, Z., Wan, Q., Ranganathan, K., Danishefsky, S. J. (2006) Mature homogeneous erythropoietin building blocks by chemical synthesis: the EPO 22-37 glycopeptide domain presenting the full *N*-linked dodecasaccharide. *Tetrahedron Lett* 47, 8009
6. Dziadek, S., Griesinger, C., Kunz, H., Reinscheid, U. M. (2006) Synthesis and structural model of an α(2,6)-sialyl-T glycosylated MUC1 eicosapeptide under physiological conditions. *Chem Eur J* 12, 4981–4993.
7. Schuberth, R., Unverzagt, C. (2005) Synthesis of a *N*-glycan nonasaccharide of the bisecting type with additional core-fucose. *Tetrahedron Lett* 46, 4201.
8. Matsuo, I., Totani, K., Tatami, A., Ito, Y. (2006) Comprehensive synthesis of ER related high-mannose-type sugar chains by convergent strategy. *Tetrahedron* 62, 8262.
9. Adams, E. W., Ratner, D. M., Bokesch, H. R., McMahon, J. B., O'Keefe, B. R., Seeberger, P. H. (2004) Oligosaccharide and glycoprotein microarrays as tools in HIV glycobiology: glycan-dependent gp120/protein interactions. *Chem Biol* 11, 875–881.
10. Ratner, D. M., Swanson, E. R., Seeberger, P. H. (2003) Automated synthesis of a protected N-linked glycoprotein core pentasaccharide. *Org Lett* 5, 4717–4720.
11. Bill, R. M., Flitsch, S. L. (1996) Chemical and biological approaches to glycoprotein synthesis. *Chem Biol* 3, 145–149.
12. Macmillan, D., Bill, R. M., Sage, K. A., Fern, D., Flitsch, S. L. (2001) Selective in vitro glycosylation of recombinant proteins: semi-synthesis of novel homogeneous glycoforms of human erythropoietin. *Chem Biol* 8, 133–145.
13. Davis, N. J., Flitsch, S. L. (1991) A novel method for the specific glycosylation of proteins. *Tetrahedron Lett* 32, 6793–6796.
14. Gamblin, D. P., Scanlan, E. M., and Davis, B. G. (2009). *Glycoprotein Synthesis: An Update. Chem Rev* 109, 131–163.
15. Buskas, T., Ingale, S., Boons, G.-J. (2006) Glycopeptides as versatile tools for glycobiology. *Glycobiology* 16, 113R–136R.
16. Baneyx, F. (1999) Recombinant protein expression in *Escherichia coli. Curr Opin Biotechnol* 10, 411–421.
17. Hollister, J. R., Jarvis, D. L. (2001) Engineering lepidopteran insect cells for sialoglycoprotein production by genetic transformation with mammalian beta 1,4-galactosyltransferase and alpha 2,6-sialyltransferase genes. *Glycobiology* 11, 1–9.
18. Jarvis, D. L., Kawar, Z. S., Hollister, J. R. (1998) Engineering *N*-glycosylation pathways in the baculovirus-insect cell system. *Curr Opin Biotechnol* 9, 528–533.
19. Wu, X., and Schultz, P.G. (2009). *Synthesis at the Interface of Chemistry and Biology. J Amer Chem Soc* 131, 12497.
20. Wacker, M., Linton, D., Hitchen, P. G., Nita-Lazar, M., Haslam, S. M., North, S. J., Panico, M., Morris, H. R., Dell, A., Wren, B. W., Aebi, M. (2002) *N*-Linked glycosylation in *Campylobacter jejuni* and its functional transfer into *E. coli. Science* 298, 1790–1793.
21. Zhang, Z., Gildersleeve, J., Yang, Y.-Y., Xu, R., Loo, J. A., Uryu, S., Wong, C.-H., Schultz, P. G. (2004) A new strategy for the synthesis of glycoproteins. *Science* 303, 371–373.
22. Dawson, P. E., Muir, T. W., Clark-Lewis, I., Kent, S. B. (1994) Synthesis of proteins by native chemical ligation. *Science* 266, 776–779.
23. Mezzato, S., Schaffrath, M., Unverzagt, C. (2005) An orthogonal double-linker resin facilitates the efficient solid-phase synthesis of complex-type N-glycopeptide thioesters suitable for native chemical ligation. *Angew Chem Int Ed* 44, 1650–1654.
24. Hojo, H., Matsumoto, Y., Nakahara, Y., Ito, E., Suzuki, Y., Suzuki, M., Suzuki, A., Nakahara, Y. (2005) Chemical synthesis of 23 kDa glycoprotein by repetitive segment condensation: a synthesis of MUC2 basal motif

carrying multiple O-GalNAc moieties. *J Am Chem Soc* 127, 13720–13725.
25. Macmillan, D., Bertozzi, C. R. (2004) Modular assembly of glycoproteins: towards the synthesis of GlyCAM-1 by using expressed protein ligation. *Angew Chem Int Ed* 43, 1355–1359.
26. Yamamoto, N., Tanabe, Y., Okamoto, R., Dawson, P. E., Kajihara, Y. (2008) Chemical synthesis of a glycoprotein having an Intact human complex-type sialyloligosaccharide under the Boc and Fmoc synthetic strategies. *J Am Chem Soc* 130, 501–510.
27. Severinov, K., Muir, T. W. (1998) Expressed protein ligation, a novel method for studying protein-protein interactions in transcription. *J Biol Chem* 273, 16205–16209.
28. Pellois, J.-P., Muir, T. W. (2006) Semisynthetic proteins in mechanistic studies: using chemistry to go where nature can't. *Curr Opin Chem Biol* 10, 487.
29. Chong, S., et al. (1997) Single-column purification of free recombinant proteins using a self-cleavable affinity tag derived from a protein splicing element. *Gene* 192(2), 271.
30. Brik, A., Ficht, S., Wong, C.-H. (2006) Strategies for the preparation of homogenous glycoproteins. *Curr Opin Chem Biol* 10(6), 638.
31. Li, B., Song, H., Hauser, S., Wang, L. X. (2006) A highly efficient chemoenzymatic approach toward glycoprotein synthesis. *Org Lett* 8(14), 3081–3084.
32. Evans, T. C., Jr., et al. (1998) Semisynthesis of cytotoxic proteins using a modified protein splicing element. *Protein Sci* 7(11), 2256.

Chapter 10

Expression of Recombinant Proteins with Uniform N-Termini

Orsolya Király, Lan Guan, and Miklós Sahin-Tóth

Abstract

Heterologously expressed proteins in *Escherichia coli* may undergo unwanted N-terminal processing by methionine and proline aminopeptidases. To overcome this problem, we present a system where the gene of interest is cloned as a fusion to a self-splicing mini-intein. This fusion construct is expressed in an engineered *E. coli* strain from which the *pepP* gene coding for aminopeptidase P has been deleted. We describe a protocol using human cationic trypsinogen as an example to demonstrate that recombinant proteins produced in this expression system contain homogeneous, unprocessed N-termini.

Key words: *Escherichia coli*, intein, human cationic trypsinogen, in vitro refolding, ecotin affinity chromatography, aminopeptidase P.

1. Introduction

The expression system presented here was developed as part of an effort to elucidate the functional effect of the p.A16V mutation in human cationic trypsinogen, which has been reported to be associated with chronic pancreatitis by several studies (1–4). This variant alters the N-terminal amino acid residue of the mature, secreted trypsinogen protein. The amino acid numbering starts with Met^1 of the pre-trypsinogen protein and the first 15 residues comprise the secretory signal peptide. It had previously been determined that other pancreatitis-associated mutations in human cationic trypsinogen increase the propensity of trypsinogen for autoactivation (5, 6). We speculated that the p.A16V mutation might have a similar effect; however, functional characterization of the recombinantly expressed mutant trypsinogen required preparations with uniform, authentic N-termini.

T.C. Evans, M.-Q. Xu (eds.), *Heterologous Gene Expression in E. coli*, Methods in Molecular Biology 705, DOI 10.1007/978-1-61737-967-3_10, © Springer Science+Business Media, LLC 2011

For high-yield heterologous expression of human cationic trypsinogen in *Escherichia coli*, an expression plasmid was constructed in which the secretory signal peptide of trypsinogen was deleted and the initiator methionine was placed immediately upstream of the mature protein. Trypsinogen expressed from this construct accumulated in the cytoplasm as inclusion bodies. The native N-terminal sequence of trypsinogen isolated from pancreatic juice is Ala^{16}-Pro^{17}-Phe^{18}. The expected N-terminal sequence of trypsinogen expressed in *E. coli* is Met- Ala^{16}-Pro^{17}-Phe^{18}. However, we found that the N-terminal sequence of recombinant trypsinogen in the inclusion bodies was heterogeneous, consisting of ~30% Pro^{17}-Phe^{18} and approximately ~70% Met-Ala^{16}-Pro^{17}-Phe^{18}. Apparently, part of the expressed trypsinogen was processed by methionine aminopeptidase and then by proline aminopeptidase (aminopeptidase P). Removal of the initiator methionine by methionine aminopeptidase is a well-documented phenomenon in *E. coli* (7) which occurs if the second amino acid residue has a small, uncharged side chain. With high-level expression of heterologous proteins, the enzyme gets saturated and only a fraction of the proteins is processed. After cleavage of the initiator methionine, proteins can be subject to cleavage by aminopeptidase P, which cleaves the N-terminal amino acid of a protein if proline is in the penultimate position (8).

To address the problem of unwanted N-terminal processing by aminopeptidases, we developed a novel expression system (9). First, the gene of interest (in this case human cationic trypsinogen) is cloned in a C-terminal fusion with the *Synechocystis* DnaB mini-intein (10). In this fusion construct (*see* **Fig. 10.1**), translation is initiated by the start codon of the intein gene and the intein moiety is subsequently removed through

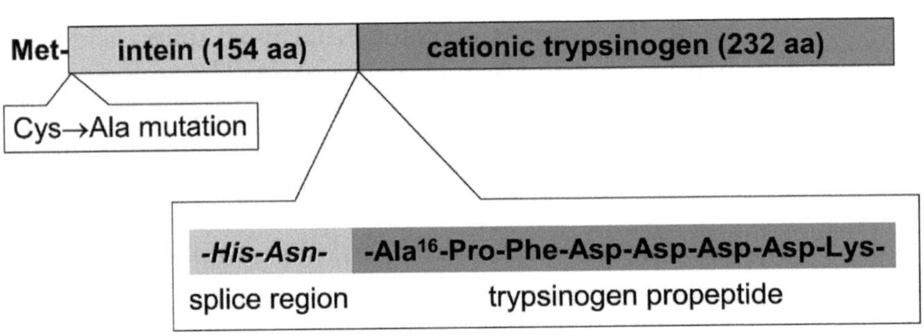

Fig. 10.1. Primary structure of the intein–trypsinogen fusion protein. An initiator methionine was placed upstream of the 154 amino acid long mini-intein, which was then fused in-frame with the 232 amino acid long human cationic trypsinogen. At the fusion junction, the C-terminal Asn of the splice region is joined to Ala^{16} of the trypsinogen propeptide. The N-terminal cysteine of the mini-intein has been mutated to Ala to disable splicing at the N-terminus. The sequence of the fusion construct has been deposited with GenBank under accession number DQ371396. Reproduced from (9) with permission from Elsevier Science.

intein self-cleavage (11). To eliminate cleavage by aminopeptidase P, the fusion construct is expressed in the aminopeptidase P deficient *E. coli* LG-3 strain. This strain was engineered by deleting the *pep*P gene coding for aminopeptidase P from the *E. coli* chromosome (12, 13), using the recombination-based method described by Datsenko and Wanner (14). The intein–trypsinogen fusion is expressed in LG-3 cells as inclusion bodies, solubilized with guanidine and re-natured in vitro followed by affinity purification on immobilized ecotin (15). Finally, we use MonoS cation-exchange chromatography to remove the small fraction of uncleaved intein fusion proteins and to obtain a pure trypsinogen preparation with uniform, authentic N-termini. The expression system described here can be useful for the heterologous expression of proteins whose N-terminal integrity is compromised by aminopeptidase activity.

2. Materials

2.1. Expression Plasmid Construction

1. Expression plasmid pTrapT7-*PRSS1* harboring the human cationic trypsinogen gene under the control of the T7 promoter (5, 6).
2. Plasmid pTWIN2 (New England Biolabs, Ipswitch, MA).
3. Oligonucleotide primers (Sigma Genosys, The Woodlands, TX).

   Primer A: 5′-CGG GAG TCC ATG GCT ATC TCT GGC GAT AGT CTG ATC AGC-3′

   Primer B: 5′ -<u>GAA AGG AGC GTT GTG TAC AAT</u> GAT GTC ATT CGC-3′

   Primer C: 5′ -<u>ATT GTA CAC AAC GCT CCT TTC</u> GAT GAT GAT GAC AAG-3′

   Primer D: 5′ -CTG CTC ATT GCC CTC AAG GAC-3′
4. Deep Vent® proofreading DNA polymerase with appropriate reaction buffer, 10 mM dNTP mix (New England Biolabs), diluted to 2 mM and stored in aliquots at –20°C.
5. Agarose gel electrophoresis equipment (e.g., Mini-Sub Cell GT system, Bio-Rad, Hercules, CA).
6. QIAquick Gel Extraction Kit (Qiagen, Valencia, CA).
7. Restriction enzymes *Nco*I and *Eco*RI, T4 DNA ligase (New England Biolabs).
8. α-Select Gold Efficiency competent *E. coli* cells (Bioline, Taunton, MA).

9. Luria–Bertani (LB) medium. Dissolve 25 g granulated LB broth, Miller (BP1426-500, Fisher Scientific) in 1 L distilled water, add 10 mL 1 M Tris-HCl, pH 7.5; sterilize by autoclaving.
10. LB agar plates containing 100 μg/mL ampicillin, stored at 4°C.
11. QIAprep Spin Miniprep kit (Qiagen).
12. DNA sequencing was performed with automated sequencing at a core facility.

2.2. Deletion of the pepP Gene from E. coli

1. Plasmids pKD4 and pCP20 (available from Dr. Barry L. Wanner, Department of Biological Sciences, Purdue University, West Lafayette, IN, *see* (14)).
2. Oligonucleotide primers (Sigma Genosys).

   P1-s: 5′-ACT CTA CAC TAA AAA CAA AAA ACG TAA GGA GAG TGT TAT GAG TGG TGT AGG CTG GAG CTG CTT C-3′

   P2-as: 5′-AGC GCC AGC GTC GCG CCC GCC ATG CCG CCA CCG ACG ATG ATT ACG CAT ATG AAT ATC CTC CTT A-3′

   T1: 5′-CCG CAA CCG ACC GCG CCA GAA G-3′

   k1: 5′-CAG TCA TAG CCG AAT AGC CT-3′

   T2: 5′-CAA ATG TAC CGG CAG CGC CC-3′

   k2: 5′-CGG TGC CCT GAA TGA ACT GC-3′
3. Restriction enzyme *Dpn*I (New England Biolabs).
4. Agarose gel electrophoresis equipment (e.g., Mini-Sub Cell GT system, Bio-Rad).
5. *E. coli* strain BW25113 harboring the helper plasmid pKD46 (14) (available from Dr. Barry L. Wanner).
6. Luria–Bertani (LB) medium. Dissolve 25 g granulated LB broth, Miller (BP1426-500, Fisher Scientific) in 1 L distilled water, add 10 mL 1 M Tris-HCl, pH 7.5; sterilize by autoclaving.
7. SOB medium: 20 g/L bacto-tryptone, 5 g/L yeast extract, 0.584 g/L NaCl, 0.186 g/L KCl, pH 7.0. Sterilize by autoclaving, store at 4°C. For complete medium, add 1 mL of 2 M Mg^{2+} solution to 99 mL SOB medium.
8. 2 M Mg^{2+} solution: 1 M $MgCl_2{\cdot}6H_2O$, 1 M $MgSO_4{\cdot}7H_2O$. Dissolve 20.33 g $MgCl_2{\cdot}6H_2O$ and 24.65 g $MgSO_4{\cdot}7H_2O$ in 100 mL distilled water, filter sterilize.
9. 2 M glucose solution: dissolve 36.04 g glucose (Sigma, St. Louis, MO) in 100 mL distilled water, filter sterilize, store at 4°C.

10. SOC medium: SOB medium with 20 mM glucose. Add 1 mL of 2 M Mg^{2+} solution and 1 mL 2 M glucose solution to 98 mL SOB medium.
11. Electroporator with 0.2 cm cuvettes (e.g., MicroPulser, Bio-Rad).
12. Kanamycin: Kanamycin sulfate (Sigma) is dissolved in sterile water at 16 mg/mL and stored at –20°C in aliquots.
13. Ampicillin: Ampicillin (Sigma) is dissolved in sterile water at 100 mg/mL and stored at –20°C in aliquots.
14. L-Arabinose: L-Arabinose (Sigma) is dissolved in sterile water at 1 M and stored at –20°C in aliquots.
15. Plasmid pGP1-2 harboring the T7 RNA polymerase under the control of a temperature-inducible λ promoter (16).
16. DNA sequencing was performed with automated sequencing at a core facility.

2.3. Expression and Purification of Trypsinogens

2.3.1. Trypsin Affinity Chromatography Column

1. Actigel ALD resin (#2701; 50 mL; Sterogene Bioseparations, Carlsbad, CA, *see* **Note 1**).
2. Coupling solution (#9704; 10 mL; 1 M sodium cyanoborohydride; Sterogene Bioseparations).
3. Bovine trypsin, TPCK treated (#3744, Worthington Biochemical, Lakewood, NJ).
4. 60 mL Buchner funnel inserted into a side-arm flask through a rubber stopper.
5. Adams Nutator mixer or shaking table.
6. Empty chromatography column, ~8 mL volume (10×100 mm, e.g., HR 10/10, Amersham-Pharmacia, Piscataway, NJ).

2.3.2. Ecotin Affinity Chromatography Column

1. Plasmid pT7-7-ecotin harboring the recombinant ecotin gene (17).
2. Competent *E. coli* BL21 (DE3) cells (Novagen, Madison, WI).
3. Luria–Bertani (LB) medium. Dissolve 25 g granulated LB broth, Miller (BP1426-500, Fisher Scientific) in 1 L distilled water, add 10 mL 1 M Tris-HCl, pH 7.5; sterilize by autoclaving.
4. Ampicillin: Ampicillin (Sigma) is dissolved in sterile water at 100 mg/mL and stored at –20°C in aliquots.

5. IPTG: Isopropyl-1-thio-β,D-galactopyranoside (Sigma) is dissolved to 1 M in sterile distilled water and stored in aliquots at –20°C.
6. Osmotic shock buffer: 30% sucrose, 20 mM Tris-HCl, pH 8.0, 5 mM Na-EDTA. Prepare fresh and store at 4°C until use.
7. 1 M Tris-HCl (pH 8.0). Dissolve 121.14 g Tris base in 1 L distilled water and adjust pH to 8.0 with HCl under a pH meter.
8. 5 M NaCl. Dissolve 292.2 g NaCl in 1 L distilled water.
9. Lyophilizer (e.g., Freezone 4.5 Freeze Dry System, Labconco Corp., Kansas City, MO).
10. Fast protein liquid chromatography (FPLC) system (e.g., LCC-501 Plus, Amersham-Pharmacia) with trypsin affinity column (*see* step 5 of **Section 3.3.1**).
11. Spectrophotometer with quartz cuvette to measure ultraviolet (UV) absorbance at 280 nm.
12. Actigel ALD resin (#2701; 50 mL; Sterogene Bioseparations).
13. Coupling solution (#9704; 10 mL; 1 M sodium cyanoborohydride, Sterogene Bioseparations).
14. Empty chromatography column with ~2 mL volume (5×100 mm; e.g., HR 5/10 or Tricorn 5/100, Amersham-Pharmacia).

2.3.3. Trypsinogen Expression and Purification Using E. coli Strain LG-3

1. *E. coli* strain LG-3: described in step 21 of **Section 3.2**.
2. Luria–Bertani (LB) medium: Dissolve 25 g granulated LB broth, Miller (BP1426-500, Fisher Scientific) in 1 L distilled water; add 10 mL 1 M Tris-HCl, pH 7.5; sterilize by autoclaving.
3. IPTG: Isopropyl-1-thio-β,D-galactopyranoside (Sigma) is dissolved to 1 M in sterile distilled water and stored in aliquots at –20°C.
4. Resuspension/wash buffer for inclusion bodies: 0.1 M Tris-HCl, pH 8.0, 5 mM K-EDTA.
5. Sonicator (e.g., Cell Disruptor Model W-200R with a microtip probe, Heat Systems Ultrasonics, Farmingdale, NY).
6. DTT: Dithiothreitol (Sigma) is dissolved in sterile distilled water to 1 M and stored in aliquots at –20°C.
7. Denaturing buffer for inclusion bodies: 4 M guanidine-HCl, 0.1 M Tris-HCl, pH 8.0, 2 mM K-EDTA, 30 mM dithiothreitol (*see* **Note 2**).

8. Refolding buffer: 0.9 M guanidine-HCl, 0.1 M Tris-HCl, pH 8.0, 2 mM K-EDTA, 1 mM L-cysteine, 1 mM L-cystine.
9. Argon gas (Grade 5).
10. Fast protein liquid chromatography (FPLC) system (e.g., LCC-501 Plus) with Mono S HR 5/5 cation-exchange column (Amersham-Pharmacia).
11. 2 mL ecotin affinity column (*see* step 18 of **Section 3.3.2**).
12. Wash buffer for ecotin affinity chromatography: 20 mM Tris-HCl, pH 8.0, 0.2 M NaCl.
13. Elution solution for ecotin affinity chromatography: 50 mM HCl.
14. Buffer A for cation-exchange chromatography: 20 mM Na-acetate, pH 5.0.
15. Buffer B for cation-exchange chromatography: 20 mM Na-acetate, pH 5.0, 0.5 M NaCl.
16. Spectrophotometer with quartz cuvette to measure ultraviolet (UV) absorbance at 280 nm.

2.3.4. SDS-PAGE

1. Running gel buffer: 1.5 M Tris-HCl, pH 8.8, 0.4% SDS.
2. Stacking gel buffer: 0.5 M Tris-HCl, pH 6.8, 0.4% SDS.
3. Acrylamide solution: 30% acrylamide with an acrylamide:bis-acrylamide ratio of 37.5:1.
4. Ammonium persulfate solution: 10% ammonium persulfate (APS, Sigma) solution. Prepared fresh in distilled water.
5. *N*,*N*,*N*′,*N*′-Tetramethylenediamine (TEMED, Sigma, electrophoresis grade).
6. 2× Laemmli sample buffer: 120 mM Tris-HCl, pH 6.8, 20% glycerol, 2% SDS, 0.003% bromophenol blue. To make 10 mL of 2× sample buffer, mix 34 mL distilled water with 10 mL glycerol, 6 mL 1 M Tris-HCl, pH 6.8, 1 g SDS, and 1.5 mg bromophenol blue (Sigma). Store the sample buffer at room temperature and add dithiothreitol to 100 mM just before use.
7. DTT: Dithiothreitol (Sigma) is dissolved in sterile distilled water to 1 M and stored in aliquots at –20°C.
8. Protein markers: MultiMark multi-colored standards (Invitrogen, Carlsbad, CA).
9. Gel staining solution: 40% methanol, 10% acetic acid, 0.25% Coomassie Brilliant Blue R-250 (*see* **Note 3**). To prepare 1 L staining solution, mix 500 mL distilled water with 400 mL methanol, 100 mL glacial acetic acid, and 2.5 g Coomassie Brilliant Blue R-250. Stir on magnetic

stirrer until the dye is completely dissolved, then filter the solution.

10. Destaining solution: 30% methanol, 10% acetic acid (*see* **Note 3**).
11. DryEase® Mini-Gel Drying System (Invitrogen, *see* **Note 4**).

3. Methods

3.1. Construction of the Expression Plasmid Harboring the Intein–Trypsinogen Fusion

1. Amplify the ~460 nt DnaB mini-intein gene from the plasmid pTWIN2 using primers A and B. Reactions contain 1× reaction buffer, 0.2 mM dNTP, 1 μM of each primer, and 0.02 units/μL of Deep Vent DNA polymerase. Cycling conditions are as follows: 35 cycles of 94°C for 20 s, 50°C for 30 s, 72°C for 1 min (RapidCycler, Idaho Technology, Salt Lake City, UT).
2. Amplify an ~230 nt portion of the recombinant human cationic trypsinogen gene from plasmid pTrapT7-*PRSS1* using primers C and D. Reaction and cycling conditions are the same as in step 1.
3. Run PCR reactions from steps 1 and 2 on an agarose gel and excise the PCR product bands from the gel using a razor blade.
4. Purify the PCR products from the gel with a gel extraction kit and elute in 10 μL elution buffer (10 mM Tris-HCl, pH 8.0).
5. Set up a new PCR reaction containing 1 μL of each product from steps 1 to 2, and primers A and D. Reaction and cycling conditions are the same as in step 1 (*see* **Note 5**).
6. Run the reaction product on an agarose gel, excise, purify, and digest with *Nco*I and *Eco*RI.
7. Digest the expression plasmid pTrapT7-*PRSS1* (5, 6) with *Nco*I and *Eco*RI and ligate the PCR product between these sites using T4 DNA ligase.
8. Transform 1 μL of the ligation reaction into chemically competent *E. coli* and spread the cells on LB agar plates containing 100 μg/mL ampicillin. Incubate the plates overnight at 37°C.
9. Pick two colonies from the plate and grow in 2 mL LB medium with 100 μg/mL ampicillin overnight with shaking.

10. Prepare plasmid DNA (miniprep) from the transformants and digest ~1 μg DNA with *Nco*I and *Eco*RI. Verify the presence of the correct insert by separating and visualizing the digestion reaction on an agarose gel.
11. Choose a positive clone and sequence the entire *Nco*I–*Eco*RI insert.

3.2. Deletion of the pepP Gene from E. coli

To remove the *pep*P gene that encodes an active proline aminopeptidase from the *E. coli* genome, the Red recombinase-based gene deletion system described by Datsenko and Wanner (14) is used. First, a recombination substrate is produced by amplifying the kanamycin resistance gene flanked by two FRT sites using primers that contain 44–45 nucleotide extensions homologous to sequences flanking the *pep*P gene on the *E. coli* chromosome. This PCR product is introduced into *E. coli* cells harboring a helper plasmid expressing the λ Red recombinase that catalyzes homologous recombination between the recombination substrate and the *E. coli* chromosome. The target gene is thus exchanged for the recombination substrate containing the kanamycin resistance gene and recombinants can be selected by kanamycin resistance. The kanamycin resistance gene is eliminated by site-specific recombination between the two FRT sites catalyzed by the FLP recombinase introduced on a second helper plasmid. Both helper plasmids are temperature sensitive and can be eliminated by growing bacteria at the non-permissive temperature. At the end of the procedure, a single FRT site remains at the locus of the *pep*P gene. This remaining FRT sequence is sometimes termed a "scar sequence":

1. Amplify the kanamycin resistance gene from plasmid pKD4 using primers P1-s and P2-as with the following cycling program: 95°C for 8 min followed by 30 cycles of 94°C for 45 s, 55°C for 45 s, 72°C for 2 min, and a final extension of 72°C for 8 min.
2. Run the PCR reaction on an agarose gel, excise and gel purify, and treat with *Dpn*I to eliminate template DNA. *Dpn*I cleaves only the methylated template plasmid DNA. Add 1 μL *Dpn*I and 5 μL 10× NEBuffer 4 to a 45 μL PCR reaction and incubate the digestion reaction at 37°C for 1 h. Run the digested product on an agarose gel and gel purify.
3. Spread the *E. coli* strain BW25113 containing the helper plasmid pKD46 (harboring the Red recombinase system genes) on a fresh LB agar plate containing 100 μg/mL ampicillin.
4. After incubating the plate at 30°C overnight, inoculate 12 mL LB medium containing 100 μg/mL ampicillin with a colony and grow overnight at 30°C.

5. To make electrocompetent cells, mix 1 mL overnight culture with 20 mL SOB, 100 μL 2 M $MgSO_4$, and 20 μL 1 M L-arabinose and grow at 30°C for 5 h.
6. Harvest cells by centrifugation (1000×*g*, 15 min at 4°C), wash twice with 40 mL ice cold water and once with 40 mL ice cold 10% glycerol, and resuspend in 0.2 mL 10% glycerol.
7. Add 3 μL *Dpn*I-treated PCR reaction (from step 2) to 100 μL of electrocompetent cells and pulse in a 0.2 cm electroporation cuvette with an electroporator setting of 2.5 kV (Ec2 setting on MicroPulser).
8. Add 0.5 mL SOC medium and incubate the cells at 37°C for 1 h.
9. Spread the cells on LB plates containing 50 μg/mL kanamycin and incubate at 37°C overnight.
10. Pick seven colonies from the plate and suspend each in 50 μL water.
11. Using 5 μL of this suspension as template, screen colonies by PCR using two primer sets: primers T-1 and k-1, and primers T-2 and k-2 (*see* **Note 6**). Reactions contain 1× Taq reaction buffer, 2.5 mM $MgSO_4$, 0.2 mM dNTP, 1 μM of each primer and 2 units of Taq polymerase in 50 μL final volume. Cycling conditions are 96°C for 4 min, followed by 25 cycles of 96°C for 45 s, 62°C for 1 min, and 72°C for 1 min, and a final extension of 72°C for 4 min.
12. Run PCR reactions on an agarose gel.
13. Streak colonies showing bands in the PCR analysis onto LB plates containing no antibiotics and incubate at 37°C overnight to cure the temperature-sensitive pKD46 plasmid. Patch all tested colonies onto LB plates with 50 μg/mL kanamycin and LB plates with 100 μg/mL ampicillin (*see* **Note 7**).
14. From colonies resistant to kanamycin but sensitive to ampicillin, inoculate one colony from the drug-free LB plate into 10 mL LB with 50 μg/mL kanamycin and grow overnight. Verify the absence of the *pep*P gene by amplifying the "scar sequence" using primers T-1 and T-2 and sequencing the PCR product. This kanamycin-resistant, aminopeptidase P-deleted strain is designated as LG-1. Prepare electrocompetent LG-1 cells as described in steps 5 and 6 (grow cells at 37°C).
15. Purify plasmid pCP20 harboring the FLP recombinase gene by miniprep after growing the appropriate cells at 30°C overnight in LB medium with 100 μg/mL ampicillin.

16. Introduce plasmid pCP20 into competent cells from step 14 by electroporation and then add 0.5 mL SOC medium and incubate the cells at 30°C for 1 h. Spread cells on LB plates containing 100 μg/mL ampicillin and incubate at 30°C overnight.
17. Pick ten colonies and streak onto drug-free LB plates, then incubate at 43°C overnight.
18. Patch streaks from step 17 on drug-free LB plates, LB plates with 50 μg/mL kanamycin, and LB plates containing 100 μg/mL ampicillin and incubate at 37°C overnight.
19. Pick one of the colonies growing in the absence of antibiotics but sensitive to both kanamycin and ampicillin, indicating excision of the kanamycin resistance gene and loss of the helper plasmid pCP20, and grow overnight in 10 mL LB. This strain is designated LG-2.
20. Prepare electrocompetent LG-2 cells as described in steps 5 and 6 (growing at 37°C).
21. Purify plasmid pGP1-2 by a miniprep protocol and introduce into electrocompetent LG-2 cells by electroporation. Spread transformants onto LB plates with 50 μg/mL kanamycin and incubate the plates at 30°C. The new strain is designated LG-3 and is used for the expression of trypsinogens.

3.3. Expression and Purification of Trypsinogens

3.3.1. Preparation of the Trypsin Affinity Chromatography Column

1. Resuspend the Actigel ALD resin (supplied in 20% ethanol) by turning the bottle upside-down a few times. Insert a 60 mL Buchner funnel into a side-arm flask through a rubber stopper and attach the side arm to the house vacuum through a hose. Pour approximately 20 mL resin into the funnel and wash several times with 50 mM Na-phosphate, pH 7.5, under suction to remove the ethanol and to equilibrate with the coupling buffer (*see* **Note 8**).
2. Disconnect the vacuum and resuspend the resin in the funnel in ~5 mL phosphate buffer. Transfer the resuspended resin from the funnel to a 50 mL Falcon tube and allow to settle.
3. Remove the supernatant with a pipette and add 14 mL 50 mM Na-phosphate, pH 7.5, and 80 mg crystalline bovine trypsin to the remaining ~16 mL wet resin. This corresponds to about 5 mg protein per mL resin immobilization ratio. Finally, add 3 mL 1 M coupling solution to 0.1 M final concentration.

4. Seal the Falcon tube's cap with Parafilm and incubate the tube at room temperature for 1–3 h. After the room temperature incubation, move the reaction to the 4°C cold room and leave overnight with gentle rocking on an Adams Nutator.
5. Pour the resin into an ~8 mL volume empty chromatography column and wash several times, alternating between 20 mM Tris-HCl, pH 8.0, 0.2 M NaCl buffer and 50 mM HCl solution. Store the column equilibrated in 20 mM Tris-HCl, pH 8.0, 0.2 M NaCl in the refrigerator.

3.3.2. Preparation of the Ecotin Affinity Chromatography Column

Ecotin is a pan-serine-protease inhibitor from *E. coli* (18) that forms a relatively tight complex with trypsinogen and thus can be used for affinity purification of the zymogen (15). Recombinant ecotin (17) is overexpressed in the periplasmic space of *E. coli* and isolated using osmotic shock (19) followed by trypsin affinity chromatography. Purified ecotin is immobilized on aldehyde-activated resin by reductive amination using cyanoborohydride (15) and loaded into a chromatography column:

1. Transform *E. coli* BL21 (DE3) cells with plasmid pT7-7-ecotin (17) and spread cells on an LB agar plate containing 100 μg/mL ampicillin to select for transformants. Incubate the plate overnight at 37°C.
2. Inoculate two flasks containing 50 mL LB medium with 100 μg/mL ampicillin (or 50 μg/mL carbenicillin; Sigma) with a streak of the pT7-7-transformed *E. coli* colonies each and incubate overnight at 37°C with shaking.
3. Inoculate two flasks of 1200 mL LB medium containing 100 μg/mL ampicillin with 50 mL overnight culture each and grow at 37°C until OD_{600} reaches 0.5 (*see* **Note 9**).
4. Induce the expression of ecotin by adding 1 mM IPTG to the cultures. Grow cultures for an additional 4 h at 37°C.
5. Harvest the cells by centrifugation at 15,000×*g* for 10 min at 4°C and resuspend in 500 mL ice cold osmotic shock buffer. Keep on ice for 10 min.
6. Centrifuge the suspension for 30 min at 15,000×*g* at 4°C. Discard the supernatant.
7. Resuspend the pellet in 250 mL ice cold water and incubate on ice for 15 min.
8. Centrifuge the suspension for 15 min at 15,000×*g* at 4°C. Remove and save the supernatant, containing the periplasmic fraction (*see* **Note 10**).
9. Add 5 mL 1 M Tris-HCl, pH 8.0, and 10 mL 5 M NaCl to the supernatant (*see* **Note 11**).

10. Centrifuge the supernatant, which is approximately 250 mL, for 15 min at 27,000×*g* at 4°C and load it on the trypsin affinity column in five runs of 50 mL each.
11. Collect the flow-through from all runs and pool. Load the pooled flow-through back onto the column.
12. Wash the column with 20 mM Tris-HCl, pH 8.0, 0.2 M NaCl. Elute ecotin with 50 mM HCl. Pool the eluates from all runs. Determine the concentration and yield of ecotin by reading the absorbance at 280 nm using a calculated molar extinction coefficient of 23,045 M^{-1} cm^{-1} and the monomeric molecular mass of 16,099.5 Da (http://ca.expasy.org/tools/protparam.html). Typical yields vary between 20 and 40 mg purified ecotin.
13. Dialyze the eluate against two changes of 3.5 L 1 mM HCl overnight. Determine its purity by subjecting samples of 1, 5, and 10 μL to SDS-PAGE (*see* **Section 3.3.4**).
14. Lyophilize the dialyzed ecotin overnight (*see* **Note 12**).
15. Store the lyophilized ecotin at –20°C until use. When needed dissolve in water to obtain ~40 mg/mL concentration.
16. Wash the Actigel ALD resin as described in step 1 of **Section 3.3.1**.
17. Mix 1 mL ecotin solution (~40 mg protein) with 8 mL wet resin (~5 mg protein per mL wet resin immobilization ratio) and 1 mL 1 M sodium cyanoborohydride. Incubate the mixture at room temperature for 1–3 h on a shaking table followed by overnight incubation in the cold room.
18. Pack the resin into a 2 mL chromatography column and wash several times by alternating 20 mM Tris-HCl, pH 8.0, 0.2 M NaCl buffer, and 50 mM HCl. Store the column equilibrated with 20 mM Tris-HCl, pH 8.0, 0.2 M NaCl in the refrigerator.

3.3.3. Expression and Purification Using E. coli Strain LG-3

Trypsinogen variants cloned into intein fusion constructs are expressed as inclusion bodies in *E. coli* LG-3. Analysis of solubilized inclusion bodies by SDS-PAGE (*see* **Section 3.3.4**) reveals that the intein moiety has been cleaved off approximately 70–80% of the fusion proteins in *E. coli* (**Fig. 10.2a**); therefore, a separate in vitro intein cleavage step is not included. After in vitro refolding, trypsinogens are purified using ecotin affinity chromatography (15). SDS-PAGE analysis of purified samples shows that eluted trypsinogens contain a small fraction of uncleaved fusion proteins (**Fig. 10.2b**) and further purification is necessary. The eluate is thus subjected to ion-exchange chromatography (**Fig. 10.2d**) which yields a homogenous preparation of recombinant trypsinogens with authentic N-termini (**Fig. 10.2c**):

1. Electroporate the plasmids containing the intein fusion constructs (from step 11 in **Section 3.1**) into the aminopeptidase P-deficient *E. coli* strain LG-3 (from step 21 in **Section 3.2**). Spread transformants onto LB agar plates containing 100 μg/mL ampicillin and 50 μg/mL kanamycin. Incubate the plates at 30°C overnight.
2. Establish starter cultures by inoculating 10 mL LB medium with 100 μg/mL ampicillin and 50 μg/mL kanamycin with a streak of colonies from each plate and incubate at 30°C overnight with shaking.
3. Inoculate 200 mL LB medium containing 100 μg/mL ampicillin and 50 μg/mL kanamycin with 10 mL starter culture and grow until culture density reaches an OD_{600} of 0.5. This will take approximately 2.5 h.
4. Transfer cultures to a 42°C incubator for 30 min and add 200 μL 1 M IPTG (*see* **Note 13**).
5. Grow cultures at 30°C for an additional 5 h (*see* **Note 14**).
6. Divide cultures into four 50 mL Falcon tubes and harvest cells by centrifugation at 2000×*g* for 10 min. Store pellets at –80°C until use.
7. Use one pellet, corresponding to 50 mL of the original culture, per round of purification. Thaw the pellet on ice and resuspend in 4 mL resuspension buffer. Keep samples on ice from this step until step 13.
8. Divide the resulting suspension into 6 aliquots of approximately 700 μL in Eppendorf tubes for more efficient sonication.
9. Sonicate the aliquots three times for 20 s using continuous mode at power setting 4.
10. Centrifuge sonicated samples in a microcentrifuge at 13,200 rpm for 5 min at 4°C.
11. Discard supernatants and resuspend the pellets in 1 mL wash buffer by pipetting. Centrifuge the resuspended samples in a microfuge at 13,200 rpm for 5 min at 4°C. Discard the supernatants, resuspend the pellets and centrifuge again (*see* **Note 15**).
12. Resuspend washed pellets, corresponding to inclusion bodies, in 500 μL denaturing buffer containing 30 mM dithiothreitol and incubate at 37°C for 30 min to reduce the disulfide bonds in trypsinogen.
13. During step 12, add L-cystine and L-cysteine to 50 mL refolding buffer and stir for 30 min on a magnetic stirrer under argon (*see* **Note 16**).

14. Microfuge samples at 13,200 rpm for 5 min at 4°C to remove any insoluble material and then dilute in 50 mL refolding buffer (containing L-cystine and L-cysteine) under argon and stir for 5 min (*see* **Note 17**).
15. Incubate refolding reactions overnight at 4°C.
16. Dilute the refolding reactions with 50 mL 0.4 M NaCl and centrifuge at 27,000×*g* for 15 min (*see* **Note 18**).
17. Load refolded trypsinogen (100 mL) onto a 2 mL ecotin affinity column (from step 18 of **Section 3.3.2**) at a flow rate of 2 mL/min.
18. Wash the column with wash buffer at a flow rate of 2 mL/min until the UV signal recedes to baseline.
19. Elute trypsinogen from the column with 50 mM HCl at a flow rate of 2 mL/min and determine the protein concentration by reading the absorbance at 280 nm and using a calculated molar extinction coefficient of 37,525 M^{-1} cm^{-1} (http://ca.expasy.org/cgi-bin/protparam). Then subject samples to SDS-PAGE (described under **Section 3.3.4**) for assessment of purity (**Fig. 10.2b**).
20. Load the ecotin eluate directly onto a Mono S HR 5/5 cation-exchange column equilibrated with 20 mM Na-acetate, pH 5.0.
21. Develop the column with a gradient of 0–0.5 M NaCl at a flow rate of 1 mL/min and collect 1 mL fractions; *see* **Fig. 10.2d**.
22. Run fractions showing the highest absorbance at 280 nm on an SDS-PAGE gel to verify purity (**Fig. 10.2c**).

3.3.4. SDS Polyacrylamide Gel Electrophoresis (SDS-PAGE)

1. The Mini-PROTEAN Gel System (Bio-Rad) is described. Rub clean glass plates with 95% ethanol to remove any grease and prepare 13% running gels by mixing (for 4 gels) 9.4 mL water, 7.5 mL running gel buffer, 13 mL 30% acrylamide solution, 100 μL APS solution, and 50 μL TEMED. Gently mix the solution, then cast the gels, overlay with water-saturated 1-butanol, and allow to set for at least 30 min.
2. Prepare stacking gels (4 gels) by mixing 11.6 mL water, 5 mL stacking gel buffer, 3.33 mL 30% acrylamide solution, 60 μL APS solution, and 50 μL TEMED.
3. Pour off the 1-butanol from the running gels, then cast the stacking gels and insert the combs. Leave stacking gels to polymerize for at least 30 min, then remove the combs and rinse the wells with distilled water (*see* **Note 19**).
4. Determine the concentration of trypsinogen samples as described in step 19 under **Section 3.3.3**. Load approximately 200 pmol trypsinogen (~5 μg) per lane.

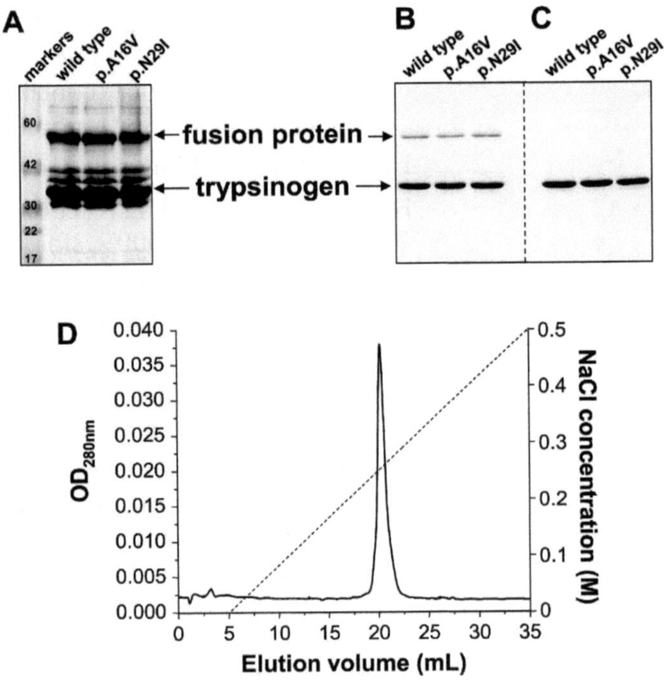

Fig. 10.2. Expression and purification of the intein–trypsinogen fusions. In the experiment shown here, wild-type cationic trypsinogen and pancreatitis-associated mutants p.A16V and p.N29I were expressed and purified. SDS-PAGE analysis of (**a**) inclusion bodies from LG-3 cells expressing the intein–trypsinogen fusions; (**b**) trypsinogens eluted from the ecotin affinity column; (**c**) trypsinogen fractions after Mono S chromatography. Inclusion body fraction prepared from 1.8 mL LG-3 culture ($OD_{600} = 1.6$) or ~5 μg purified protein was loaded per lane. (**d**) Mono S chromatography of cationic trypsinogen eluted from the ecotin affinity column. The column was developed with a 0–0.5 M gradient of NaCl at a flow rate of 1 mL/min. Under these conditions only pure trypsinogen is eluted, while unprocessed fusion proteins are not recovered. Reproduced from (9) with permission from Elsevier Science.

5. Precipitate trypsinogens by adding trichloroacetic acid at 10% final concentration and incubate on ice for 10 min.
6. Centrifuge the samples at 13,200 rpm for 10 min in a microfuge at 4°C and discard the supernatants.
7. Resuspend the pellets in 30 μL reducing Laemmli sample buffer, briefly vortex, and heat-denature by incubating at 90°C for 5 min.
8. Load the samples onto a gel using a 20 μL pipet tip or Hamilton syringe. One lane on each gel should be loaded with protein markers.
9. Run the gel by applying a 30 mA current until the blue dye front exits the gel.
10. After electrophoresis, remove the gel from the unit and stain in Coomassie blue staining solution for 20–30 min on a rocking table (*see* **Note 20**).

11. Remove excess staining by incubating the gel in several changes of destaining solution on a rocking table until the gel background is clear (*see* **Note 20**).
12. Preserve gels by soaking in Gel-Dry^{TM} solution for 20 min and then stretching on a drying frame between two cellophane sheets and air-drying overnight (*see* **Note 21**).

4. Notes

1. There are several newer generation forms of Actigel ALD available (e.g., ultraflow, superflow) which claim to exhibit better flow rates. In our application we experienced no benefit from using these resins and we obtained the best results with the least expensive, regular Actigel ALD resin.
2. Add dithiothreitol immediately before use from a 1 M solution kept at –20°C in aliquots.
3. Solutions containing methanol are hazardous and should be disposed of accordingly.
4. Gel-Dry^{TM} solution contains ethanol, polyethylene glycol, methanol, and isopropanol and should be disposed of as hazardous waste.
5. The underlined sequences in primers B and C overlap and thus allow the hybridization of PCR products obtained in the first round of amplifications (*see* (20)).
6. In both primer sets, one primer anneals within the kanamycin resistance gene and the other anneals to a nearby sequence in the *E. coli* chromosome. Amplification of PCR products indicates that recombination between the *pep*P gene and the recombination substrate has occurred.
7. Colonies resistant to kanamycin have undergone homologous recombination and have lost the *pep*P gene, while resistance to ampicillin indicates that the helper plasmid has not been lost.
8. The coupling buffer must be free of amines; therefore, phosphate-based buffers are recommended. Do not use Tris buffer or other amine-containing buffers.
9. Growing the bacterial culture followed by the isolation of the periplasmic fraction is a full-day procedure, so it is practical to start the culture as early as possible.
10. At this step, the supernatant has to be clear. Centrifuge again if necessary before proceeding to step 9.

11. The procedure can be paused here and the supernatant stored on ice overnight.
12. Distribute the dialyzed sample to 50 mL Falcon tubes (about 15 mL in each). Using adequate cold protection, hold the tubes tilted in a Dewar flask with liquid nitrogen and slowly rotate the tubes so that ecotin freezes onto the tube wall as a layer. Cover the tubes with Parafilm and punch holes in the Parafilm to let water evaporate. Place tubes in a larger freeze-dry bottle or flask and lyophilize overnight.
13. Addition of IPTG is necessary because the region upstream of the intein–trypsinogen fusion contains the lac operator (5, 6).
14. It is useful to check expression levels before starting the purification procedure. After approximately 4 h of culturing, remove a 1 mL sample from each culture and centrifuge for 5 min at 12,300 rpm. Discard the supernatant, resuspend the pellet in 1 mL resuspension buffer and sonicate three times for 20 s (continuous mode, power setting 5). Microfuge the samples for 5 min at 12,300 rpm and discard the supernatants. Resuspend the pellets in 30 μL reducing Laemmli sample buffer, vortex briefly, and incubate at 90°C for 5 min. Load these samples on an SDS polyacrylamide gel (*see* **Section 3.3.4**). Purified trypsinogen should be loaded as positive control and samples with a strong band at the same position are processed further.
15. Pellets can be combined to produce a single pellet at the end of the washing procedure.
16. First, add L-cystine in powder form to 50 mL refolding buffer and dissolve by vigorous stirring at room temperature on a magnetic stirrer for about 30 min. There is no need for argon at this step. L-Cystine is poorly soluble in water and undissolved crystals may still be visible after 30 min. This has no impact on the refolding procedure. Add L-cysteine after the refolding solution has been equilibrated with argon for a few minutes. L-Cysteine should dissolve readily.
17. Attach a 200 μL pipet tip to the tubing from the gas cylinder and punch it through the Parafilm covering the flask with the refolding buffer. Place the flask on a magnetic stirrer and after a few minutes of argon flow, add the denatured trypsinogen sample dropwise to the buffer under moderate stirring. Then stir the solution for another 5 min under argon.
18. This step is included to dilute the guanidine content of the sample. Centrifugation is necessary to remove any precipi-

tated proteins, which can clog up the chromatography system and the column.

19. Gels can be stored for 2 weeks at 4°C wrapped in wet paper towels in airtight plastic bags.
20. To minimize exposure to methanol vapor, cover the dishes tightly with aluminum foil during these steps. Solutions containing methanol are hazardous and should be disposed of accordingly. To reduce the amount of hazardous waste, slightly stained destaining solutions can be re-used at the beginning of a destaining procedure. Speed and efficiency of destaining can be increased by placing a small sheet of Kimwipe in the dish which adsorbs colloidal stain particles.
21. Care should be taken to avoid air bubbles between the cellophane sheets as these can result in white spots on dried gels. Gel-Dry™ solution contains ethanol, polyethylene glycol, methanol, and isopropanol and should be covered during soaking of gels and disposed of as hazardous waste.

Acknowledgments

We thank Barry L. Wanner (Department of Biological Sciences, Purdue University, West Lafayette, IN) for sharing the plasmids and bacterial strains of the Red recombinase-based gene deletion system. The valuable support and contributions of Ronald Kaback, Edit Szepessy, Zoltán Kukor, and Miklós Tóth are gratefully acknowledged. This work was supported by NIH Grant DK058088 to M.S.-T.

References

1. Witt, H., Luck, W., Becker, M. (1999) A signal peptide cleavage site mutation in the cationic trypsinogen gene is strongly associated with chronic pancreatitis. *Gastroenterology* 117, 7–10.
2. Pfützer, R. H., Whitcomb, D. C. (1999) Trypsinogen mutations in chronic pancreatitis. *Gastroenterology* 117, 1507–1508.
3. Chen, J. M., Raguenes, O., Ferec, C., Deprez, P. H., Verellen-Dumoulin, C., Andriulli, A. (1999) The A16V signal peptide cleavage site mutation in the cationic trypsinogen gene and chronic pancreatitis. *Gastroenterology* 117, 1508–1509.
4. Teich, N., Bauer, N., Mössner, J., Keim, V. (2002) Mutational screening of patients with nonalcoholic chronic pancreatitis: identification of further trypsinogen variants. *Am J Gastroenterol* 97, 341–346.
5. Sahin-Tóth, M. (2000) Human cationic trypsinogen. Role of Asn-21 in zymogen activation and implications in hereditary pancreatitis. *J Biol Chem* 275, 22750–22755.
6. Sahin-Tóth, M., Tóth, M. (2000) Gain-of-function mutations associated with hereditary pancreatitis enhance autoactivation of human cationic trypsinogen. *Biochem Biophys Res Commun* 278, 286–289.
7. Ben-Bassat, A., Bauer, K., Chang, S. Y., Myambo, K., Boosman, A., Chang, S. (1987) Processing of the initiation methionine from proteins: properties of the *Escherichia coli* methionine aminopeptidase and its gene structure. *J Bacteriol* 169, 751–757.

8. Yaron, A., Mlynar, D. (1968) Aminopeptidase-P. *Biochem Biophys Res Commun* 32, 658–663.
9. Király, O., Guan, L., Szepessy, E., Tóth, M., Kukor, Z., Sahin-Tóth, M. (2006) Expression of human cationic trypsinogen with an authentic N terminus using intein-mediated splicing in aminopeptidase P deficient *Escherichia coli*. *Protein Expr Purif* 48, 104–111.
10. Wu, H., Xu, M. Q., Liu, X. Q. (1998) Protein trans-splicing and functional mini-inteins of a cyanobacterial dnaB intein. *Biochim Biophys Acta* 1387, 422–432.
11. Evans, T. C., Jr., Benner, J., Xu, M. Q. (1999) The cyclization and polymerization of bacterially expressed proteins using modified self-splicing inteins. *J Biol Chem* 274, 18359–18363.
12. Yoshimoto, T., Murayama, N., Honda, T., Tone, H., Tsuru, D. (1988) Cloning and expression of aminopeptidase P gene from *Escherichia coli* HB101 and characterization of expressed enzyme. *J Biochem* 104, 93–97.
13. Yoshimoto, T., Tone, H., Honda, T., Osatomi, K., Kobayashi, R., Tsuru, D. (1989) Sequencing and high expression of aminopeptidase P gene from *Escherichia coli* HB101. *J Biochem* 105, 412–416.
14. Datsenko, K. A., Wanner, B. L. (2000) One-step inactivation of chromosomal genes in *Escherichia coli* K-12 using PCR products. *Proc Natl Acad Sci USA* 97, 6640–6645.
15. Lengyel, Z., Pál, G., Sahin-Tóth, M. (1998) Affinity purification of recombinant trypsinogen using immobilized ecotin. *Protein Expr Purif* 12, 291–294.
16. Tabor, S., Richardson, C. C. (1985) A bacteriophage T7 RNA polymerase/promoter system for controlled exclusive expression of specific genes. *Proc Natl Acad Sci USA* 82, 1074–1078.
17. Pál, G., Sprengel, G., Patthy, A., Gráf, L. (1994) Alteration of the specificity of ecotin, an *E. coli* serine proteinase inhibitor, by site directed mutagenesis. *FEBS Lett* 342, 57–60.
18. Chung, C. H., Ives, H. E., Almeda, S., Goldberg, A. L. (1983) Purification from *Escherichia coli* of a periplasmic protein that is a potent inhibitor of pancreatic proteases. *J Biol Chem* 258, 11032–11038.
19. Neu, H. C., Heppel, L. A. (1965) The release of enzymes from *Escherichia coli* by osmotic shock and during the formation of spheroplasts. *J Biol Chem* 240, 3685–3692.
20. Aiyar, A., Xiang, Y., Leis, J. (1996) Site-directed mutagenesis using overlap extension PCR. *Methods Mol Biol* 57, 177–191.

Chapter 11

Recent Developments in Difficult Protein Expression: A Guide to *E. coli* Strains, Promoters, and Relevant Host Mutations

James C. Samuelson

Abstract

Escherichia coli is a versatile and popular tool for heterologous protein production. Some of the reasons for its popularity include rapid growth, a variety of portable vectors, relatively simple genetics, and the potential for high-density cultivation. In addition, the extensive laboratory use of *E. coli* has resulted in technologies to target protein overexpression to various intracellular compartments. This is advantageous because these compartments have different environments that may facilitate folding of particular proteins of interest. This chapter discusses the properties of many of the *E. coli* strains available for protein expression in order to facilitate the choice of the best expression host for a particular protein of interest.

Key words: *Escherichia coli*, protein expression, T7 promoter, *tac* promoter, membrane protein expression, expression strain.

1. Introduction

Escherichia coli is a versatile tool for the production of recombinant protein. Attractive features include portable vector systems, relatively simple genetics, rapid growth, and the potential for high-density cultivation. Furthermore, each of the well-characterized cellular compartments may serve as a reservoir for the overexpressed protein of interest. The following sections offer comprehensive guidance in using *E. coli* as a laboratory tool for the expression of well-behaved or difficult proteins.

T.C. Evans, M.-Q. Xu (eds.), *Heterologous Gene Expression in E. coli*, Methods in Molecular Biology 705,
DOI 10.1007/978-1-61737-967-3_11,

2. Protein Expression Using IPTG-Regulated Promoters

Prior to 1990, recombinant protein expression in *E. coli* was accomplished primarily with plasmids encoding a promoter derived from the *lac* operon (e.g., *Plac*, *PlacUV5*, *Ptac*, and *Ptrc*). Each of these promoters includes the binding site for the LacI repressor. In wild-type *E. coli*, the lac repressor is present at approximately ten molecules per cell (1). Thus, efficient repression of multicopy plasmid expression requires that the plasmid also carry its own *lacI* gene or *lacI^q^* gene. The *lacI^q^* mutant was reported by Calos in 1978 and this mutation is simply an "up" promoter mutation resulting in a tenfold enhancement of lacI repressor expression (2). In cases where the plasmid does not carry the *lacI* gene or repression is not sufficient, host strains carrying the *lacI^q^* gene are recommended. For example, the NEB Express Iq strain is a BL21 derivative carrying the *lacI^q^* gene on a single-copy mini-F plasmid. Preexistence of the Lac repressor protein within a competent cell may be necessary to obtain transformants of so-called toxic clones, even when the plasmid clone carries the *lacI* gene.

The *lacUV5* promoter is a variant of the *lac* promoter with two mutations within the –10 region of the promoter. In addition, a mutation is present at –66 within the CAP (catabolite gene activator protein) binding site. These mutations increase the promoter strength relative to the wild-type *lac* promoter and expression from *lacUV5* is less subject to catabolite repression (3); yet the *lacUV5* promoter is stimulated two- to threefold by the presence of cAMP according to in vitro studies performed by Eron and Block (4).

The *tac* promoter was first described by DeBoer et al. (5). This strong promoter is a hybrid of the –10 region of the *lacUV5* promoter and the –35 region of the *trp* promoter. Amann et al. reported that the *tac* promoter is at least five times more efficient than the *lacUV5* promoter (6). The *trc* promoter should be considered equivalent to the *tac* promoter since the one bp difference in spacing between the –35 and –10 consensus sequences does not affect promoter strength (7). Note that the *tac* and *trc* promoters are not subject to catabolite repression as the CAP binding site is missing. Even so, *Ptac* and *Ptrc* systems are generally well controlled by LacI repression.

Note: In practice, we have observed multiple cases where protein expression from a *tac* promoter is as equally robust as expression from a T7 promoter (with other factors being equal: host strain, rbs, etc.). However, one advantage of the T7 system is the potential for lower basal expression when T7 lysozyme is employed to inhibit the basal activity of T7 RNA polymerase.

For the last 15 plus years, recombinant protein expression has been dominated by the use of BL21(DE3) and derivative strains. DE3 refers to the lambda prophage carrying phage T7 gene 1, which encodes the T7 RNA polymerase. T7 gene 1 expression is controlled by the *lacUV5* promoter in DE3 strains. The DE3 fragment (present within the chromosomal *att* site) also carries a *lacI* gene with its native promoter. Thus, induction of the T7 RNA polymerase in DE3 strains is mediated by IPTG or allolactose in the case of auto-induction protocols (8). Early on, DE3 strains were recognized to be "leaky" in terms of basal-level protein expression. This phenomenon is the result of incomplete repression of T7 RNA polymerase synthesis from the *lacUV5* promoter. Any basal expression of T7 RNA polymerase may lead to a significant level of the target protein due to the processivity of the T7 RNA polymerase and the presence of multiple plasmid-encoded copies of the target gene in each cell. In fact, Pan and Malcolm (9) demonstrated that basal target protein expression may be extremely high in BL21(DE3) grown in rich media (Terrific Broth). These researchers also showed that basal expression may be reduced by the addition of 1% glucose, which results in lower cAMP levels. In turn, cAMP/CAP stimulation of the *lacUV5* promoter is minimized. Pan and Malcolm recommended switching from glucose to a poor carbon source for the final growth cycles to achieve maximal IPTG-induced expression.

Certainly, the best way to counteract basal T7 expression of a target protein is to add T7 lysozyme to the system. T7 lysozyme naturally inhibits T7 RNA polymerase function by a 1:1 protein–protein interaction (10). T7 lysozyme control may be provided by three vehicles. Plasmids pLysS and pLysE produce "active" lysozyme meaning that this protein exhibits amidase activity when given access to the peptidoglycan layer of the *E. coli* cell wall, whereas lysY strains express a K128Y variant of T7 lysozyme that lacks amidase activity. The lysY variant retains full function in inhibiting T7 RNA polymerase activity (11) and we find that lysY strains provide the same level of performance as pLysS strains with respect to expression control. Use of pLysS is most common as the constitutive expression of lysozyme from this plasmid is much lower than the level expressed by pLysE. Strains carrying pLysE may exhibit a growth defect (12) and in extreme cases culture lysis has been observed when the protein of interest is targeted to the cell envelope in a pLysE host (personal observation).

Note: If an expression plasmid does not yield transformants when using BL21(DE3), the first response should be to test transformation into a lysY or pLysS strain. Even mildly toxic gene products may be lethal to BL21(DE3) upon transformation. In contrast, lysY or pLysS strains may yield normal colonies and express the protein of interest at moderate to high levels. Finally, it should be noted that the choice of lysY or pLysS should take into account downstream processing of cells. Strains expressing active lysozyme from pLysS may lyse spontaneously upon one freeze–thaw cycle. If this feature is not desired, lysY strains are a good alternative.

3. Other T7 Expression Strains

This section will cover other T7 expression strains introduced to the US market in the last 5 years. The BL21-AI strain (Invitrogen) was constructed to encode a chromosomal copy of the T7 RNA polymerase gene under control of the *ParaBAD* promoter (AI = arabinose-inducible). This promoter is tightly repressed by glucose and this strain is an option for the expression of toxic proteins.

KRX protein expression strain (Promega) is a T7 expression strain where the T7 RNA polymerase is controlled by the *PrhaBAD* promoter. This is a relatively new strain and we expect it will perform well for the controlled expression of toxic proteins. Glucose repression of the rhamnose promoter (*PrhaBAD*) is reported to be even better than repression of the *ParaBAD* promoter (13).

The T7 Express strains (New England Biolabs) are BL21 derivatives with unique features: T7 Express is an enhanced BL21 derivative with the T7 RNA polymerase expressed from the wild-type *lac* promoter from within the *lac* operon (in contrast to DE3 strains). This means that basal T7 expression is lower than that observed in DE3 strains. Added features in other T7 Express strains are *lacI^q* control, *lysY* control, or a combination of both *lysY* and *lacI^q*. These additional features are carried by a single-copy mini-F plasmid that is stably maintained without antibiotic supplementation. All T7 Express strains are phage T1 resistant (*fhuA2* mutation), and each is suitable for cloning (*endA1* mutation results in high-quality plasmid yields and transformation efficiencies range up to 10^9 per μg plasmid DNA).

The OverExpress™ C41(DE3) and C43(DE3) strains were first described by Miroux and Walker in 1996 (14) and are now offered in the United States by Lucigen. These strains were isolated by transforming BL21(DE3) with plasmid clones expressing a toxic membrane protein. The few colonies that survived this transformation step were analyzed for the ability to tolerate expression of the toxic protein. Two isolates able to tolerate expression of the toxic proteins were designated C41(DE3) and C43(DE3) but the mutations in these host strains were apparently not characterized. Only recently, Wagner et al. provided information to explain the tolerance of these two strains to T7 expression of otherwise toxic protein (15). The key host mutations in the

"Walker" strains appear to simply moderate the expression level of the T7 RNA polymerase. Both strains have GT to AA mutations in the –10 region of the chromosomal T7 RNA polymerase promoter that effectively change the sequence from *lacUV5* to the weaker wild-type *lac* promoter. The importance of a third mutation in the *lac* operator sequence is uncertain, but it may in fact also lessen T7 expression since Wagner and colleagues demonstrated by Western blotting that the level of T7 RNA polymerase in both strains does not accumulate to an appreciable level even at 120 min post-IPTG addition. Thus, the Walker strains tolerate toxic proteins because less of the protein is produced per cell per unit time. In the context of membrane protein expression, this can actually equate to a greater culture yield of protein since individual cells are more likely to remain vital and maintain expression of the protein of interest.

The study of the Walker strains led to the construction of the Lemo21(DE3) strain. In this BL21(DE3) derivative, the *activity* of T7 RNA polymerase is modulated by the user by coexpression of the T7 lysozyme variant lysY using the *PrhaBAD* promoter. Wagner et al. demonstrate that this T7 expression system is very tunable in the range of 1–1,000 μM L-rhamnose and the Lemo21(DE3) strain is particularly useful for membrane protein expression. Importantly, this strain allows the user to test various expression levels to find the optimal level where the membrane protein of interest is maximally produced without inflicting toxicity to the host cells. The Lemo21(DE3) strain may provide a solution for producing other difficult proteins where tunable expression is desired.

4. Other New Approaches in Protein Expression

The Single Protein Production system (SPP™ system) was developed by Masayori Inouye (16) and is marketed by Takara Bio Inc. This is a two-vector system suitable for use in most *E. coli* strains. The protein of interest is expressed from a vector with the cold-inducible *E. coli cpsA* promoter. However, the unique enabling feature is the induction of a site-specific mRNA interferase (MazF) from the second plasmid, which degrades endogenous mRNA by acting at ACA sites. Accordingly, the gene of interest must be designed and synthesized to lack ACA sequences, so that the mRNA of interest persists and becomes the primary substrate for the translation machinery. The elimination of most host-derived mRNA is reported to create a quasi-dormant cell where expression of the protein of interest may be sustained for up to several days. As the cost of gene synthesis decreases, this system should gain increasing popularity.

Giacalone et al. have shown recently that vectors employing the *rhaT* promoter are useful for direct L-rhamnose-inducible protein expression (17). In this work, GFP expression from a *PrhaT* vector was compared to expression from a *ParaBAD* vector (L-arabinose induction) and pUC19 (IPTG induction). Flow cytometry analyses indicated that L-rhamnose induction is tunable (meaning that protein expression per cell is proportional to inducer concentration), whereas the other promoters tend to operate by on/off mode. This study was conducted in *E. coli* strain MG1655 containing an intact rhamnose operon. So apparently, metabolism of the inducer does not affect the tunable nature of this *PrhaT* expression system.

Arabinose-inducible protein expression is a method of choice for studies where near-complete repression is desired to avoid cytotoxicity. Initially, expression from *ParaBAD* was thought to be tunable, but careful studies of this promoter by Siegele et al. (18) and Giacalone et al. (17) both agreed that at subsaturating levels of L-arabinose, protein expression cultures contain a mixed population of cells with only some of the cells expressing protein. Thus, the following solutions have been presented to make *araBAD* expression more tunable: (1) Khlebnikov et al. (19) constructed strains where the AraE low-affinity, high-capacity transporter is controlled by a constitutive promoter and (2) Marc Better (of Xoma Technology Ltd.) described the application of strains deficient in either AraE or the high-affinity transport system AraFGH (20).

Numerous expression studies have concluded that colder induction temperatures (15–20°C) can be beneficial for obtaining a greater yield of soluble, properly folded protein. The ArcticExpress™ strains from Stratagene are reported to aid in even lower temperature (10–13°C) folding of heterologous proteins. This is mediated by coexpression of cold-adapted chaperonins Cpn10 and Cpn60 from *Oleispira antarctica*. The development of these strains resulted from a study demonstrating that the *O. antarctica* chaperonins display protein refolding activities at 4–12°C (21).

Auto-induction of T7 expression has gained popularity since the detailed examination of this method by William Studier (8). Cells are cultured in defined media containing a mix of carbon sources: generally glucose, lactose, and glycerol. When glucose is depleted, lactose serves to induce expression of the T7 RNA polymerase upon conversion to allo-lactose by beta-galactosidase (the *lacZ* gene product). Thus, Studier points out that auto-induction should be performed in strains encoding an intact *lac* operon.

Note: The T7 Express strains (NEB) are not suitable for auto-induction protocols as the T7 RNA polymerase gene is located within the *lac* operon.

5. Disulfide Bond Formation

In wild-type *E. coli*, a set of enzymes (DsbA, DsbB, DsbC, DsbD) mediates disulfide bond formation within proteins in the periplasm. DsbA acts as the primary oxido-reductase in this process, while periplasmic DsbC is thought to act as a disulfide bond isomerase to help proteins attain the correct disulfide linkages. Thus, if the expressed protein of interest requires disulfide bonds, one option to obtain an active protein is to target it to the periplasm of *E. coli*. However, the yield of the expressed protein will be limited by the throughput capacity of inner membrane transport and the volumetric capacity of the periplasmic compartment. Research in the lab of Jon Beckwith (22, 23) has led to a line of *E. coli* strains that enable disulfide bond formation within proteins expressed in the cytoplasm. These strains are marketed by Novagen using the names Origami, Rosetta-gami, and variations thereof. These strains have deletions in the glutathione reductase (*gor*) and thioredoxin reductase (*trxB*) loci that alter the reducing potential of the cytoplasm, which makes the cytoplasm a more favorable environment for disulfide bond formation. A suppressor mutation within the *ahpC* gene also appears to be necessary in these strains. A recent development in the expression of proteins requiring disulfide bonds is a new set of strains offered by New England Biolabs. The SHuffle strains also carry *gor*, *trxB*, and *ahpC* mutations but an added feature is the cytoplasmic expression of DbsC. The SHuffle strains have been engineered by Mehmet Berkmen of NEB to express an optimal level of the DsbC isomerase to assist expressed proteins in achieving their correct structure.

6. Protein Export

Export of proteins to the *E. coli* periplasm is most commonly carried out using a signal sequence (ss) from a Sec translocase-dependent protein. For example, OmpA and PelB signal sequences are included in many pET vectors. This export strategy typically works best for proteins that are normally exported in their native host. Newer strategies utilize DsbA or DsbC signal sequences or fusions to periplasmic DsbA or DsbC (pET 39 or pET40 vectors). Export of passenger proteins fused to the DbsA signal sequence was shown to be mediated by the signal recognition particle (24). Utilization of the cotranslational SRP pathway is a complementary approach for proteins that do not get exported efficiently by the SecB-mediated posttranslational pathway.

Export to the periplasm by fusion to maltose binding protein (MBP) is a well-established expression method. Fusion proteins are isolated by affinity purification on amylose resin and we have found that many fused proteins retain the activity of the native protein. If desired, the protein of interest may be cleaved from MBP using Enterokinase, Factor Xa, or Genenase^{TM} I. The newest version of the New England Biolabs MBP fusion vector for periplasmic expression is pMAL-p5. This vector encodes a variant of MBP that has a higher affinity for amylose resin and the polylinker has been altered to match other NEB protein expression vectors: pKLAC2, pTYB21, and pTYB22. During the development of the pMALp5 vector, protein export was compared between the K-12 strain TB1 and the BL21 derivative NEB Express. Expression and protein export were clearly superior in the Lon minus, OmpT minus NEB Express strain (P. Riggs, unpublished results).

7. Membrane Protein Expression

Membrane protein overexpression within a bacterial cell presents an entirely distinct set of problems. In contrast to the overexpression of well-behaved water-soluble proteins, the expression of an alpha-helical membrane protein must be precisely controlled in order to maximize yield. The most practical approach is to express the desired membrane protein to the *E. coli* inner membrane to promote proper folding within the host cell. To that end, it may be necessary or advantageous to modify the protein of interest if it does not possess a typical N-terminal signal peptide that will be recognized by the *E. coli* signal recognition particle. The presence of a native N-terminal signal peptide (and the topology of other membrane alpha-helices) may be analyzed by submitting the protein sequence to the SignalP (25) or Phobius (26) predictors. Even if these algorithms predict a signal peptide, membrane targeting may be improved by replacing the native signal with an N-terminal fusion partner that is known to navigate the SRP/Sec translocase pathway. Grisshammer et al. showed that membrane proteins fused to pre-MBP make their way to the inner membrane as the MBP domain is exported through the Sec translocase and into the periplasm (27). However, many studies have demonstrated that MBP export is primarily a posttranslational process that is mediated by the SecB chaperone. Thus, one might predict that MBP-membrane protein fusions would be especially prone to cytoplasmic aggregation once the overexpressed membrane protein exceeds the capacity of the SecB chaperone pool. To those who study bacterial membrane protein biogenesis, an ideal N-terminal fusion partner would travel the cotranslational SRP pathway into the inner membrane, thus taking the same path as most endogenous membrane proteins. This, in theory, would minimize the opportunity for accumulation of toxic cytoplasmic protein aggregates.

New approaches for inner membrane targeting include employing the SRP-dependent, rationally designed P8CBD fusion partner (28), fusion to the *E. coli* GlpF protein (29), or simple replacement of the native signal with an *E. coli* SRP-dependent signal peptide such as the DsbA or TorT signal sequences (30).

Some researchers may prefer to overproduce the desired membrane protein within cytoplasmic inclusion bodies since this material can be easily isolated. However, even if the researcher is willing to work out optimal conditions for in vitro folding of the protein, this approach may not yield quality protein since it is well established that cytoplasmic protease expression is upregulated to deal with protein aggregation.

Producing an excess of the mRNA of interest (e.g., uncontrolled T7 expression) is not recommended when attempting to overexpress membrane proteins to the inner membrane. The SPP system is one exception to this advice since this system is based on elimination of endogenous mRNA by the MazF ribonuclease and thus, the mRNA for the membrane protein of interest may be present at a great excess over cellular mRNAs. The net result should be a scenario where the Sec translocase is focused on integrating the protein of interest into the inner membrane and early reports on the application of the SPP system for membrane protein expression appear to be favorable. The SPP vectors (pColdI-V) utilize the *E. coli cspA* promoter, which is cold-inducible, so expression optimization may require sampling different induction temperatures.

8. Notes on Membrane Protein Expression

1. Test many inducer concentrations (especially within the low range) as the critical concern during overexpression of membrane proteins in *E. coli* is saturation of the Sec translocation channel, the major bottleneck in membrane protein expression. Tightly regulated and tunable expression systems appear to be best for the expression of membrane proteins. When basal expression is completely repressed during the initial outgrowth phase, aggregated protein does not

accumulate in the cytoplasm, stress responses are not triggered, plasmid loss is minimized, and cell growth rate is less affected.

2. Consider altering the Shine–Delgarno sequence: One overlooked means of optimizing membrane protein expression is the alteration of ribosomal binding site (rbs) strength. When using a *Ptac* vector expressing a protein that utilizes the SRP cotranslational pathway, we have found that a weak rbs may improve results significantly (28). This finding was reached after all other measures of optimization were attempted. We confirmed that the weaker rbs resulted in lower basal expression and upon induction with 400 μM IPTG the cells continued to grow and the culture achieved a high cell density after 20 h at 20°C. In this case, expression toxicity was avoided by moderating the expression level by simply altering translation initiation efficiency of the mRNA of interest. We found that the yield of protein was increased up to tenfold per liter of shake flask and the yield of protein per cell was also improved. In this experiment, the 5′-untranslated region (uppercase) was altered from TGGAAACTTCCTCatg to AAAGGACGGCCGGatg.

   Note: For a list of rbs sequences that are ranked in order of their translational efficiency in *E. coli*, *see* supplemental Fig. 2b in Gardner et al. (31).

3. Use a low to medium copy plasmid. High-copy number plasmids (pUC origins) will almost certainly result in clone instability and/or loss of plasmid or expression by a fraction of the cells within the culture. Plasmid loss is often a significant problem with high copy clones expressing ampicillin resistance since the resistance-conferring protein (beta-lactamase) is exported in excess causing the ampicillin concentration in the growth medium to be depleted. This creates an opportunity for plasmid-free cells to multiply and outgrow the slower growing protein expressing cells. Plasmids carrying KanR and CamR genes are generally recommended for the expression of membrane proteins. Other extreme measures may be applied to counteract plasmid loss. For example, plasmid maintenance systems may be engineered into membrane protein expression vectors. One such example (the hok/sok gene cassette) is in use by groups expressing GPCRs in *E. coli* (32, 33).

4. Always determine how much of the protein is membrane integrated vs. aggregated. Analyze the expression level by Coomassie-stained protein gel, Western blot, or activity assay. Most of the overexpressed target membrane protein should be in the low-speed spin supernatant after cell breakage by French Press, cell disruption, or sonication

(in combination with EDTA-lysozyme treatment). If a significant fraction of the target protein is insoluble (low-speed pellet), repeat expression at a lower temperature (20–25°C). Expression of protein into the membrane fraction is also generally improved by early induction (OD600 = 0.35–0.45).

9. Beneficial Host Strain Mutations/ Phenotypes

The *ΔompT* deletion eliminates the outer membrane protease OmpT, which cleaves polypeptides primarily between pairs of basic amino acids. If not deleted, this activity may cleave membrane proteins as well as soluble proteins of interest during processing of cell lysates.

The *ompP* gene is encoded by the F′ episome of *E. coli*. Therefore, deletion of the ompP gene, *ΔompP*, is relevant only for strains with an F′ episome. It should be noted that the Promega KRX strain carries an F′ where the *ompP* gene is deleted. OmpP is 71% identical to the outer membrane protease OmpT. Thus, OmpP exhibits an enzymatic activity similar to OmpT (34).

Note: The features carried by F′ episomes are generally not necessary for protein overexpression. In fact, typical F′ episomes encode the potentially deleterious OmpP protease (exception: Promega KRX F′ strain). F′-containing strains are not recommended for overexpression of proteins targeted to the cell envelope (inner membrane, periplasm, outer membrane). Strains carrying mini-F plasmids must be distinguished from F′ strains. Mini-F strains (NEB T7 Express lysY, T7 Express lys/Iq, and T7 Express Iq) do not express an F pilus for conjugation and do not carry the *ompP* gene.

B strains are naturally deficient in Lon protease. Until recently, the absence of Lon protease was widely viewed as beneficial for protein overexpression. Lon is an ATP-dependent cytoplasmic protease that plays a significant role in protein quality control. Lon is known to degrade misfolded or truncated proteins that might otherwise aggregate. Recent publications suggest that the activity of quality control proteases (e.g., FtsH) may be important in some situations (35). Cytoplasmic protein aggregation is thought to be a major factor in the "toxic effect" observed upon membrane protein overexpression (15). Perhaps the activity of Lon, in addition to the activity of the Clp family proteases, would help to reduce the toxicity associated with membrane protein overexpression.

The BL21 Star strains carry a mutation in the gene encoding RNaseE, *rne131*, one of the primary enzymes involved in mRNA degradation. Since T7 RNA polymerase synthesizes mRNA faster than *E. coli* RNA polymerase and transcription from the T7 promoter is not coupled to translation, the transcript of interest may be left unprotected and may be susceptible to enzymatic degradation. The *rne131* mutation is found in the BL21(DE3) Star strains marketed by Invitrogen. Invitrogen's product manual reports that Star strains grow slightly slower than *rne* plus strains. Although we have not tested the Star strains, we expect that the Star strains may provide a higher protein yield in the following situation: where maximal mRNA level directly correlates to maximal protein production (e.g., well-behaved soluble proteins).

The *dcm* mutation is not necessarily relevant to protein expression, but is mentioned since all BL21 derivatives lack methylation at the internal cytosine of the 5′-CCWGG-3′ Dcm recognition sequence (in contrast to most K-12 strains). Thus, plasmids isolated from BL21 strains or T7 Express strains (NEB) will be amenable to cleavage by Dcm-sensitive restriction enzymes.

The *endA1* mutation eliminates expression of Endonuclease I, the major nonspecific endonuclease present within the periplasm of *E. coli.* Strains with this mutation yield higher quality plasmid preparations. Most BL21 derivatives contain this problematic activity; however, all T7 Express strains (NEB) and Single Step KRX (Promega) are examples of expression strains with the *endA1* mutation.

Homologous recombination is abolished and DNA repair processes are compromised in strains carrying a *recA* mutation. This mutation is desired in cloning strains but its utility is questionable in protein expression strains. Note that *recA* minus strains should not be employed to overexpress proteins that have the potential to modify or damage DNA or to induce an SOS response indirectly. Another consideration is that strains lacking RecA function grow more slowly.

SulA is a natural substrate of Lon protease. In the absence of Lon protease (e.g., BL21 derivative strains) SulA is expressed and persists during an SOS response and functions to inhibit cell division. Therefore, cells may not recover from an SOS response. In Lon minus strains, the *sulA11* mutation enables cells to better recover from an SOS response and proceed through cell division (36).

The *cya* mutation is reported to significantly reduce basal expression and improve plasmid maintenance when present in BL21(DE3) or other strains where T7 RNA pol expression is controlled by the *lacUV5* promoter (37). The *cya* mutation eliminates adenylate cyclase activity and cAMP production. Grossman et al. reported that cAMP, low pH, and acetate act synergistically to promote leaky lac expression. In *cya+* strains, addition of 1% glucose to agar plates and liquid cultures is expected to control basal expression mediated by *lac* as well as *lacUV5* promoters.

The *lacY1* mutation eliminates the activity of lactose permease resulting in a more uniform uptake of IPTG. This mutation is present in Tuner(DE3), Origami B(DE3), and Rosetta-gami B(DE3) available from Novagen.

Ara^- strains are preferred for the expression of protein from *ParaBAD* so that the inducer (L-arabinose) is not metabolized. Two recommended *ara* minus strains are LMG194 (*Δara714*) described by Guzman et al. (38) and MC1061 which carries the (*araA-leu)7697* deletion of the *araBAD* operon. Ara^+ strains may also be used for expression from *ParaBAD* plasmids with only a slight effect on protein yield when using rich media (38). Therefore, the more commonly used protease minus (Ara^+) strains are suitable hosts for *ParaBAD* expression.

Acknowledgments

The author would like to thank Marion Sibley, Elisabeth Raleigh, Paul Riggs, Mehmet Berkmen, Bryce Causey, and Jianying Luo of New England Biolabs for many valuable discussions and for sharing unpublished results. NIH-SBIR grant 1 R43 GM 083413 – 01 has supported work related to the overexpression of membrane proteins.

References

1. Gilbert, W., Müller-Hill, B. (1966) Isolation of the lac repressor. *Proc Natl Acad Sci USA* 56, 1891–1898.
2. Calos, M. P. (1978) DNA sequence for a low-level promoter of the *lac* repressor gene and an 'up' promoter mutation. *Nature* 274, 762–765.
3. Hirschel, B. J., Shen, V., Schlessinger, D. (1980) Lactose operon transcription from wild-type and L8-UV5 lac promoters in *Escherichia coli* treated with chloramphenicol. *J Bacteriol* 143, 1534–1537.
4. Eron, L., Block, R. (1971) Mechanism of initiation and repression of in vitro transcription of the *lac* operon of *Escherichia coli*. *Proc Natl Acad Sci USA* 68, 1828–1832.
5. deBoer H. A., Comstock, L. J., Vasser, M. (1983) The *tac* promoter: a functional hybrid derived from *trp* and *lac* promoters. *Proc Natl Acad Sci USA* 80, 21–25.
6. Amann, E., Brosius, J., Ptashne, M. (1983) Vectors bearing a hybrid *trp-lac* promoter useful for regulated expression of cloned genes in *Escherichia coli*. *Gene* 25, 167–178.
7. Mulligan, M. E., Brosius, J., Clure, W. R. (1985) Characterization in vitro of the effect of spacer length on the activity of *Escherichia coli* RNA polymerase at the *tac* promoter. *J Biol Chem* 260, 3529–3538.
8. Studier, F. W. (2005) Protein production by auto-induction in high-density shaking cultures. *Protein Expr Purif* 41, 207–234.
9. Pan, S. H., Malcolm, B. A. (2000) Reduced background expression and improved plasmid stability with pET vectors in BL21(DE3). *Biotechniques* 29, 1234–1237.
10. Zhang, X., Studier, W. F. (1997) Mechanism of inhibition of bacteriophage T7 RNA polymerase by T7 lysozyme. *J Mol Biol* 269, 10–27.
11. Cheng, X., Zhang, X., Pflugrath, J. W., Studier, W. F. (1994) The structure of bacteriophage T7 lysozyme, a zinc amidase and an inhibitor of T7 RNA polymerase. *Proc Natl Acad Sci USA* 91, 4034–4038.

12. Studier, W. F., Rosenberg, A. H., Dunn, J. J., Dubendorff, J. W. (1990) Use of T7 RNA polymerase to direct expression of cloned genes. *Methods Enzymol* 185, 60–89.
13. Haldimann, A., Daniels, L. L., Wanner, B. L. (1998) Use of new methods for construction of tightly regulated arabinose and rhamnose promoter fusions in studies of the *Escherichia coli* phosphate regulon. *J Bacteriol* 180, 1277–1286.
14. Miroux, B., Walker, J. E. (1996) Over-production of proteins in *Escherichia coli*: mutant hosts that allow synthesis of some membrane proteins and globular proteins at high levels. *J Mol Biol* 260, 289–298.
15. Wagner, S., Klepsch, M. M., Schlegel, S., Appel, A., Draheim, R., Tarry, M., Högbom, M., van Wijk, K. J., Slotboom, D. J., Persson, J. O., de Gier, J. W. (2008) Tuning *Escherichia coli* for membrane protein over-expression. *Proc Natl Acad Sci USA* 105, 14371–14376.
16. Suzuki, M., Mao, L., Inouye, M. (2007) Single protein production (SPP) system in *Escherichia coli. Nat Protocols* 2, 1802–1810.
17. Giacalone, M. J., Gentile, A. M., Lovitt, B. T., Berkley, N. L., Gunderson, C. W., Surber, M. W. (2006) Toxic protein expression in *Escherichia coli* using a rhamnose-based tightly regulated and tunable promoter system. *Biotechniques* 40, 355–364.
18. Siegele, D. A., Hu J. C. (1997) Gene expression from plasmids containing the araBAD promoter at subsaturating inducer concentrations represents mixed populations. *Proc Natl Acad Sci USA* 94, 8168–8172.
19. Khlebnikov, A., Datsenko, K. A., Skaug, T., Wanner, B. L., Keasling, J. D. (2001) Homogeneous expression of the P(BAD) promoter in *Escherichia coli* by constitutive expression of the low-affinity high-capacity AraE transporter. *Microbiology* 147, 3241–3247.
20. Better, M. D. (2004) Methods and cells for expression of recombinant protein products under the transcriptional control of an inducible promoter. *U.S. Patent* 6,803,210 Assignee: Xoma Technology Ltd., Berkeley, CA, October 12, 2004
21. Ferrer, M., Chernikova, T. N., Yakimov, M., Golyshin, P. N., Timmis, K. N. (2003) Chaperonins govern growth of *Escherichia coli* at low temperatures. *Nat Biotechnol* 21, 1266–1267.
22. Derman, A. I., Prinz, W. A. Belin, D., Beckwith, J. (1993) Mutations that allow disulfide bond formation in the cytoplasm of *Escherichia coli. Science* 262, 1744–1747.
23. Stewart, E. J., Aslund, F., Beckwith, J. (1998) Disulfide bond formation in the *Escherichia coli* cytoplasm: an in vivo role reversal for the thioredoxins. *EMBO J* 17, 5543–5550.
24. Schierle, C.F., Berkmen, M., Huber, D., Kumamoto, C., Boyd, D., Beckwith, J. (2003) The DsbA signal sequence directs efficient, cotranslational export of passenger proteins to the *Escherichia coli* periplasm via the signal recognition particle pathway. *J Bacteriol* 185, 5706–5713.
25. Emanuelsson, O., Brunak, S., von Heijne, G., Nielsen, H. (2006) Locating proteins in the cell using TargetP, SignalP and related tools. *Nat Protocols* 2, 953–971.
26. Käll, L., Krogh, A., Sonnhammer, E. L. (2007) Advantages of combined transmembrane topology and signal peptide prediction – the Phobius web server. *Nucleic Acids Res* 35, W429–W432.
27. Grisshammer, R., Duckworth, R., Henderson, R. (1993) Expression of a rat neurotensin receptor in *Escherichia coli. Biochem J* 295, 571–576.
28. Luo, J., Choulet, J., Samuelson, J. C. (2009) Rational design of a fusion partner for membrane protein expression in *E. coli. Protein Sci* 18, 1735–1744.
29. Neophytou, I., Harvey, R., Lawrence, J., Marsh, P., Panaretou, B., Barlow, D. (2007) Eukaryotic integral membrane protein expression utilizing the *Escherichia coli* glycerol-conducting channel protein (GlpF). *Appl Microbiol Biotechnol* 77, 375–381.
30. Huber, D., Boyd, D., Xia, Y., Olma, M. H., Gerstein M., Beckwith J. (2005) Use of thioredoxin as a reporter to identify a subset of *Escherichia coli* signal sequences that promote signal recognition particle-dependent translocation. *J Bacteriol* 187, 2983–2991.
31. Gardner, T. S., Cantor, C. R., Collins, J. J. (2000) Construction of a genetic toggle switch in *Escherichia coli. Nature* 403, 339–342.
32. Tucker, J., Grisshammer, R. (1996) Purification of a rat neurotensin receptor expressed in *Escherichia coli. Biochem J* 317, 891–899.
33. Yeliseev, A. A., Wong, K. K., Soubias. O., Gawrisch, K. (2005) Expression of human peripheral cannabinoid receptor for structural studies. *Protein Sci* 14, 2638–2653.
34. Hwang, B. Y., Varadarajan, N., Li, H., Rodriguez, S., Iverson, B. L., Georgiou, G. (2007) Substrate specificity of the *Escherichia coli* outer membrane protease OmpP. *J Bacteriol* 189, 522–530.
35. Link, A. J., Skretas, G., Strauch, E.-M., Chari, N. S., Georgiou, G. (2008) Efficient production of membrane-integrated

and detergent-soluble G protein-coupled receptors in *Escherichia coli*. *Protein Sci* 17, 1857–1863.
36. Lewis, K. (2000) Programmed death in bacteria. *Micro Mol Biol Rev* 64, 503–514.
37. Grossman, T. H., Kawasaki, E. S., Punreddy, S. R., Osburne, M. S. (1998) Spontaneous cAMP-dependent derepression of gene expression in stationary phase plays a role in recombinant expression instability. *Gene* 209, 95–103.
38. Guzman, L. M., Belin, D., Carson, M. J., Beckwith, J. (1995) Tight regulation, modulation, and high-level expression by vectors containing the arabinose *PBAD* promoter. *J Bacteriol* 177, 4121–4130.

Chapter 12

Periplasmic Chaperones Used to Enhance Functional Secretion of Proteins in *E. coli*

Martin Schlapschy and Arne Skerra

Abstract

While *Escherichia coli* is in wide use as a host organism for preparative protein production, problems with the folding of the recombinant gene product as well as protein aggregation, i.e., formation of inclusion bodies, are frequently encountered. This is particularly true for proteins that carry structural disulfide bonds, including antibody fragments, cytokines, growth factors, and extracellular fragments of eukaryotic cell surface receptors. In these cases, secretion into the oxidizing milieu of the bacterial periplasm in principle enables disulfide bond formation, resulting in a correctly folded and soluble protein. However, this process often occurs at low efficiency, depending on the nature of the recombinant gene product. Therefore, we have developed the helper plasmid pTUM4, which effects overexpression of four established periplasmic chaperones and/or folding catalysts: the thiol-disulfide oxidoreductases DsbA and DsbC, which catalyze the formation and isomerization of disulfide bridges, and two peptidyl-prolyl *cis/trans* isomerases with chaperone activity, FkpA and SurA. Here, we present a detailed protocol how to use this system for the bacterial secretion of recombinant proteins, including human EGF as a new example, and we give hints on optimization of the expression procedure.

Key words: Chaperone, disulfide isomerase, DsbA, DsbC, FkpA, folding catalyst, peptidyl-prolyl *cis/trans* isomerase, SurA.

1. Introduction

The periplasm of the Gram-negative bacterium *Escherichia coli* is the compartment of choice for the heterologous expression of extracellular eukaryotic proteins in a correctly folded and soluble state as its oxidizing milieu favors the formation of structural disulfide bonds (1). Thus, the method of periplasmic secretion has been utilized for the functional production of many

T.C. Evans, M.-Q. Xu (eds.), *Heterologous Gene Expression in E. coli*, Methods in Molecular Biology 705,
DOI 10.1007/978-1-61737-967-3_12, © Springer Science+Business Media, LLC 2011

recombinant proteins that carry structural disulfide bridges (2) such as, for example, antibody fragments. The broad application of recombinant antibody fragments in the life sciences and the inherent difficulty in their production by means of conventional cytoplasmic expression make them the protein class that is probably the most widely expressed using periplasmic secretion (3, 4).

However, the folding efficiency of such proteins in the bacterial periplasm varies and can significantly depend on their individual properties. Whereas some proteins are almost quantitatively recovered in the active state, others are to a large extent deposited as aggregated material, which limits the yield of easily purified product (2, 5). Provided that appropriate bacterial signal peptides – e.g., OmpA, PhoA, PelB, STII, and the like (6) – are correctly fused to the mature part of the heterologous gene product, this effect is usually not due to inefficient membrane translocation or processing by the signal peptidase (7). Rather, the cause lies in the limited solubility of the assembled recombinant protein itself or in the appearance of labile folding intermediates, in particular those with incompletely formed disulfide bonds, which are prone to nonspecific association. Thus, "periplasmic aggregates" have to be distinguished from cytoplasmic "inclusion bodies": the latter are typically obtained if proteins are overproduced without a signal peptide, and these two types of protein deposition are morphologically clearly different when investigated by electron microscopy (8).

Inefficient protein folding in the periplasm is often accompanied by a toxic effect on the bacterial host cell, even though there seems to be no strict correlation (9). This leads to reduced growth after induction of foreign gene expression and to an increased tendency of lysis, thus hampering the efficiency of periplasmic cell fractionation. Mechanistic studies on protein folding in the periplasm of *E. coli* received early attention (10), but only more recently the presence of molecular chaperones and several folding catalysts that support disulfide bond formation or catalyze peptidyl-prolyl *cis/trans* isomerization has been recognized (6, 7).

Folding of proteins that carry more than one disulfide bridge involves two distinct reactions, and either one may be rate limiting: (i) introduction of disulfide bonds into the nascent polypeptide chain via pairwise oxidation of Cys residues and (ii) isomerization of intermediary nonphysiological disulfide bonds to attain the native connectivity. Both processes are catalyzed by the thiol-disulfide oxidoreductases DsbA (11) and DsbC (12, 13), respectively.

Another well-known step along the folding pathway of proteins that often requires catalysis in vivo is the *cis/trans* isomerization of prolyl-iminopeptide bonds (14). While the *trans* configuration of Xaa-Pro bonds is favored in newly synthesized polypeptides via stereospecific peptide bond formation by the ribosome, on average ca. 5% of all prolyl-peptide bonds occur in *cis* configuration in natively folded proteins. The isomerization is catalyzed by peptidyl-prolyl *cis/trans* isomerases (PPIases, also termed rotamases) (15). So far, four PPIases have been found in the periplasm of *E. coli*: the parvulins SurA (16) and PpiD (17), the cyclophilin PpiA (alias CypA or RotA) (18), and the bacterial ortholog of the FK506-binding protein, FkpA (19).

Among these, the homodimeric FkpA is known to further exhibit generic folding enhancer activity with a broad substrate range, apart from its PPIase function (7). The "survival" protein SurA is known to be involved in the folding and assembly of endogenous outer membrane porins (20, 21). Although exerting PPIase activity, it was recently shown that the major function of SurA is that of a chaperone, an activity apparently confined to a separate structural domain (22).

Few other true periplasmic chaperones have been described for *E. coli*, in contrast to a series of cytoplasmic chaperones and chaperonins whose coexpression is increasingly utilized to suppress the formation of inclusion bodies by "directly" expressed recombinant proteins (7). For example, the periplasmic "seventeen kilodalton protein," Skp (23), possesses chaperone activity for the biogenesis of bacterial outer membrane proteins, but examples of successfully employing Skp to enhance the efficiency of soluble protein secretion in *E. coli* have remained rare.

Thus, SurA and FkpA appear to be most promising in their combination of chaperone with PPIase activity to support the proper periplasmic folding of recombinant proteins together with DsbA and DsbC to ensure disulfide bond formation. pTUM4 (*see* **Fig. 12.1**) was designed as a novel helper vector permitting the constitutive overexpression of these four folding catalysts (24). This plasmid carries two artificial dicistronic operons, one encoding FkpA and SurA under common control of the constitutive *fkp*A promoter and the other one encoding DsbA and DsbC under common control of the constitutive *dsb*A promoter followed by the strong *lpp* transcription terminator. pTUM4 has a p15a origin of replication (25) and a chloramphenicol resistance gene (26), which makes it compatible with commonly used bacterial secretion vectors for recombinant proteins (27–29) that often carry an ampicillin resistance gene and a ColEI origin of replication.

In addition to the proper choice of a bacterial signal peptide that has to be fused to the protein of interest to direct it to the periplasm (see above), the use of an appropriate *E. coli* strain with beneficial secretion properties is instrumental for periplasmic protein biosynthesis in a functional state. For example, JM83 (30), KS272 (31), HM125 (32), MC4100Δskp (33), W3110 (34), and BL21 (35) were successfully tested in our lab for various

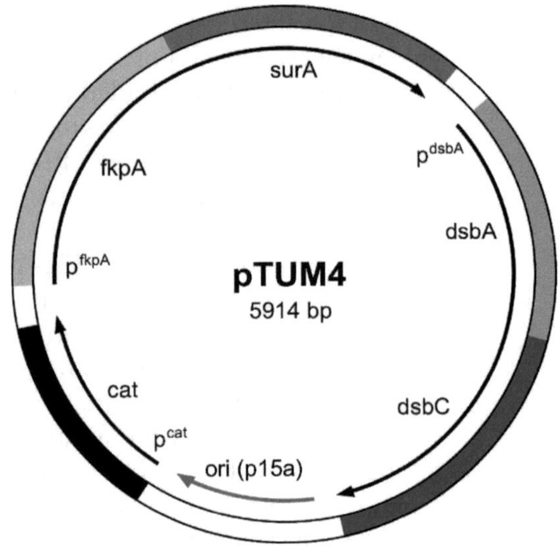

Fig. 12.1. Helper plasmid pTUM4 for the concomitant overexpression of DsbA, DsbC, FkpA, and SurA in *E. coli*. pTUM4 carries a p15a origin of replication (ori) and the chloramphenicol resistance gene (cat) with its own constitutive promoter (p^{cat}). In addition, this vector carries two artificial dicistronic operons, one with the structural genes for the periplasmic thiol/disulfide oxidoreductases DsbA and DsbC under common control of the constitutive *dsbA* promoter (p^{dsbA}) and a second one with the structural genes for the PPIases FkpA and SurA under common control of the constitutive *fkpA* promoter (p^{fkpA}).

proteins in combination with pTUM4. Even "leaky" strains of *E. coli*, which release periplasmic proteins into the culture supernatant (2), may be used.

pTUM4 complements existing cytosolic chaperone coexpression plasmids (e.g., the chaperone plasmid set from Takara Bio, Shiga, Japan), and it was demonstrated in several cases to be useful for the functional secretion of diverse recombinant proteins with hampered folding efficiency. Examples include various lipocalins (36–38), extracellular receptor domains (e.g., DC-SIGN (24), Langerin (24, 39)), and several Fab fragments (40). Furthermore, a strongly improved yield was reported for the recombinant Pfs48/45 protein and fragments thereof, a protein specifically expressed in sexual stages of *Plasmodium falciparum* which represents a promising transmission-blocking vaccine candidate (41). pTUM4 was also described to boost the secretion of the *Pseudozyma antarctica* lipase B (PalB) into the periplasm of *E. coli* (42, 43). Finally, an improved bacterial production in the presence of pTUM4 has been demonstrated for the epididymis-specific antimicrobial peptide BIN1b/SPAG11E (44).

Hence, by enabling the simultaneous overexpression of DsbA, DsbC, FkpA, and SurA, pTUM4 constitutes a generally useful tool to enhance the periplasmic secretion of functional recombinant proteins in *E. coli*. While the individual influence of the different encoded folding helper proteins may vary, depending on the nature of the heterologous gene product, this system is useful to (i) prevent inefficient disulfide bond formation, (ii) raise the yield of correctly folded, soluble recombinant protein, and (iii) abolish toxic effects on the bacterial host that may lead to premature cell lysis during expression.

2. Materials

2.1. Helper Plasmid

pTUM4 is available from the authors upon request.

2.2. *E. coli* Cultivation

1. Luria Bertani (LB) medium: 10 g/l Bacto Tryptone, 5 g/l Bacto Yeast Extract, 5 g/l NaCl.
   Stir until a clear solution is formed, adjust to pH 7.5 with 5 M NaOH (in the case of culture plates add 15 g/l Bacto Agar) and sterilize by autoclaving at 121°C for 20 min.
2. Chloramphenicol (Cam): 30 mg/ml in 70% (v/v) ethanol. This is a 1,000× stock solution; therefore use by adding 1:1,000 to the autoclaved medium.
3. Ampicillin (Amp): 100 mg/ml in water, filter sterilize. This is a 1,000× stock solution; therefore use by adding 1:1,000 to the autoclaved medium.

2.3. Preparation of the Periplasmic Extract

Periplasma extraction (PE) buffer: 500 mM sucrose, 1 mM EDTA, 100 mM Tris–HCl pH 8.0.

3. Methods

The gene encoding the protein of interest should be cloned on a suitable expression vector in a way that it is targeted to the *E. coli* periplasmic space. To effect secretion across the inner cell membrane (typically via a type II secretion mechanism (2)), the recombinant protein has to be equipped at its N-terminus with a bacterial signal peptide (see above). Many examples for this strategy (*see* **Fig. 12.2**) have been described in the literature together with suitable expression vectors (6, 28, 36, 45), which are also commercially available (e.g., from IBA, Göttingen, Germany; http://www.iba-go.de). Preferably, the bacterial signal peptide should be precisely fused – using appropriate PCR primers for subcloning or, afterwards, trimming by site-directed mutagenesis – with the N-terminal amino acid of the mature polypeptide

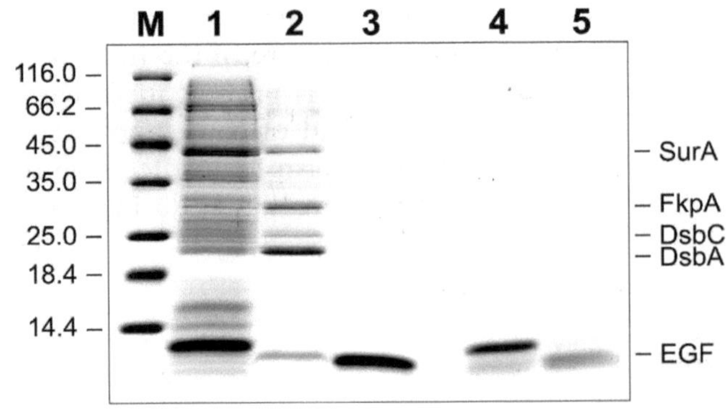

Fig. 12.2. Bacterial secretion and purification of human EGF in the presence of pTUM4 as analyzed by SDS-PAGE. *E. coli* JM83 harboring pTUM4 and pASK-IBA4-EGF, which encodes a recombinant protein comprising the OmpA signal peptide, the *Strep*-tag II (48), and the mature part of human EGF (49), was cultivated at 22°C in LB/Amp/Cam medium. Protein expression was induced at $OD_{550} = 0.5$ by addition of anhydrotetracycline (aTc) (28). After 3 h induction the cells were harvested and the periplasmic cell extract was prepared. *Lane M*, molecular size marker; *Lane 1*, sample of the whole cell extract at the time point of harvest; *Lane 2*, periplasmic cell fraction revealing prominent bands for the four folding helper proteins as well as the recombinant EGF (see legend on the right); *Lanes 3 and 4*, recombinant EGF affinity purified from the periplasmic cell fraction; *Lane 5*, for comparison: recombinant EGF affinity purified from the periplasmic cell fraction of *E. coli* JM83 harboring only pASK-IBA4-EGF. In the latter case, three different disulfide isomers give rise to diffuse bands. The samples in lanes 1–3 were reduced with 2-mercaptoethanol before application to the gel whereas samples in lanes 4 and 5 were not reduced. In addition to an increase in yield of purified protein by a factor of 5 in this example (cf. lane 4 vs. 5), the presence of pTUM4 led to the formation of just one predominant disulfide bond isomer of EGF with retarded electrophoretic mobility (note the prominent band in lane 4), representing the correctly folded protein.

such that after its processing by the bacterial periplasmic signal peptidase the authentic primary sequence is obtained.

The expression vector for the recombinant protein of interest should carry an origin of replication that is compatible with pTUM4, for example the widely used ColEI ori, and confer an antibiotic resistance different from Cam, preferably Amp. For the preparation of the expression strain, two alternative strategies are recommended. First, a suitable *E. coli* strain may be transformed both with the expression vector for the recombinant protein and with pTUM4 at the same time. In this case, which is described in further detail below, transformation efficiencies are typically much lower than when using one plasmid at a time.

Second, competent *E. coli* cells may first be transformed with pTUM4. Then a culture – using Cam for selection – is grown from one of the resulting colonies and competent cells are again prepared. Aliquots of these cells may be stored at –80°C for later use. These cells are then transformed with the expression vector for the recombinant protein in the second step, using both Cam and Amp (or another antibiotic as appropriate) for selection.

For both strategies, competent cells prepared according to the $CaCl_2$ method (46) are usually sufficient in order to obtain a satisfactory number of colonies when using 1 μl from a conventional DNA minipreparation for each of the two plasmids as detailed below. Alternatively, other procedures for the preparation of chemically competent *E. coli* cells or electroporation may be used for transformation (*see* **Note 1**).

The following protocol describes periplasmic protein expression at the shaker flask scale. However, pTUM4 is also convenient for laboratory fermentation (37, 38). A generally suitable protocol from our laboratory has been published (47). Care must be taken that the two antibiotics are present both during the preculture and fermentation stages.

3.1. Preparing Chemically Competent E. coli

1. Plate the *E. coli* expression strain of choice onto agar plates. Expression strains that we have successfully used for periplasmic protein production are JM83 (30), KS272 (31), HM125 (32), MC4100Δskp (33), W3110 (34), and BL21 (35).
2. Pick a single bacterial colony (1–2 mm in diameter) from a plate that has been incubated for 14–16 h at 37°C. Transfer the colony into 4 ml LB broth in a 13 ml culture tube and incubate 12–14 h at 37°C with vigorous agitation, resulting in an OD_{550} between 2 and 3, depending on the strain.
3. Use 500 μl of this preculture to inoculate 50 ml LB broth in a 100 ml Erlenmeyer flask and incubate at 37°C with vigorous agitation, monitoring the growth of the culture.
4. For efficient transformation, it is essential that the number of viable cells does not exceed 10^8 cells/ml, which for most strains of *E. coli* is equivalent to an OD_{550} of approximately 0.4. To ensure that the culture does not grow to higher density and approaches the stationary phase, measure the OD_{550} of the culture every 10–15 min. Harvest the culture at OD_{550} of 0.4.
5. Transfer the culture to two sterile ice-cold 50 ml polypropylene tubes. Recover the cells by centrifugation at 5,087×*g* for 10 min at 4°C in a Sigma 4K15 centrifuge (5,000 rpm using a swinging bucket rotor no. 11156, for example).
6. Decant the medium from the cell pellets. Keep the tubes on ice for 1 min and then remove traces of media with a microtip pipette.
7. Resuspend the drained cell pellet by gentle vortexing or repeated suction with a glass pipette in 40 ml ice-cold 100 mM $MgCl_2$ and centrifuge again as before.

8. Decant the medium and drain the cell pellet as above. Resuspend the bacterial cells in 20 ml ice-cold 50 mM $CaCl_2$ solution and incubate 30 min on ice.
9. Recover the cells by centrifugation as before and resuspend the drained pellet in 2 ml ice-cold 50 mM $CaCl_2$ containing 15% (v/v) glycerol. Aliquot the cells in 200 μl portions. Use them immediately for transformation or store them at –80°C.

3.2. Cotransformation with pTUM4 and a Compatible Expression Vector

1. For cotransformation, add typically 50–100 ng plasmid DNA of both pTUM4 and your expression vector (each in a total volume of 5 μl or less) to 200 μl of the chemically competent *E. coli* cells from above in a 1.5 ml Eppendorf tube and mix gently.
2. Incubate for 30 min on ice.
3. Transfer the tube into an equilibrated 37°C heating block. Store the tube in the block for exactly 5 min (without shaking).
4. Rapidly pipette the cells into a culture tube with 2 ml LB medium. Incubate the tube for 45 min at 37°C with shaking at 200 rpm to allow the bacteria to recover and to express the antibiotic resistance markers encoded on both plasmids.
5. Then, plate an appropriate volume (100–150 μl per 82 mm petri dish) of this culture onto an LB agar plate containing 30 μg/ml Cam as well as the appropriate antibiotic for your expression vector (e.g., 100 μg/ml Amp).

3.3. Preparation of the Expression Culture

1. A colony from the previous step is used to grow an expression culture. We recommend a growth temperature lower than 37°C, e.g., 30°C (*see* **Note 2**). A fresh single colony of *E. coli* transformed with pTUM4 and the compatible expression vector is used to inoculate 50 ml LB medium in a 100 ml flask containing 30 μg/ml Cam and the appropriate second antibiotic (e.g., 100 μg/ml Amp). The preculture is incubated at 30°C upon shaking at 200 rpm overnight.
2. The preculture (40 ml) is added to 2l of LB medium containing the same antibiotics in a 5 l Erlenmeyer flask. Smaller expression cultures may be used but should be inoculated with less volume of the preculture as appropriate. The culture is incubated at 22°C upon shaking at 200 rpm. Bacterial growth should be monitored by measuring the OD_{550}.
3. Expression is induced at $OD_{550} = 0.5$ (after correction with an LB blank value) by adding the appropriate inducer for the expression plasmid, e.g., IPTG (46) or aTc (28). The optimal induction period varies (from 2 to 16 h) and may depend on toxic effects caused by the heterologous protein

that can ultimately lead to culture lysis instead of a stationary phase. The best time for harvest is when the growth curve has just reached a plateau (*see* **Note 3**).

4. The culture is quickly distributed to large centrifuge tubes (e.g., six Sorvall SLA3000 tubes) and centrifuged at 4,400×*g* for 15 min at 4°C. Ensure that the tubes and rotor are prechilled to 4°C.
5. After discarding the culture supernatant the tubes are put on ice in a tilted position, with the pellets elevated, and residual culture medium – which accumulates at the opposite side on the bottom of the centrifuge vessel – is removed with a pipette.

3.4. Preparation of the Periplasmic Extract

Usually, the expression of foreign proteins in the periplasm of *E. coli* leads to a fragile outer membrane such that addition of EDTA (to remove divalent cations that stabilize the lipopolysaccharide) in combination with high osmotic pressure is sufficient to trigger the release of the soluble periplasmic content (*see* **Note 4**). Solutions should be prechilled to 4°C before use. Note that time is a critical factor because cells may lyse if the pellet is stored for too long.

1. The sedimented and drained bacterial cell pellet from the 2 l expression culture is carefully resuspended in 20 ml of ice-cold PE buffer, transferred to a 50 ml Falcon tube, and incubated for 30 min on ice. Use a 25 ml glass pipette for repeated suction with 10 ml portions of the buffer (for each three of the centrifugation cups). The goal is to quickly obtain a homogeneous cell suspension, without visible flakes, but to avoid excessive shear stress and air bubbles. If an expression culture with less volume is prepared, all steps described here must be appropriately scaled down.
2. The spheroplasts are sedimented by centrifugation at 5,087× *g* for 15 min at 4°C in a Sigma 4K15 bench top centrifuge (5,000 rpm using a swinging bucket rotor no. 11156, for example) and the supernatant is carefully recovered as the crude periplasmic cell fraction.
3. In order to clear the supernatant from residual debris it is transferred to fresh centrifuge tubes (e.g., Sorvall SS34) and submitted to a second centrifugation step at 27,000×*g* for 15 min at 4°C.
4. The periplasmic protein extract can usually be frozen at –20°C for storage or directly dialysed against 2 l of a suitable buffer overnight at 4°C prior to chromatographic purification.
5. The His_6 tag (45) and the *Strep*-tag II (48) provide convenient affinity purification methods. If an affinity tag is to be

used it must be included during cloning of the gene encoding the target protein. The periplasmic production and affinity purification of recombinant human epidermal growth factor, EGF (49), which carries three disulfide bonds, according to this procedure is shown in **Fig. 12.2**.

6. It is advisable to check the success of protein expression and the quality of the periplasmic cell extract by Coomassie-stained SDS-PAGE and, if necessary, by Western blotting. An appropriate periplasmic extract should show a distinct band pattern when compared to a sample of the whole cell protein extract as only ca. 5% of all *E. coli* proteins are secreted into this compartment (*see* **Notes 5** and **6**). Generally, all four folding catalysts cloned on pTUM4, i.e. DsbA, DsbC, FkpA, and SurA, are strongly overexpressed and give rise to prominent bands (24).

4. Notes

1. If the number of colonies obtained from the double transformation of *E. coli* is too small, electroporation according to published protocols (46) is recommended. Use 5 ng of each plasmid obtained from a typical minipreparation kit in a total volume of 1–2 μl for 40 μl of the electrocompetent cell suspension. Make sure that the plasmid solution is free of salt or buffer (TE/10 buffer may be allowed).
2. Often, an elevated growth temperature leads to reduced folding efficiency for foreign proteins in the bacterial periplasm as well as early onset of cell lysis. An expression temperature of 22°C is usually a good compromise between folding efficiency and cell growth rate. If a shaker with a cooling device is not available the culture may be grown at ambient temperature. Note, however, that results often vary if the temperature conditions differ by just 1–2°C between experiments. The practice of growing the culture first at 37°C, until the OD_{550} for induction is reached, and then lowering the temperature is not recommended as the cells do not have sufficient time to adjust their physiology. This will cost yield, especially for proteins that are difficult to express. Yet, in the case of less demanding recombinant proteins, conditions may be handled in a more flexible way, e.g., by inducing the promoter at a higher cell density, such as an $OD_{550} = 1$, and by continuing expression overnight.

3. It is advisable first to grow a culture after induction of gene expression – without cell harvest – for a prolonged time while monitoring the cell density. In some cases there may be an onset of culture lysis rather soon after reaching a plateau phase or even a peak (50). This time point should not be missed because otherwise protein may be lost into the supernatant of the culture (1) and the periplasmic fractionation can be less efficient. Under these circumstances, cell lysis may liberate many contaminating cytoplasmic proteins as well as chromosomal DNA, thus leading to high viscosity of the extract. After even longer induction time the cells may be overgrown by other cells that have lost their plasmid and/or no longer produce the recombinant protein of interest.
4. If the periplasmic extraction appears inefficient, i.e., too few bands are visible by SDS-PAGE, more stringent conditions may be applied. A convenient method is the addition of lysozyme up to a final concentration of 200 μg/ml (from a fresh 10 mg/ml stock solution in PE buffer). The lysozyme enters the periplasmic space via fractures in the cell envelope and hydrolyzes the peptidoglycan, which destabilizes the outer membrane. Alternatively, some harsh osmotic shock procedures are known, involving dilution of the spheroplasts into distilled water (51). However, such protocols should be applied with caution as extended cell lysis is a frequent problem. If the viscosity of the periplasmic extract is too high – but it is still possible to pellet the spheroplasts – the protein solution can be rescued for purification trials by adding Benzonase (Merck, Darmstadt, Germany; purity grade I, 250 U/μl; recommended concentration: 25 U/ml; incubate at 4°C for 16 h with the addition of 5 mM $MgCl_2$). Generally, if the conditions for periplasmic cell fractionation have to be optimized it is recommended to use periplasmic and cytoplasmic marker enzymes (50) – e.g., β-lactamase and β-galactosidase, respectively – and determine their specific activities both in the periplasmic extract, prepared under varying conditions, and in a whole cell lysate obtained from an equivalent aliquot of the cell suspension via mechanical lysis using either a French pressure cell or a sonifier.
5. For SDS-PAGE or Western blot analysis, 10–20 μl samples of the undiluted periplasmic extract should be applied to the gel. To prepare a whole cell extract, remove a 1 ml sample from the bacterial culture – at time points of induction and harvest, respectively – spin them down in a 1.5 ml tube for 5 min at 14,000 rpm in a bench top centrifuge, and remove the supernatant. The cell pellet is then resuspended in 80 μl of a solution of 12.5 U/ml Benzonase in 100 mM

Tris–HCl pH 8.0, 5 mM $MgCl_2$, and gently mixed with 20 μl of reducing 5× SDS-PAGE sample buffer [7.5% (w/v) SDS, 25% (v/v) glycerol, 0.25 M Tris–HCl pH 8.0, 12.5% (v/v) 2-mercaptoethanol, 0.25 mg/ml bromophenol blue]. After 1 h incubation on ice for degradation of the chromosomal DNA the sample should be directly applied – after heating at 95°C for 5 min – to the gel or stored frozen until SDS-PAGE is performed.

6. In certain cases it may happen that the secretion of the recombinant protein itself is hampered (2, 6), leading to its accumulation in the bacterial cytoplasm. A clear sign of this is if no soluble protein can be recovered with the periplasmic extract but the protein band detected in the whole cell lysate after Coomassie-stained SDS-PAGE or Western blotting is by ca. 2 kDa larger than expected, indicating that the signal peptide has not been processed by the periplasmic signal peptidase. This may result from overloading the bacterial secretion system, for example, by using very strong promoters such as the T7 system. In such a case weaker promoters or conditions of partial induction are recommended. In addition, one can try to exchange the signal peptide. This may also help if the cleavage site itself, at the junction of the signal peptide and the mature recombinant protein, constitutes a poor substrate of signal peptidase. Finally, one should be aware that several commonly used cloning strains, e.g., XL1-Blue and DH5α, but also *E. coli* B strains such as BL21 and its derivatives, have inferior protein secretion capability.

Acknowledgments

The authors wish to thank S. Grimm, A. Holla, A. Richter, and K. Wachinger for their help in the construction of pTUM4 and related plasmids and their support in testing this system with various proteins of interest.

References

1. Georgiou, G., Segatori, L. (2005) Preparative expression of secreted proteins in bacteria: status report and future prospects. *Curr Opin Biotechnol* 16, 538–545.
2. Mergulhao, F. J., Summers, D. K., Monteiro, G. A. (2005) Recombinant protein secretion in *Escherichia coli*. *Biotechnol Adv* 23, 177–202.
3. Skerra, A. (1993) Bacterial expression of immunoglobulin fragments. *Curr Opin Immunol* 5, 256–262.
4. Humphreys, D. P. (2003) Production of antibodies and antibody fragments in *Escherichia coli* and a comparison of their functions, uses and modification. *Curr Opin Drug Discov Develop* 6, 188–196.

5. Skerra, A., Plückthun, A. (1991) Secretion and in vivo folding of the Fab fragment of the antibody McPC603 in *Escherichia coli*: influence of disulphides and cis-prolines. *Protein Eng* 4, 971–979.
6. Choi, J. H., Lee, S. Y. (2004) Secretory and extracellular production of recombinant proteins using *Escherichia coli*. *Appl Microbiol Biotechnol* 64, 625–635.
7. Baneyx, F., Mujacic, M. (2004) Recombinant protein folding and misfolding in *Escherichia coli*. *Nat Biotechnol* 22, 1399–1408.
8. Chan, W., Helms, L. R., Brooks, I., Lee, G., Ngola, S., McNulty, D., Maleeff, B., Hensley, P., Wetzel, R. (1996) Mutational effects on inclusion body formation in the periplasmic expression of the immunoglobulin VL domain REI. *Fold Des* 1, 77–89.
9. Knappik, A., Krebber, C., Plückthun, A. (1993) The effect of folding catalysts on the in vivo folding process of different antibody fragments expressed in *Escherichia coli*. *Biotechnology* 11, 77–83.
10. Wülfing, C., Plückthun, A. (1994) Protein folding in the periplasm of *Escherichia coli*. *Mol Microbiol* 12, 685–692.
11. Bardwell, J. C., McGovern, K., Beckwith, J. (1991) Identification of a protein required for disulfide bond formation in vivo. *Cell* 67, 581–589.
12. Missiakas, D., Georgopoulos, C., Raina, S. (1994) The *Escherichia coli* dsbC (xprA) gene encodes a periplasmic protein involved in disulfide bond formation. *EMBO J* 13, 2013–2020.
13. Shevchik, V. E., Condemine, G., Robert-Baudouy, J. (1994) Characterization of DsbC, a periplasmic protein of *Erwinia chrysanthemi* and *Escherichia coli* with disulfide isomerase activity. *EMBO J* 13, 2007–2012.
14. Fischer, G., Tradler, T., Zarnt, T. (1998) The mode of action of peptidyl prolyl *cis/trans* isomerases in vivo: binding vs. catalysis. *FEBS Lett* 426, 17–20.
15. Fanghanel, J., Fischer, G. (2004) Insights into the catalytic mechanism of peptidyl prolyl *cis/trans* isomerases. *Front Biosci* 9, 3453–3478.
16. Rouviere, P. E., Gross, C. A. (1996) SurA, a periplasmic protein with peptidyl-prolyl isomerase activity, participates in the assembly of outer membrane porins. *Genes Develop* 10, 3170–3182.
17. Dartigalongue, C., Raina, S. (1998) A new heat-shock gene, ppiD, encodes a peptidyl-prolyl isomerase required for folding of outer membrane proteins in *Escherichia coli*. *EMBO J* 17, 3968–3980.
18. Liu, J., Walsh, C. T. (1990) Peptidyl-prolyl *cis-trans*-isomerase from *Escherichia coli*: a periplasmic homolog of cyclophilin that is not inhibited by cyclosporin A. *Proc Natl Acad Sci USA* 87, 4028–4032.
19. Horne, S. M., Young, K. D. (1995) *Escherichia coli* and other species of the Enterobacteriaceae encode a protein similar to the family of Mip-like FK506-binding proteins. *Arch Microbiol* 163, 357–365.
20. Lazar, S. W., Kolter, R. (1996) SurA assists the folding of *Escherichia coli* outer membrane proteins. *J Bacteriol* 178, 1770–1773.
21. Hennecke, G., Nolte, J., Volkmer-Engert, R., Schneider-Mergener, J., Behrens, S. (2005) The periplasmic chaperone SurA exploits two features characteristic of integral outer membrane proteins for selective substrate recognition. *J Biol Chem* 280, 23540–23548.
22. Bitto, E., McKay, D. B. (2002) Crystallographic structure of SurA, a molecular chaperone that facilitates folding of outer membrane porins. *Structure* 10, 1489–1498.
23. Korndörfer, I. P., Dommel, M. K., Skerra, A. (2004) Structure of the periplasmic chaperone Skp suggests functional similarity with cytosolic chaperones despite differing architecture. *Nat Struct Mol Biol* 11, 1015–1020.
24. Schlapschy, M., Grimm, S., Skerra, A. (2006) A system for concomitant overexpression of four periplasmic folding catalysts to improve secretory protein production in *Escherichia coli*. *Protein Eng Des Sel* 19, 385–390.
25. Selzer, G., Som, T., Itoh, T., Tomizawa, J. (1983) The origin of replication of plasmid p15A and comparative studies on the nucleotide sequences around the origin of related plasmids. *Cell* 32, 119–129.
26. Alton, N. K., Vapnek, D. (1979) Nucleotide sequence analysis of the chloramphenicol resistance transposon Tn9. *Nature* 282, 864–869.
27. Skerra, A., Pfitzinger, I., Plückthun, A. (1991) The functional expression of antibody Fv fragments in *Escherichia coli*: improved vectors and a generally applicable purification technique. *Biotechnology* 9, 273–278.
28. Skerra, A. (1994) Use of the tetracycline promoter for the tightly regulated production of a murine antibody fragment in *Escherichia coli*. *Gene* 151, 131–135.
29. Sϕrensen, H. P., Mortensen, K. K. (2005) Advanced genetic strategies for recombinant protein expression in *Escherichia coli*. *J Biotechnol* 115, 113–128.

30. Yanisch-Perron, C., Vieira, J., Messing, J. (1985) Improved M13 phage cloning vectors and host strains: nucleotide sequences of the M13mp18 and pUC19 vectors. *Gene* 33, 103–119.
31. Strauch, K. L., Beckwith, J. (1988) An *Escherichia coli* mutation preventing degradation of abnormal periplasmic proteins. *Proc Natl Acad Sci USA* 85, 1576–1580.
32. Meerman, H. J., Georgiou, G. (1994) Construction and characterization of a set of *E. coli* strains deficient in all known loci affecting the proteolytic stability of secreted recombinant proteins. *Biotechnology* 12, 1107–1110.
33. Schäfer, U., Beck, K., Müller, M. (1999) Skp, a molecular chaperone of Gram-negative bacteria, is required for the formation of soluble periplasmic intermediates of outer membrane proteins. *J Biol Chem* 274, 24567–24574.
34. Jensen, K. F. (1993) The *Escherichia coli* K-12 "wild types" W3110 and MG1655 have an rph frameshift mutation that leads to pyrimidine starvation due to low pyrE expression levels. *J Bacteriol* 175, 3401–3407.
35. Studier, F. W., Moffatt, B. A. (1986) Use of bacteriophage T7 RNA polymerase to direct selective high-level expression of cloned genes. *J Mol Biol* 189, 113–130.
36. Breustedt, D. A., Schönfeld, D. L., Skerra, A. (2006) Comparative ligand-binding analysis of ten human lipocalins. *Biochim Biophys Acta* 1764, 161–173.
37. Nasreen, A., Vogt, M., Kim, H. J., Eichinger, A., Skerra, A. (2006) Solubility engineering and crystallization of human apolipoprotein D. *Protein Sci* 15, 190–199.
38. Schönfeld, D. L., Ravelli, R. B., Mueller, U., Skerra, A. (2008) The 1.8-Å crystal structure of α1-acid glycoprotein (Orosomucoid) solved by UV RIP reveals the broad drug-binding activity of this human plasma lipocalin. *J Mol Biol* 384, 393–405.
39. Chatwell, L., Holla, A., Kaufer, B. B., Skerra, A. (2008) The carbohydrate recognition domain of Langerin reveals high structural similarity with the one of DC-SIGN but an additional, calcium-independent sugar-binding site. *Mol Immunol* 45, 1981–1994.
40. Friedrich, L., Stangl, S., Hahne, H., Küster, B., Köhler, P., Multhoff, G., Skerra, A. (2010) Bacterial production and functional characterization of the Fab fragment of the murine IgG1/λ monoclonal antibody cmHsp70.1, a reagent for tumour diagnostics. *Protein Eng Des Sel* 23, 161–168.
41. Outchkourov, N. S., Roeffen, W., Kaan, A., Jansen, J., Luty, A., Schuiffel, D., van Gemert, G. J., van de Vegte-Bolmer, M., Sauerwein, R. W., Stunnenberg, H. G. (2008) Correctly folded Pfs48/45 protein of *Plasmodium falciparum* elicits malaria transmission-blocking immunity in mice. *Proc Natl Acad Sci USA* 105, 4301–4305.
42. Xu, Y., Lewis, D., Chou, C. P. (2008) Effect of folding factors in rescuing unstable heterologous lipase B to enhance its overexpression in the periplasm of *Escherichia coli. Appl Microbiol Biotechnol* 79, 1035–1044.
43. Xu, Y., Yasin, A., Wucherpfennig, T., Chou, C. P. (2008) Enhancing functional expression of heterologous lipase in the periplasm of *Escherichia coli. World J Microbiol Biotechnol* 24, 2827–2835.
44. Guo, C., Diao, H., Lian, Y., Yu, H., Gao, H., Zhang, Y., Lin, D. (2009) Recombinant expression and characterization of an epididymis-specific antimicrobial peptide BIN1b/SPAG11E. *J Biotechnol* 139, 33–37.
45. Skerra, A. (1994) A general vector, pASK84, for cloning, bacterial production, and single-step purification of antibody Fab fragments. *Gene* 141, 79–84.
46. Sambrook, J., Fritsch, E. F., Maniatis, T. (2001). *Molecular Cloning: A Laboratory Manual*, 3rd ed. Cold Spring Harbor Laboratory Press, Cold Spring Harbor, NY.
47. Schiweck, W., Skerra, A. (1995) Fermenter production of an artificial Fab fragment, rationally designed for the antigen cystatin, and its optimized crystallization through constant domain shuffling. *Proteins* 23, 561–565.
48. Schmidt, T. G., Skerra, A. (2007) The *Strep*-tag system for one-step purification and high-affinity detection or capturing of proteins. *Nat Protoc* 2, 1528–1535.
49. Bell, G. I., Fong, N. M., Stempien, M. M., Wormsted, M. A., Caput, D., Ku, L. L., Urdea, M. S., Rall, L. B., Sanchez-Pescador, R. (1986) Human epidermal growth factor precursor: cDNA sequence, expression in vitro and gene organization. *Nucleic Acids Res* 14, 8427–8446.
50. Plückthun, A., Skerra, A. (1989) Expression of functional antibody Fv and Fab fragments in *Escherichia coli. Methods Enzymol* 178, 497–515.
51. Ewis, H. E., Lu, C. D. (2005) Osmotic shock: a mechanosensitive channel blocker can prevent release of cytoplasmic but not periplasmic proteins. *FEMS Microbiol Lett* 253, 295–301.

Chapter 13

Engineering Unusual Amino Acids into Peptides Using Lantibiotic Synthetase

Jun-ichi Nagao, Kouki Shioya, Yoshitaka Harada, Ken-ichi Okuda, Takeshi Zendo, Jiro Nakayama, and Kenji Sonomoto

Abstract

Alteration of protein structure and function by introducing unusual amino acids has great potential to develop new biological tool and to produce novel therapeutic agents. Lantibiotics produced by Gram-positive bacteria are ribosomally synthesized and post-translationally modified antimicrobial peptides. The modification enzyme involved in lantibiotic biosynthesis can catalyze the formation of unusual amino acids in the nascent lantibiotic prepeptide. Here, a novel methodology on the lantibiotic nukacin ISK-1 is described for engineering unusual amino acid residues into hexa-histidine-tagged (His-tagged) prepeptide NukA by the modification enzyme NukM in *Escherichia coli*. Co-expression of His-tagged NukA and NukM, purification of the resulting His-tagged prepeptide by affinity chromatography, and subsequent mass spectrometry analysis show that the prepeptide is converted into a postulated peptide with decrease in mass which results from the formation of unusual amino acids such as dehydrated amino acid, lanthionine, or 3-methyl lanthionine at the expected positions. The modified prepeptide can be readily obtained by one-step purification. This strategy will thus be a simple and powerful tool for introducing unusual amino acid residues aimed at peptide engineering.

Key words: Lantibiotic, unusual amino acid, post-translational modification, peptide engineering.

1. Introduction

Recent advances in peptide chemistry, molecular biology, and genetics have increased the possibility of the engineering of peptides and proteins. Amino acid substitution, D-amino acid replacement, chemical modification, glycosylation, and backbone cyclization are the present technologies that can ameliorate the potentiality of antimicrobial peptides and proteins (1). In

T.C. Evans, M.-Q. Xu (eds.), *Heterologous Gene Expression in E. coli*, Methods in Molecular Biology 705,
DOI 10.1007/978-1-61737-967-3_13, © Springer Science+Business Media, LLC 2011

post-genomic era, the molecular design and alteration of a protein structure and function by introducing unusual amino acids have attracted much interest and is highly forthcoming to develop a new biological tool for producing novel therapeutic agents. In particular, the engineering of a new peptide shows promise for producing potential antibiotics, since effective traditional antibiotics are decreasing in number because of the development of antibiotic resistance among microbial populations.

Lantibiotics are ribosomally synthesized and post-translationally modified antimicrobial peptides that contain unusual amino acids such as lanthionine (Lan) and/or 3-methyl lanthionine (MeLan), and dehydrated amino acids (2–4). The lantibiotic prepeptide consists of an N-terminal leader peptide followed by a C-terminal propeptide moiety that undergoes several post-translational modification events including unusual amino acids formation, cleavage of leader peptide, and export. Serine and threonine residues at specific positions in the propeptide moiety are dehydrated to dehydroalanine (Dha) and dehydrobutyrine (Dhb) residues, respectively and the thioether rings of Lan and MeLan are formed by the respective cyclization of Dha and Dhb residues with the thiol group of cysteine residues. Various biological activities and significant structural stability of lantibiotics have been reported and all the activities depend on the presence of unusual amino acids. Thus, enzymes involved in lantibiotic biosynthesis may have high potentiality for peptide engineering by introduction of unusual amino acids into desired peptides, which might envisage the founding of a universal approach to advance for structural design of novel peptides, termed lantibiotic engineering (2).

A lantibiotic nukacin ISK-1 is produced by *Staphylococcus warneri* ISK-1 (5–7). Nukacin ISK-1 prepeptide (NukA) is post-translationally modified by the modification enzyme NukM (5). The overall strategy developed is to explore the Gram-negative bacterium *Escherichia coli* as a heterologous host for the unusual amino acid formation. Here we describe a novel methodology for engineering unusual amino acids into hexa-histidine-tagged (His-tagged) NukA by NukM (5). Co-expression of His-tagged NukA and NukM, purification of modified prepeptides by affinity chromatography, and subsequent mass spectrometry analysis demonstrate that considerable amounts of the modified His-tagged NukA containing unusual amino acids are readily obtained. The lanthionine-introducing system in *E. coli* has many advantages such as easy manipulation of genetic engineering and rapid and instant usages. Therefore, the strategy developed will be a novel and powerful tool for the peptide engineering by introduction of unusual amino acids.

2. Materials

2.1. Construction of Plasmids

1. KOD plus DNA polymerase (Toyobo, Osaka, Japan) is used for PCR.
2. Restriction enzyme *Bam*HI (Toyobo) is used according to manufacturer's instructions.
3. *E. coli* JM109 is used as a cloning host.
4. *E. coli* BL21(DE3) (Novagen, Madison, WI, USA) is used as an expression strain.
5. Plasmid pET-14b (Novagen) is used as an expression vector.
6. Mag extractor plasmid extraction kit (Toyobo) is used to isolate plasmids from *E. coli.*
7. Gene pulser apparatus (Bio-Rad, Hercules, CA, USA).

2.2. Bacterial Culture and Cell Disruption

1. 2xYT media: 1.6% trypton, 1.0% yeast extract, 0.5% NaCl.
2. Ampicillin (Nacalai Tesque, Kyoto, Japan) is dissolved at 50 mg/mL in sterilized water and stored in aliquots at –30°C.
3. Isopropyl 1-thio-β-D-galactoside (IPTG) (Nacalai Tesque) is dissolved at 0.5 M in sterilized water and stored in aliquots at –30°C.
4. Tris-HCl buffer (pH 7.5) is prepared at 1 M and stored at room temperature.
5. SONIFIER cell disruptor 350 (Branson Ultrasonics, Danbury, CT, USA).

2.3. Purification by Affinity Chromatography and Desalting

1. HiTrap Chelating HP 1-mL column (Amersham Biosciences, Uppsala, Sweden).
2. Binding buffer: 20 mM Tris-HCl, 0.5 M NaCl, 10 mM imidazole. Stored at 4°C.
3. Washing buffer: 20 mM Tris-HCl, 0.5 M NaCl, 100 mM imidazole. Stored at 4°C.
4. Elution buffer: 20 mM Tris-HCl, 0.5 M NaCl, 300 mM imidazole. Stored at 4°C.
5. Reverse-phase column, PepRPC HR 5/5 (Amersham Biosciences) integrated in LC-10A HPLC system (Shimadzu, Kyoto, Japan).
6. Acetonitrile (HPLC grade, Kanto Chemical, Japan) containing 0.1% trifluoroacetic acid (TFA, Nacalai Tesque). Make fresh as required.
7. MilliQ water containing 0.1% TFA. Make fresh as required.

2.4. Electrospray Ionization-Mass Spectrometry (ESI–MS) Analysis

1. ESI-MS (Accutof T100LC, JEOL, Tokyo, Japan).
2. 95% acetonitrile (HPLC grade, Kanto Chemical) containing 0.05% TFA (Nacalai Tesque). Make fresh as required.

2.5. Characterization of the Modified His-Tagged Prepeptide

2.5.1. Amino Acid Sequencing

1. Lysylendopeptidase (Lys-C, Wako, Osaka, Japan) dissolved at 0.1 mg/mL in sterilized water and stored at 4°C.
2. NH_4HCO_3 buffer (pH 8.5) is prepared at 100 mM and stored at room temperature.
3. Evaporator (Speed Vac Plus SC110A, Savant, Farmingdale, NY, USA)
4. PPSQ-21 gas-phase automatic protein sequence analyzer (Shimadzu).

2.5.2. Amino Acid Composition Analysis

1. D,L-Lanthionine (Tokyo Kasei, Tokyo, Japan) is used as a standard for Lan or MeLan.
2. LC-500/V amino acid analyzer (JEOL).
3. 6 N HCl. Make fresh as required.

2.5.3. Characterization with 1-Cyano-4-dimethylaminopyridinium Tetrafluoroborate (CDAP)

1. CDAP (Sigma Chemical, St. Louis, MO, USA) is dissolved at 0.1 M in citrate buffer at pH 3.0. CDAP is unstable; therefore, make fresh as required.
2. Tris(2-carboxyethyl) phosphate-HCl (TCEP, PIERCE, Rockford, IL, USA) is dissolved at 1 M in citrate buffer at pH 3.0 as a stock solution and stored under N_2 at –30°C.
3. LC/MS device [LC: Agilent HP1100, column: SOURCE 5RPC ST 2.1/150 (Amersham Biosciences), MS: JEOL Accutof T100LC].

3. Methods

To obtain reliable and reproducible results, it is important to maintain the culture and induction condition for co-expression of His-tagged NukA and NukM. In this co-expression system, *nukA* gene is in frame with hexa-histidine tag at N-terminus to purify the modified His-tagged NukA easily after cell lysis. The molecular mass data of the obtained peptides are not enough to identify the presence of Lan or MeLan, because the intramolecular cyclization of Dha or Dhb residue with cysteine residue does not produce a change in mass. Physico-chemical analyses such as amino acid sequencing and composition analysis are therefore required to determine the unusual amino acids. Besides, CDAP, which cyanylates the thiol group of free cysteine residue, is used for rapid and sensitive determination of the Lan formation.

3.1. Construction of Plasmid

General molecular biology techniques are basically carried out following the established protocols (8).

1. The gene fragment encoding NukA and NukM is successively amplified from the template plasmid pInuk (6) by PCR with the following set of primers: 5′-TAGGAGGATCCAAACATGGAAAATTCTAAAGTT-3′ and 5′-TCATGGATCCCTTCCAAACTAAATGTTTG-3′.
2. The amplified fragment is cloned under T7 promoter in pET-14b at *Bam*HI site. Resultant plasmid pETnukAM is in-frame translation with hexa-histidine tag sequence at N-terminus of NukA.
3. The gene fragment encoding NukA is amplified from the template plasmid pInuk (6) by PCR with the following set of primers, 5′-TTTAGGAGGATCCAAACATGGAAAA-3′ and 5′-TATTAAGGATCCAATACTCTCCA-3′. The amplified fragment is cloned into pET-14b at *Bam*HI site. Resultant plasmid pETnukA is in-frame translation with hexa-histidine tag sequence at N-terminus of NukA.
4. *E. coli* BL21(DE3) is transformed with the plasmid pET-nukAM or pETnukA. Resultant bacterial strains can be stored in final 15% (v/v) glycerol at –80°C. *E. coli* strains used are listed in **Table 13.1** (*see* **Note 1**).

Table 13.1
***E. coli* strains used in this study**

Bacterial strains	Description[a]	Reference or source
E. coli BL21(DE3)	F^-, *ompT*, *hsd*$S_B(r_B^- m_B^-)$ *gal*, *dcm* (DE3)	Novagen
E. coli BL21(DE3) recombinant[b]		
(pETnukA)	Ap^r, BL21(DE3) carrying pETnukA	(7)
(pETnukAM)	Ap^r, BL21(DE3) carrying pETnukAM	(5)

[a] Ap^r, ampicillin resistance
[b] Plasmids introduced in strains are indicated in parentheses

3.2. Culture Condition for Co-expression

1. To recover the bacterial strain from –80°C glycerol stock, streak it onto an LB agar plate containing 50 μg/mL of ampicillin and grow overnight at 37°C.
2. Pick a single colony from the plate and grow overnight into 5 mL of 2xYT media containing 50 μg/mL of ampicillin at 37°C with shaking (120 rpm).
3. About2 mL of overnight culture is inoculated into 200 mL of 2×YT media containing 50 μg/mL of ampicillin and grown at 37°C with shaking (120 rpm) until OD_{600} reaches 0.6.
4. Prior to addition of IPTG, incubate the culture at 20°C for 15 min with shaking (120 rpm) for adaptation of the bacteria to subsequent temperature change for induction (*see* **Note 2**).
5. Add IPTG at a final concentration of 0.5 mM and further culture at 20°C for 20 h with shaking (120 rpm).
6. Harvest cells by centrifugation at 10,000×*g* for 15 min at 4°C.
7. Resuspend cells in 30 mL of 20 mM Tris-HCl buffer (pH 7.5) and centrifuge at 6000×*g* for 15 min at 4°C.
8. Resuspend cells in 5 mL of 20 mM Tris-HCl buffer (pH 7.5).
9. Disrupt cells on the cold ice by sonication with a SONIFIER cell disruptor 350 (output 4, 40% duty, 1 min × 6 cycles).
10. Centrifuge at 6000×*g* for 20 min at 4°C to prepare cell-free extract (*see* **Note 3**).

3.3. Purification of His-Tagged Prepeptide by Affinity Chromatography

HiTrap chelating HP 1-mL column is used for purification of His-tagged NukA. Generally, all operations can be achieved at room temperature by a syringe at flow rate of 1 mL/min according to the following protocol:

1. Equilibrate the column with 10 mL of binding buffer.
2. Apply 5 mL of the cell-free extract.
3. Wash with 10 mL of binding buffer.
4. Wash with 5 mL of washing buffer to remove non-specific host proteins.
5. Elute with 5 mL of elution buffer (*see* **Note 4**).
6. Re-equilibrate the column with 10 mL of binding buffer.
7. Desalt His-tagged NukA in elution fraction by subjecting to the reverse-phase HPLC (RP-HPLC). The absorbance is

measured at 220 nm. Peptide is eluted from the column at flow rate of 1 mL/min with a linear gradient of acetonitrile/0.1% TFA (15–80% for 30 min).

3.4. ESI-MS Analysis of His-Tagged Prepeptide

His-tagged peptides eluted from the RP-HPLC are directly analyzed by ESI-MS (*see* **Notes 5** and **6**).

1. About 5 μL of column elutes are applied to ESI-MS.
2. The mass spectrometer is operated under the following conditions: positive polarity; capacity temperature, 250°C; needle voltage, 2.0 kV; orifice voltage, 92 V; and ring voltage, 10 V.
3. After scanning for molecular ions in the *m/z* range of 500–3000, ion chromatograms extracted are plotted with detector counts at the indicated *m/z*. The results obtained are shown in **Fig. 13.1a** (*see* **Notes** 7 and **8**).

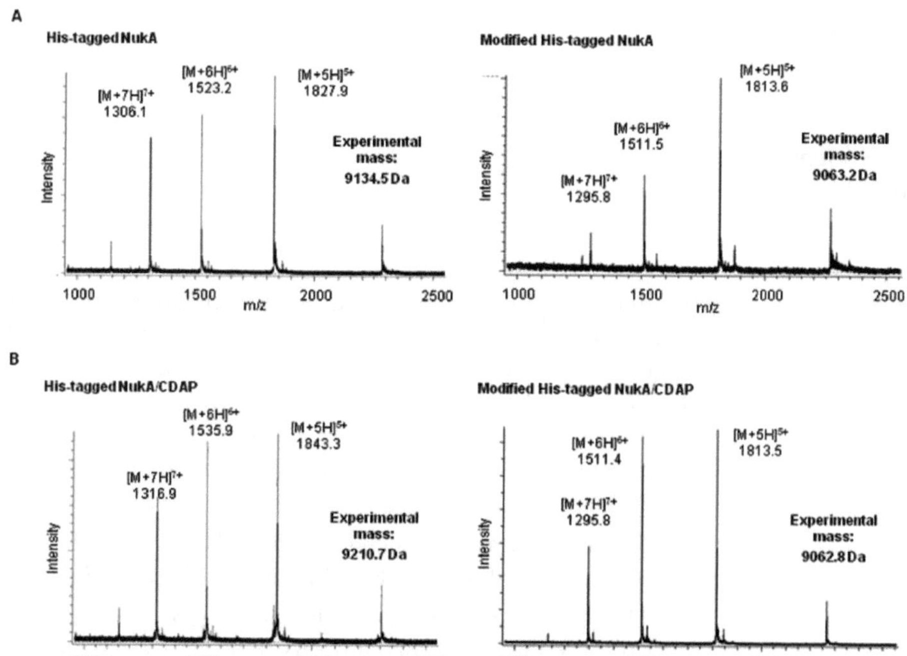

Fig. 13.1. ESI-MS spectra of the purified His-tagged NukA and the modified His-tagged NukA (**a**), and their masses cyanylated with CDAP (**b**). (**a**) From the recorded multiply charged ions, experimental masses of 9134.5 Da and 9063.2 Da could be calculated for His-tagged NukA and the modified His-tagged NukA, respectively. These results indicate that His-tagged NukA (9134.5 Da) co-expressed with NukM for 20 h at 20°C is converted into the peptide (the modified His-tagged NukA, 9063.2 Da) with decrease in mass of 71.3 Da. This mass reduction is in good agreement with the formation of unusual amino acids at the expected positions (*see* **Fig. 13.2**). (**b**) His-tagged NukA containing three Cys residues is used as a control. A total mass shift with +76.2 Da is observed by CDAP treatment, indicating complete CDAP addition to three free cysteine residues. In contrast, the modified His-tagged NukA shows no mass change by CDAP treatment. These results suggest the presence of three thioether bridges in the modified His-tagged NukA.

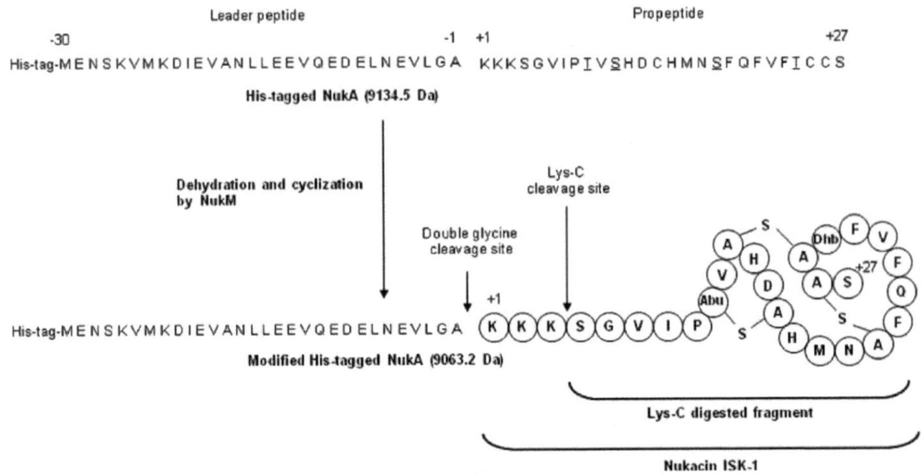

Fig. 13.2. Engineering unusual amino acids into His-tagged NukA by the modification enzyme NukM. NukA consists of an N-terminal leader peptide and a C-terminal propeptide moiety. NukM catalyzes post-translational modification in C-terminal moiety to form unusual amino acids such as lanthionine, 3-methyl lanthionine, and dehydrobutyrine. Nukacin ISK-1 is obtained after cleavage of the leader peptide at the double glycine site. C-Terminal propeptide moiety of nukacin ISK-1 is obtained by the digestion with Lys-C at an appropriate site. A-S-A, lanthionine; Abu-S-A, 3-methyl lanthionine; Dhb, dehydrobutyrine. Underlined Ser and Thr residues in His-tagged NukA can be dehydrated by NukM.

3.5. Characterization of the Modified His-Tagged Prepeptide

3.5.1. Amino Acid Sequencing

To determine the presence and location of unusual amino acids, the modified His-tagged NukA with decrease in mass of 71.3 Da is digested with lysylendopeptidase (Lys-C) (*see* **Fig. 13.2**).

1. The modified His-tagged NukA (42 μg) is dissolved in 1 mL of 100 mM NH_4HCO_3 buffer (pH 8.5) containing the protease Lys-C (0.5 μg) and then incubated at 25°C for 20 h.
2. The digests are subjected to RP-HPLC and the obtained peaks are analyzed by ESI-MS as described in **Section 3.4**.
3. The Lys-C fragment of the modified His-tagged NukA (2574.8 Da) corresponding to that of nukacin ISK-1 is collected and concentrated with evaporator.
4. Lys-C-digested fragment of the modified His-tagged NukA (1 nmol) is analyzed via Edman degradation using a PPSQ-21 gas-phase automatic protein sequence analyzer. (*see* **Note 9**).

3.5.2. Amino Acid Composition Analysis

1. The modified or no modified His-tagged NukA (10 nmol) is hydrolyzed in 1 mL of 6 N HCl under a vacuum at 110°C for 24 h.

2. Amino acid analysis of the hydrolyzate is performed with LC-500/V amino acid analyzer.
3. The sum of the Lan and MeLan residues is determined using D,L-lanthionine as a standard. The result is shown in **Table 13.2** (*see* **Note 10**).

Table 13.2
Amino acid composition of His-tagged NukA and the modified His-tagged NukA[a]

Amino acid	His-tagged NukA		Modified His-tagged NukA	
	Predicted[b]	Detected	Predicted[b]	Detected
Ala	2	2.2	2	2.0
Leu	6	6.0	6	6.0
Cys[c]	3	ND[e]	0	ND[e]
Ser	10	8.8	8	7.1
Thr	2	2.1	0	0.3
Lan/MeLan[d]	0	0	3	3.2

[a]The number of amino acid residues is determined for the molar ratio relative to Leu
[b]Predicted from the gene sequences
[c]Cys is destroyed upon acid hydrolysis
[d]Lan, lanthionine; MeLan, 3-methyl lanthionine
[e] *ND*, not detected

3.5.3. Rapid Evaluation of Unusual Amino Acids Formation with CDAP

To rapidly evaluate the cyclization extent of the modified His-tagged NukA with high sensitivity, CDAP is used to cyanylate free cysteine residues (9). Cyanylated cysteine residue causes a mass shift of +25 Da, whereas cysteine residues involved in ring structure do not (*see* **Note 11**).

1. The peptide dried with evaporator [fewer than 10 μg (1 nmol)] is dissolved in 20 μL of sterilized water.
2. Add 2 μL of 100 mM TCEP buffer and incubate at room temperature for 15 min to reduce disulfide bonds.
3. Add 6 μL of 100 mM CDAP and incubated for an additional 15 min.
4. The reaction products are applied to an LC/MS device at a flow rate of 0.2 mL/min with a linear gradient of acetonitrile/0.05% TFA (15–100% for 40 min). The column elutes are directly loaded into ESI-MS under the same condition as described in **Section 3.4**. The results are shown in **Fig. 13.1b**.

4. Notes

1. Another co-expression system is also tried as follows. Stable maintenance of two plasmids in the same host requires compatible origins of replication and a complementary set of selection markers, so both genes, *nukA* and *nukM*, are independently cloned under IPTG-inducible T7 promoter of pET-14b and pACYC184, respectively. Resulting plasmids, pETnukA and pACYCnukM, are co-introduced into *E. coli* BL21(DE3). However, His-tagged NukA induced for 20 h at 20°C or 3 h at 37°C in this system shows no post-translational modification reaction in His-tagged NukA under induction conditions used (5).
2. Incubator is placed in cold room (4°C) and set it up at 20°C.
3. Quite low amount of mature nukacin ISK-1 (estimated to be below 15 ng per 200 μg protein of cell-free extract) is detected in cell-free extract by ESI-MS analysis (5). This might suggest that the absence of the processing enzyme of nukacin ISK-1 might be partially complemented by a protease encoded by a gene of *E. coli*.
4. Most of the modified His-tagged NukA is present in elution fraction with high purity which shows a single band on SDS-polyacrylamide gel electrophoresis.
5. Expression of NukM is detected only by the induction for 20 h at 20°C not for 3 h at 37°C. Accordingly, no mass change is observed in His-tagged prepeptide induced for 3 h at 37°C (5).
6. We cannot find any non-dehydrated and partially dehydrated peptides other than the obtained peptide (5). This result indicates that post-translational modification by NukM is completely occurred in *E. coli* under the condition used.
7. The modified His-tagged NukA shows no antimicrobial activity against *Lactobacillus sakei* subsp. *sakei* JCM 1157^T used as an indicator strain even by 100-fold of minimum inhibitory concentration (MIC) for nukacin ISK-1 (5). Leader peptide might keep the modified His-tagged NukA inactive that has also been found in other lantibiotics (10).
8. The modified His-tagged NukA is constantly obtained over 1.5 mg/L of culture.
9. Unusual amino acids are generally identified as unidentified residues. Furthermore, dehydrated amino acid leads to the block in Edman degradation. In case of the

modified His-tagged NukA, following N-terminal sequence is obtained: S-G-V-I-P-X-V-X-H-D-X-H-M-N-X-F-Q-F-V-F-X. Except for residues shown as X, unidentified residues, this sequence is identical to that of the Lys-C-digested fragment of nukacin ISK-1. After 21 cycles cause the blockage of subsequent sequence due to the dehydration of the Thr at position 22 to Dhb (residue 24 of the modified His-tagged NukA) (*see* **Fig. 13.2**).

10. If the post-translational modification by NukM is successfully occurred, we can observe loss of two each of Ser and Thr residues and also appearance of three Lan or MeLan residues (**Table 13.2**). There is no commercial standard for MeLan available, and it is not clear from the literature whether Lan and MeLan can be differentiated by RP-HPLC.
11. Chemical modification of free cysteine residue in the peptide with CDAP is rapid and highly sensitive compared to amino acid sequencing and composition analysis. However, we should pay attention that this is not direct method to determine the presence of unusual amino acids.

Acknowledgments

We are grateful to K. Ogawa (Asahi Kasei Pharma Corporation, Ohito, Japan) for amino acid analysis and professor K. Furukawa of Kyushu University, Japan, for giving us kind access to amino acid sequencer.

This work was partially supported by grants from the following organizations: the Japan Society for the Promotion of Science (JSPS), the Japan Science Society, the Novartis Foundation (Japan) for the Promotion of Science, the Novozymes Japan Research Fund, and the Nagase Science and Technology Foundation.

References

1. Lee, K. H. (2002) Development of short antimicrobial peptides derived from host defense peptide or by combinatorial libraries. *Curr Pharm* 8, 795–813.
2. Nagao, J., Asaduzzaman, S. M., Aso, Y., Okuda, K., Nakayama, J., Sonomoto, K. (2006) Lantibiotics: Insight and foresight for new paradigm. *J Biosci Bioeng* 102, 139–149.
3. Chatterjee, C., Paul, M., Xie, L., van der Donk, W. A. (2005) Biosynthesis and mode of action of lantibiotics. *Chem Rev* 105, 633–684.
4. Dufour, A., Hindre, T., Haras, D., Le Pennec, J. P. (2007) The biology of lantibiotics from the lacticin 481 group is coming of age. *FEMS Microbiol Rev* 31, 134–167.
5. Nagao, J., Harada, Y., Shioya, K., Aso, Y., Zendo, T., Nakayama, J., Sonomoto, K. (2005) Lanthionine introduction into nukacin ISK-1 prepeptide by co-expression

with the modification enzyme NukM in *Escherichia coli*. *Biochem Biophys Res Commun* 336, 507–513.

6. Aso, Y., Nagao, J., Koga, H., Okuda, K., Kanemasa, Y., Sashihara, T., Nakayama, J., Sonomoto, K. (2004) Heterologous expression and functional analysis of the gene cluster for the biosynthesis of and immunity to the lantibiotic, nukacin ISK-1. *J Biosci Bioeng* 98, 429–436.
7. Nagao, J., Aso, Y., Sashihara, T., Shioya, K., Adachi, A., Nakayama, J., Sonomoto, K. (2005) Localization and interaction of the biosynthetic proteins for the lantibiotic, nukacin ISK-1. *Biosci Biotechnol Biochem* 69, 1341–1347.
8. Sambrook, J., Fritsch, E. F., Maniatis, T. (1989) *Molecular Cloning: a Laboratory Manual*, 2nd edn. Cold Spring Harbor Laboratory, Cold Spring Harbor, NY.
9. Wu, J., Watson, J. T. (1997) A novel methodology for assignment of disulfide bond pairing in proteins. *Protein Sci* 6, 391–398.
10. Xie, L., Miller, L. M., Chatterjee, C., Averin, O., Kelleher, N. L., van der Donk W. A. (2004) Lacticin 481: In vitro reconstitution of lantibiotic synthetase activity. *Science* 303, 679–681.

Chapter 14

The Targeted Expression of Nucleotide Sugar Transporters to the *E. coli* Inner Membrane

Joe Tiralongo and Andrea Maggioni

Abstract

The heterologous expression of functional mammalian integral membrane proteins still represents a significant hurdle towards evaluating the relationship between their structure and function. We have therefore utilised the OmpA signal sequence to deliberately target the expression of a mammalian nucleotide sugar transporter, the murine CMP-sialic acid transporter, to the *E. coli* inner membrane. The functionality of the recombinant CMP-sialic acid transporter could then be evaluated either following the spheroplasting of *E. coli* cells or through the isolation of the *E. coli* inner membrane and the formation of mixed phosphatidylcholine-inner membrane proteoliposomes.

Key words: Membrane proteins, nucleotide sugar transporter, OmpA signal sequence, protein expression.

1. Introduction

Membrane proteins represent approximately 30% of the total prokaryotic and eukaryotic proteome. However, integral membrane proteins represent only about 0.3% of all the protein structures elucidated, with an even smaller percentage if only the structures of eukaryotic membrane protein are considered (1, 2). Despite the vast array of expression systems currently available, the main bottleneck in eukaryotic membrane protein crystallisation and elucidation of structure–function relationships remains the quantitative expression and purification of functional protein (1, 2). The expression of eukaryotic integral membrane proteins in prokaryotic expression systems (e.g. *E. coli*) typically leads to the sequestering of these recombinant proteins as inclusion

T.C. Evans, M.-Q. Xu (eds.), *Heterologous Gene Expression in E. coli*, Methods in Molecular Biology 705, DOI 10.1007/978-1-61737-967-3_14, © Springer Science+Business Media, LLC 2011

bodies, from which functional membrane protein is often difficult to recover ((3) and references therein).

The CMP-sialic acid transporter (CST) is a Golgi resident hydrophobic protein with ten putative transmembrane domains that catalyses the transport of CMP-sialic acid (the universal donor substrate for sialyltransferases) into the Golgi apparatus of eukaryotic cells (4). The CST is a member of a highly conserved family of multiple membrane spanning proteins collectively referred to as nucleotide sugar transporters (5). Therefore, our interest (6, 7) in not only probing the relationship between CST structure and function, but also in generating protein suitable for structural elucidation, led us to develop an efficient system for the heterologous expression of integral membrane proteins in *E. coli*. This system utilises the OmpA signal sequence to target integral membrane proteins to the *E. coli* inner membrane.

2. Material

2.1. Protein Expression and E. coli Inner Membrane Isolation

1. pFLAG-mCST (8) (*see* **Note 1** and **Fig. 14.1**).
2. Electrocompetent *E. coli* BL21 (Merck Biosciences, Darmstadt, Germany) (*see* **Note 2**).
3. 50 mg/mL Ampicillin: Dissolve 1 g ampicillin sodium salt in 20 mL Milli Q H_2O (*see* **Note 3**) and sterilise by passing through a 0.2 μm filter. Aliquot and freeze at –20°C.

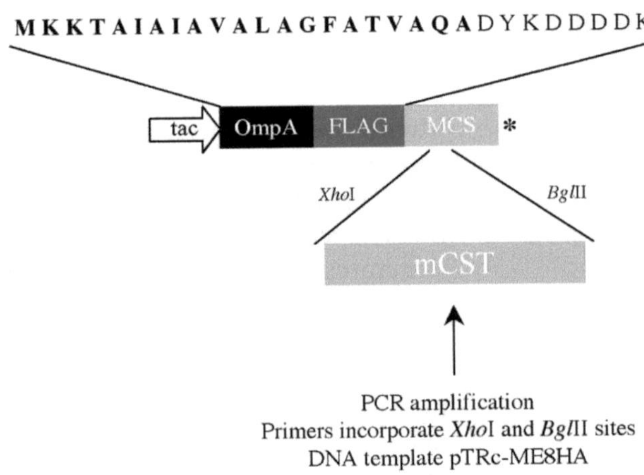

Fig. 14.1. pFLAG-mCST was constructed as described in (8). The coding sequence of the full-length mCST was amplified by PCR using pTRc-ME8HA (7) as the template and a forward primer that introduced an *XhoI* site and a reverse primer that introduced a *BglII* site. The resultant PCR product was digested with *XhoI* and *BglII* and ligated into the corresponding sites of pFLAG-ATS to generate pFLAG-mCST. Using this plasmid the mCST can be expressed in *E. coli* BL21 fused to an N-terminal OmpA signal sequence, shown in bold, and the eight amino acid residue FLAG-tag under the control of the IPTG inducible *tac* promoter (a hybrid of the *E. coli trp* and *lac* promoters).

4. LB (Luria-Bertani) medium: Dissolve 10 g Tryptone (Oxoid, Cambridge, UK), 5 g Yeast Extract (Oxoid) and 10 g NaCl in 1 L Milli Q H_2O and sterilise by autoclaving at 121°C for 20 min. Store at room temperature.
5. LB agar-ampicillin plates: Add 15 g Agar Bacteriological to LB medium described above prior to autoclaving. Allow medium to cool to approximately 55°C and add ampicillin (50 mg/mL) to a final concentration of 100 μg/mL. Immediately pour (approximately 25–30 mL) into 10 cm diameter Petri dishes and allow to set at room temperature. Store at 4°C.
6. 1 M IPTG: Dissolve 2.83 g isopropyl-β-D-thio-galactopyranoside (IPTG) in 8 mL Milli Q H_2O. Bring to 10 mL with additional Milli Q H_2O and sterilise by passing through a 0.2 μm filter. Freeze at –20°C.
7. Phosphate buffer saline (PBS): Dissolve 8 g NaCl, 0.2 g KCl, 1.44 g Na_2HPO_4 and 0.24 g KH_2PO_4 in approximately 800 mL of Milli Q H_2O and adjust pH to 7.4. Bring to 1 L and store at room temperature.
8. PBS-Protease inhibitor cocktail: Dissolve 1 Complete Protease Inhibitors Cocktail Tablet (Roche Applied Science, Castle Hill, NSW, Australia) in 50 mL PBS and supplement with 0.5 M EDTA (pH 8) to a final concentration of 1 mM.
9. Sucrose solutions (w/w): For all sucrose concentrations the indicated amounts are dissolved in 120–150 mL Milli Q H_2O. Ten mL of 0.1 M EDTA (pH 7.5) is then added and the solution is adjusted to 200 mL by the addition of Milli Q H_2O.
   55% sucrose, 138.3 g; 50% sucrose, 123.0 g; 45% sucrose, 108.2; 40% sucrose, 94.1 g; 35% sucrose, 80.6 g; 30% sucrose, 67.6 g.

2.2. Evaluation of Recombinant Membrane Protein Functionality

2.2.1. Spheroplasting of *E. coli* Cells

1. 200 mM Tris–HCl (pH 8): Dissolve 24.23 g Tris base in 900 mL of Milli Q water and adjust pH to 8 with HCl. Bring to 1 L and store at room temperature.
2. 200 mM Tris–HCl (pH 8) containing 1 M sucrose: Dissolve 24.23 g Tris base and 342.3 g sucrose in 500 mL of Milli Q water, bring volume to 900 mL and adjust pH to 8 with HCl. Bring to 1 L and store at room temperature.
3. Egg white lysozyme (*see* **Note 4**).
4. 1 M $MgCl_2$ solution: Dissolve 10.2 g magnesium chloride hexahydrate in 20 mL of Milli Q water and sterilise by passing through a 0.2 μm filter. Aliquot and freeze at –20°C.

5. 100 mM EDTA (pH 7.6) solution: Dissolve 3.73 g of EDTA-Na_2 dihydrate in 70 mL Milli Q water and while stirring adjust to pH 7.6 with NaOH. Bring to 100 mL and store at room temperature.

2.2.2. Evaluation of Protein Functionality: Generation of Mixed Phosphatidylcholine-Inner Membrane Proteoliposomes

1. Phosphatidylcholine (Type XI-E, 100 mg/mL in chloroform).
2. 10 mM Tris–HCl (pH 7) containing 2 mM $MgCl_2$: Prepare 1 M Tris–HCl (pH 7) by dissolving 12.11 g Tris base in 70 mL of Milli Q water and adjust pH to 7 with HCl. Bring to 100 mL and store at room temperature. Add 1 mL 1 M Tris–HCl (pH 7) and 0.2 mL 1 M $MgCl_2$ solution to 98.8 mL of H_2O, mix and store a 4°C.

3. Methods

3.1. Protein Expression and Inner Membrane Isolation

The protocol provided is for the targeted expression of the mouse CMP-sialic acid transporter (mCST) to the *E. coli* inner membrane using the OmpA leader sequence under optimal conditions (15°C, 0.1 mM IPTG, 3 h). These conditions were determined by evaluating various temperatures (15, 20, 25 and 37°C), IPTG concentrations (0.1–1 mM) and induction times (up to 4 h). Therefore, the same would need to be determined for each protein expressed using our system; however, the basic methodology described is in essence the same regardless of temperature, IPTG concentration and induction times used.

The OmpA signal sequence targets recombinant proteins to the Sec translocase (the *E. coli* translocation machinery in the inner membrane) (3). The isolation of *E. coli* inner membrane following the targeted expression of membrane proteins using the OmpA leader sequence serves two purposes. First, it is used to verify the localisation of the recombinant CST to the inner membrane. Second, the isolated inner membrane fraction can subsequently be used to generate mixed phosphatidylcholine-inner membrane proteoliposomes (described in **Section 3.2.2**) for the evaluation of recombinant CST functionality.

1. Transform electrocompetent *E. coli* BL21 cells with pFLAG-mCST (*see* **Note 2**) and select for positive transformants by plating on LB agar-ampicillin plates (*see* **Note 5**).
2. An overnight culture of *E. coli* BL21 transformed with pFLAG-mCST is prepared by inoculating 3 mL LB medium containing 100 μg/mL ampicillin with a single

transformant (colony) and incubating at 37°C for 16 h with shaking (225 rpm).

3. Two mL of the resulting overnight culture is used to inoculate 1 L LB medium containing 100 μg/mL ampicillin and the culture is incubated at 37°C with shaking until an OD_{600} (*see* **Note 6**) of 0.4–0.5 is reached. At this stage the culture is then shifted to an incubator set at 15°C (*see* **Note 7**) and incubated with shaking until an OD_{600} of 0.6–0.7 is reached (generally takes 45–60 min).
4. At an OD_{600} of 0.6–0.7 protein expression is induced by the addition of IPTG to a final concentration of 0.1 mM (induced cells) (*see* **Note 8**).
5. After 3 h incubation at 15°C, cells are harvested by centrifugation (4,000×*g*, 15 min, 4°C) and suspended in PBS-Protease inhibitor cocktail supplemented with 1 mM EDTA.
6. Induced *E. coli* cells while on ice are lysed by sonication using an ultrasonic processor fitted with a tapered microtip probe set at 40% maximum output for a 30 s pulse followed by a 30 s pause on ice. This is repeated for four cycles. The resulting total cell lysate is pre-cleared by centrifugation (20,000×*g*, 30 min, 4°C); this removes un-lysed cells, cell debris and inclusion bodies (*see* **Note 9**).
7. A total membrane fraction (*see* **Note 10**) is subsequently obtained by further centrifugation of the supernatant following pre-clearing by ultracentrifugation at 100,000×*g*, 4°C, 1 h.
8. The total membrane fraction is layered on a 30–55% sucrose gradient (Ultra Clear Tubes, Beckman, Part number 344058) essentially as described by Osborn and Munson (9) and shown in **Fig. 14.2**, and centrifuged at 100,000×*g* for 18 h at 4°C using a swing-out bucket rotor (e.g. SW32 Ti, Beckman Coulter, Fullerton, CA, USA).
9. Following centrifugation, 1 mL fractions are carefully removed beginning at the top of the tube and set on ice.
10. The protein concentration of each fraction is determined using the BCA protein assay (Pierce, Rockford, IL, USA) or similar protein assay kit.
11. Protein is separated by SDS-PAGE (*see* **Note 11**) and detected by Western blot analysis (*see* **Note 12**). The result from a typical isolation using a sucrose density gradient of the *E. coli* inner membrane expressing recombinant CST is shown in **Fig. 14.3**.
12. Fractions identified containing recombinant CST are pooled and used immediately.

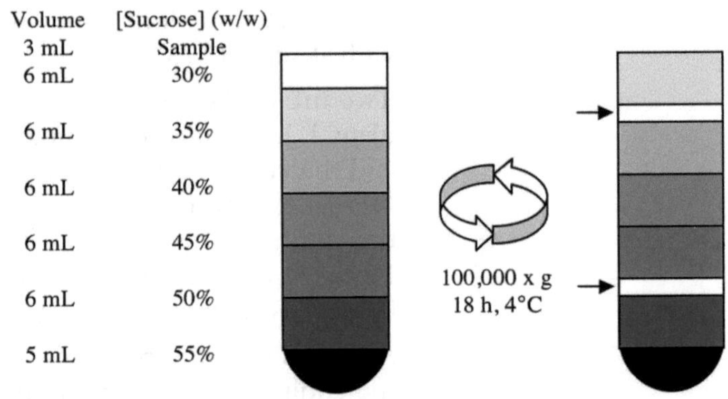

Fig. 14.2. The *E. coli* inner and outer membrane can be separated by layering a total membrane fraction onto a 30–55% sucrose gradient as shown. The volumes required to generate this gradient in a 38 mL Ultra Clear centrifuge tube are indicated and centrifugation is performed at 100,000×*g* for 18 h at 4°C using a swing-out bucket rotor. The outer membrane accumulates at the 30–35% sucrose interface and the inner membrane accumulates at the 45–50% sucrose interface. The membrane fractions are indicated with arrows.

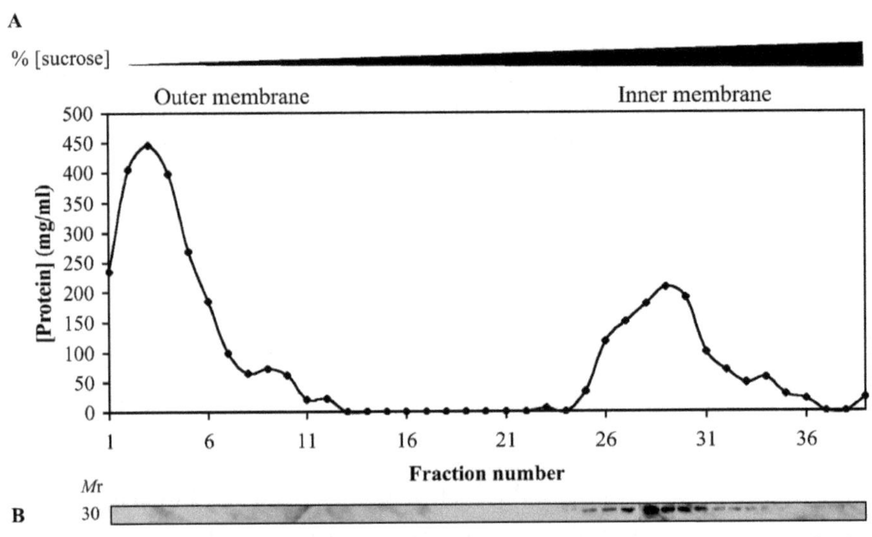

Fig. 14.3. A typical separation of the *E. coli* inner membrane incorporating the recombinant mCST from the outer membrane using a sucrose density gradient. (**a**) Fractions of 1 mL were carefully removed beginning at the top of the centrifuge tube and analysed for protein content. (**b**) Recombinant mCST content in the fractions was verified by Western blotting using anti-FLAG mAb M2 as the primary antibody. The region of the blot corresponding to 30 kDa, the expected molecular mass of mCST, is shown. Twenty micrograms of protein was loaded per lane.

3.2. Evaluation of Recombinant Membrane Protein Functionality

The following protocols describe two alternative methods that permit the evaluation of recombinant CST functionality after targeted expression to the *E. coli* inner membrane. The first utilises spheroplasted *E. coli* cells (described in **Section 3.2.1**), where the peptidoglycan network connecting the inner and outer membrane is destabilised using lysozyme leading to the almost complete removal of the outer membrane without cell lysis. This treatment results in the *E. coli* inner membrane becoming exposed, allowing the functionality of the recombinant protein localised to the membrane to be evaluated. The use of spheroplasts to evaluate recombinant membrane protein functionality has a number of advantages: (i) they are relatively quick to prepare, thus minimising sample handling; (ii) they are suitable for small-scale pilot experiments; (iii) they can be easily collected by centrifugation or filtration because of their size; and (iv) based on the positive-inside rule (10), the orientation of the recombinant protein within the membrane should be uniform. Therefore, the use of spheroplasted cells represents an ideal starting point for determining recombinant protein functionality. However, the integration of a recombinant membrane protein can, if expression is not tightly regulated, destabilise the inner membrane making spheroplasting difficult due to cell lysis. If spheroplasting cannot be achieved, we recommend isolation of the inner membrane as described in **Section 3.1**, and the generation of mixed phosphatidylcholine-inner membrane proteoliposomes (described in **Section 3.2.2**).

The generation of mixed proteoliposomes is particularly necessary if the transport activity of a recombinant solute transporter, such as the CST, is to be assessed. A significant advantage of using mixed proteoliposomes, particularly when evaluating recombinant solute transporter functionality, is the ability to pre-load the proteoliposomes with substrates or inhibitors. That is, many solute transporters function via an antiporter mechanism where the influx of one molecule is driven by the efflux of a counter-molecule; such a mechanism is utilised by the CST (7). Therefore, the ability to pre-load proteoliposomes with substrates or inhibitors allows detailed evaluation of recombinant transporter functionality.

3.2.1. Spheroplasting of *E. coli* Cells

Spheroplasting was performed essentially according to Witholt et al. (11, 12).

1. Following induction with 0.1 mM IPTG (**Section 3.1**, Step 5), collect the *E. coli* cells by centrifugation (4,000×*g*, 15 min, 4°C). Resuspend the cells at 40 mg/mL in 200 mM Tris–HCl buffer (pH 8.0) (*see* **Note 13**).
2. Dilute the cell suspension with an equal volume of 200 mM Tris–HCl (pH 8.0) containing 1 M sucrose (final concentration: 200 mM Tris–HCl (pH 8.0) containing 0.5 M sucrose).
3. Add EDTA (pH 7.6) to a final volume equivalent to 0.5% of the cell suspension obtained at Step 2.
4. Add lysozyme to a final concentration of 60 μg/mL.

5. Induce osmotic shock by diluting the cell suspension with an equal volume of Milli Q water under constant and gentle stirring.
6. Monitor spheroplasting progression by light microscopy (*see* **Note 14**).
7. When spheroplasting efficiency reaches 80–85%, add $MgCl_2$ to a final concentration of 20 mM to stabilise the spheroplasts.
8. *E. coli* spheroplasted in this manner can now be collected by centrifugation, resuspended in an appropriate buffer system and used for evaluation of functionality (e.g. binding, enzyme activity or transport assays).

3.2.2. Generation of Mixed Phosphatidylcholine-Inner Membrane Proteoliposomes

1. An appropriate amount (*see* **Note 15**) of the 100 mg/mL phosphatidylcholine solution is transferred to a glass round bottom flask and the bulk of the chloroform removed quickly by rotary evaporation. The lipid cake is then transferred to a high vacuum pump and left to dry overnight.
2. The resulting lipid cake is rehydrated with 10 mM Tris–HCl (pH 7) containing 2 mM $MgCl_2$ to give a final phosphatidylcholine concentration of 30 mg/mL.
3. The lipid suspension is extruded 11 times with the Avanti Lipid Mini-Extruder using a polycarbonate filter with 200 nm diameter (*see* **Note 16**).
4. The phosphatidylcholine unilamellar vesicle suspension is then diluted in 10 mM Tris–HCl (pH 7) containing 2 mM $MgCl_2$ to a final concentration of 3 mg/mL and stored on ice until required.
5. The inner membrane fraction purified on sucrose gradient as described above (**Section 3.1**) is mixed with the phosphatidylcholine vesicles at a ratio of 1:10 (w/w).
6. Fusion of purified inner membranes with unilamellar phosphatidylcholine vesicle is induced by snap-freezing in liquid nitrogen followed by thawing at room temperature. This freeze–thaw cycling is repeated five times (*see* **Note 17**).
7. After the last freeze–thaw cycle, the mixed phosphatidylcholine-inner membrane proteoliposomes are once again extruded through a 200 nm polycarbonate filter 11 times and immediately applied to the determination of membrane protein functionality (e.g. binding, enzyme activity or transport assays).

4. Notes

1. pFLAG-mCST was constructed as described in (8). The coding sequence of the full-length mCST incorporating *Xho*I and *Bgl*II sites was ligated into the corresponding sites of pFLAG-ATS (summarised in **Fig. 14.1**). Protein expression is controlled by the IPTG inducible *tac* promoter, a hybrid of the *E. coli trp* and *lac* promoters. All genetic manipulations are performed in the *rec*A *E. coli* strain DH5α.
2. Due to the use of a *tac* promoter, *E. coli* DE3 lysogenic strains (required when using the T7 promoter) are not required for protein expression. We use *E. coli* BL21 (*F⁻ompT hsdS*$_B$ *(r⁻*$_B$ *m⁻*$_B$*) gal dcm*) because, as is the case with all other *E. coli* B strains, *E. coli* BL21 is deficient in the *Ion* protease and lacks the *ompT* outer membrane protease. There are a number of methods available for preparing electrocompetent *E. coli* cells (*see* Sambrook and Russell (13)), or electrocompetent *E. coli* BL21 can be purchased from Merck Biosciences.
3. All solutions should be prepared in water that has a resistivity of 18.2 MΩ cm and is of the highest possible purity. This is referred to as Milli Q H_2O in this text.
4. Be aware that different lysozyme preparations might have very different specific activities. It is therefore necessary to empirically identify optimal lysozyme concentrations depending on the preparation used.
5. It is recommended not to use *E. coli* BL21 or similar expression strains for the preparation of long-term glycerol stocks of the plasmid since they are not *rec*A strains. In order to identify plasmid and/or protein toxicity, upon transformation it is recommended that a toxicity test be performed. This is performed by plating the same dilution of transformed cells onto LB agar, LB agar supplemented with 100 μg/mL ampicillin (plasmid toxicity), LB agar supplemented with both 100 μg/mL ampicillin and 1 mM IPTG (protein toxicity) and comparing the number of cells obtained on each plate (14, 15).
6. OD_{600} refers to the optical density of a bacterial suspension read at a wavelength of 600 nm using a standard spectrophotometer. That is, light passing through a bacterial suspension is scattered, and the amount of scatter is an indication of the biomass present in the suspension.

7. Maintaining an incubator temperature of 15°C is nearly impossible to achieve in standard laboratories; however, an easy solution is to place an incubator in a 4°C cold room and use the incubator's temperature control to set the temperature to 15°C. To avoid shock responses, whenever possible it is preferable to reduce the temperature stepwise rather than shifting abruptly from 37 to 15°C, thus allowing the cell culture to equilibrate and reach the desired OD_{600} at 15°C before inducing protein expression by the addition of IPTG.
8. *E. coli* cells (approximately 50 mL) can be removed prior to IPTG induction (referred to as un-induced cells) and analysed by Western blotting. This analysis allows the investigator to assess the stringency of the induction process; that is, if any "leaky" protein expression occurred in the absence of IPTG.
9. It is critical that the *E. coli* suspension is kept cold during sonication in order to avoid heat denaturation of proteins, and that foaming of the sample be minimised. This can be achieved by keeping the tube containing the *E. coli* suspension in a beaker of ice during the entire sonication process, and by placing the microtip probe greater than 10 mm below the surface of the sample. The sonication settings described have been optimised for an ultrasonic processor with a net power output of 750 W (e.g. Sonics Vibra Cell VC 750). Therefore, sonication settings will need to be adjusted depending on energy output of a given sonicator. To monitor cell lysis, a small aliquot of the total cell lysate is removed following each sonication cycle and centrifuged (13,000 × *g*, 30 s, 4°C). The protein content in the resulting supernatant is determined using a standard protein assay (e.g. Bradford Assay). The corresponding sonication cycles required to give maximum protein content can then be determined.
10. This fraction contains both the *E. coli* inner and outer membrane.
11. There are numerous standard protocols available for performing SDS-PAGE. We perform SDS-PAGE essentially as described by Laemmli in 1970 on a 10% acrylamide gel using the Mini-PROTEAN 3 Cell Electrophoresis System (Bio-Rad). However, samples are prepared for SDS-PAGE by dilution 1:1 with a modified 2× SDS-PAGE sample buffer (0.125 M Tris–HCl (pH 6.8), 4% SDS, 20% glycerol, 0.2 M dithiotheritol (DTT), 0.02% bromophenol blue, 7 M urea and 2 M thiourea) and incubated at 30°C for 15 min. To avoid carbamylation of proteins, it is recommended not to heat samples containing urea at

temperatures in excess of 37°C (16). Moreover, at higher temperature hydrophobic proteins, such as the mCST, tend to aggregate in the presence of SDS, resulting in the appearance of high molecular weight smears rather than discrete protein bands (17).

12. There are numerous standard methodologies available for performing Western blotting. We transfer protein following separation by SDS-PAGE to a PVDF membrane using the Trans-Blot SD Semi-Dry Electrophoretic Transfer Cell (Bio-Rad) for 1 h instead of the 15 min described by the manufacturer. Blocking of PVDF membrane is performed using 2% (w/v) skim milk powder in Tris buffered saline (TBS, pH 7.4) containing 0.1% Tween-20. The primary antibodies that can be utilised for Western blot detection are anti-FLAG M1 (recognises N-terminal DYKDDDDK; dilution 1:10,000) or anti-FLAG M2 (recognises DYKDDDDK dilution; 1:10,000). We commonly use horseradish peroxidase-conjugated anti-mouse IgG (Bio-Rad, Hercules, CA, USA; diluted 1:10,000) and the SuperSignal® West Pico Chemiluminescent Substrate (Pierce, Rockford, IL, USA) as the substrate for horseradish peroxidase. All antibody solutions are made up in 2% skim milk powder in TBS containing 0.1% Tween-20. Immunoreactive bands are visualised by exposing X-ray film (Kodak, Australia) to the PVDF membrane in an X-ray film cassette (Kodak) for up to 30 min (the exact exposure time will depend on expression levels).
13. It is recommended that freshly induced *E. coli* cells be used to prepare spheroplasts and not cells that have been frozen.
14. To follow spheroplasting efficiency by light microscopy, transfer a small aliquot (5 μL) of the *E. coli* suspension onto a glass slide and cover with a cover slip. Over time, as spheroplasting progresses, the shape of the *E. coli* cells will change from a rod-like to a round morphology. When 80–85% of the cells appear round spheroplasting can be terminated.
15. The amount of phosphatidylcholine and therefore the volume of rehydration buffer to be used are dependent on the number of samples to be prepared. When pre-loaded proteoliposomes are required, substrates or inhibitors can be added to the rehydration buffer at this stage. Pre-loaded proteoliposomes can subsequently be retrieved either by gel filtration chromatography (e.g. using PD-10 column) or by ultracentrifugation.
16. Several extrusion apparatus can be purchased from different suppliers depending on the specific applications,

from hand-driven small-scale analytical extruders for semi-preparative applications to fully automated devices suitable for handling large sample volumes. We use a small manual extrusion apparatus from Avanti (The Mini-Extruder). When using this device make sure to extrude the lipid suspension an odd number of times. By doing so it is possible to ensure that contaminants present in the donating syringe are not passed into the receiving syringe and therefore carried through to subsequent phases of the experimental procedure. When critical, proteoliposomes size distribution can be monitored either by gel filtration chromatography or by dynamic light scattering (18–20).

17. This freeze–thawing method for the generation of proteoliposomes takes advantage of the spontaneous fusing of lipid vesicles when the temperature is lowered to below their phase transition temperature (21).

References

1. Grisshammer, R., Tate, C. G. (2003) Overexpression of integral membrane proteins. *Biochim Biophys Acta* 1610, 1.
2. Link, A. J., Georgiou, G. (2007) Advances and challenges in membrane protein expression. *AIChE J* 53, 752–756.
3. Drew, D., Froderberg, L., Baars, L., de Gier, J. W. (2003) Assembly and overexpression of membrane proteins in *Escherichia coli*. *Biochim Biophys Acta* 1610, 3–10.
4. Eckhardt, M., Muhlenhoff, M., Bethe, A., Gerardy-Schahn, R. (1996) Expression cloning of the Golgi CMP-sialic acid transporter. *Proc Natl Acad Sci USA* 93, 7572–7576.
5. Caffaro, C. E., Hirschberg, C. B. (2006) Nucleotide sugar transporters of the Golgi apparatus: from basic science to diseases. *Acc Chem Res* 39, 805–812.
6. Tiralongo, J., Abo, S., Danylec, B., Gerardy-Schahn, R., von Itzstein, M. (2000) A high-throughput assay for rat liver Golgi and *Saccharomyces cerevisiae*-expressed murine CMP-*N*-acetylneuraminic acid transport proteins. *Anal Biochem* 285, 21–32.
7. Tiralongo, J., Ashikov, A., Routier, F., Eckhardt, M., Bakker, H., Gerardy-Schahn, R., von Itzstein, M. (2006) Functional expression of the CMP-sialic acid transporter in *Escherichia coli* and its identification as a simple mobile carrier. *Glycobiology* 16, 73–81.
8. Maggioni, A., von Itzstein, M., Gerardy-Schahn, R., Tiralongo, J. (2007) Targeting the expression of functional murine CMP-sialic acid transporter to the *E. coli* inner membrane. *Biochem Biophys Res Commun* 362, 779–784.
9. Osborn, M. J., Munson, R. (1974) Separation of the inner (cytoplasmic) and outer membranes of Gram-negative bacteria. *Methods Enzymol* 31, 642–653.
10. White, S. H., von Heijne, G. (2004) The machinery of membrane protein assembly. *Curr Opin Struct Biol* 14, 397–404.
11. Witholt, B., Boekhout, M., Brock, M., Kingma, J., Heerikhuizen, H. V., Leij, L. D. (1976) An efficient and reproducible procedure for the formation of spheroplasts from variously grown *Escherichia coli*. *Anal Biochem* 74, 160–170.
12. Witholt, B., Heerikhuizen, H. V., De Leij, L. (1976) How does lysozyme penetrate through the bacterial outer membrane? *Biochim Biophys Acta* 443, 534–544.
13. Sambrook, J., Russell, D. W. (2001) *Molecular Cloning: A Laboratory Manual*, Vol. 1. Cold Spring Harbor Laboratory Press, Cold Spring Harbor, NY.
14. Dumon-Seignovert, L., Cariot, G., Vuillard, L. (2004) The toxicity of recombinant proteins in *Escherichia coli*: a comparison of overexpression in BL21(DE3), C41(DE3), and C43(DE3). *Protein Expr Purif* 37, 203–206.
15. Miroux, B., Walker, J. E. (1996) Overproduction of proteins in *Escherichia coli*: mutant hosts that allow synthesis of some membrane proteins and globular proteins at high levels. *J Mol Biol* 260, 289–298.
16. Stark, G. R., Stein, W. H., Moore, S. (1960) Reactions of the cyanate present in aqueous

urea with amino acids and proteins. *J Biol Chem* 235, 3177–3181.
17. Ragan, C. I. (1986) Analysis of membrane protein composition by gel electrophoresis. In (Ragan, C. I., Cherry, R. J. eds.), *Techniques for the Analysis of Membrane Proteins.* Chapman and Hall, London, pp. 1–25.
18. Mui, B., Chow, L., Hope, M. J. (2003) Extrusion technique to generate liposomes of defined size. *Methods Enzymol* 367, 3–14.
19. Vojta, A., Scheuring, J., Neumaier, N., Mirus, O., Weinkauf, S., Schleiff, E. (2005) Determination of liposome size: a tool for protein reconstitution. *Anal Biochem* 347, 24–33.
20. Olson, F., Hunt, C. A., Szoka, F. C., Vail, W. J., Papahadjopoulos, D. (1979) Preparation of liposomes of defined size distribution by extrusion through polycarbonate membranes. *Biochim Biophys Acta* 557, 9–23.
21. Kasahara, M., Hinkle, P. C. (1976) Reconstitution of D-glucose transport catalyzed by a protein fraction from human erythrocytes in sonicated liposomes. *Proc Natl Acad Sci USA* 73, 396–400.

Chapter 15

Detection of Protein–Protein Interactions in Bacteria by GFP-Fragment Reconstitution

Akira Kanno, Takeaki Ozawa, and Yoshio Umezawa

Abstract

Protein–protein interaction is one of the most pivotal roles of proteins in living organisms. Association/dissociation of proteins reflects responses to intrinsic or extrinsic perturbations of signaling pathways, involved in gene expression, cell division, cell differentiation, and apoptosis. For further understanding of the biological processes, it is important to monitor protein–protein interactions in model organisms. In particular, *Escherichia coli*-based methods are suitable to assess large libraries of proteins. Many of these proteins cannot be used in yeast due to toxicity or poor expression. Herein we describe a general method based on an intein-mediated protein reconstitution system (PRS) to detect protein–protein interactions in bacterial cells. The PRS-based approach requires no other agents including enzymes, substrates, and ATP. Another advantage is that matured green fluorescent protein (GFP) accumulates in a targeted cell till degraded. These allow highly sensitive screening of protein–protein interactions.

Key words: Intein, protein splicing reaction, green fluorescent protein, protein–protein interaction.

1. Introduction

Complete sequences of genomes are available for several organisms such as *Escherichia coli* (*E. coli*) (1), *Saccharomyces cerevisiae* (2), *Caenorhabditis elegans* (3), *Arabidopsis thaliana* (4), *Drosophila melanogaster* (5), and *Homo sapiens* (6). However, the complete genome sequences do not necessarily reveal the mechanisms of biological phenomena within each organism. It is because the molecular functions of predicted open reading frames have remained experimentally uncharacterized (7). Therefore, it is essential to observe functions of proteins in vivo. Protein–protein interaction is one of the most pivotal roles of proteins

T.C. Evans, M.-Q. Xu (eds.), *Heterologous Gene Expression in E. coli*, Methods in Molecular Biology 705,
DOI 10.1007/978-1-61737-967-3_15,

in living organisms: Association/dissociation of proteins reflects responses to intrinsic or extrinsic perturbations of signaling pathways, involved in gene expression, cell division, cell differentiation, and apoptosis.

In order to increase understanding of these biological processes, several techniques have so far been developed for monitoring protein–protein interactions in model organisms (8–22). In particular, *E. coli*-based methods are suitable to assess large libraries of proteins, many of which cannot be used in yeast due to toxicity or poor expression (23). Bacterial two-hybrid systems are useful for selection and screening of $>10^8$ library proteins (24, 25). The protein-fragment complementation assay/bimolecular fluorescence complementation analysis (PCA/BiFC analysis) is a method that allows insights into drug screening beyond the static representations of protein–protein interactions (26, 27).

Herein we describe a general method based on an intein-mediated protein reconstitution system (PRS) for monitoring protein–protein interactions in bacterial cells (28, 29). "Intein" is an intervening protein sequence that triggers autocatalytic cleavage of its flanking proteins (exteins) and simultaneous ligation of the exteins by a peptide bond (so-called protein splicing reaction; *see* **Fig. 15.1a**). A schematic diagram in **Fig. 15.1b** shows the basic concept of the present technique. Green fluorescent protein (GFP) is rationally dissected into the N- and C-terminal fragments (N-GFP and C-GFP). The N- and C-GFP are fused to the N- and C-terminal inteins (N-Int and C-Int). Each fusion protein is linked to a protein of interest (protein X) and its target (protein Y). N- and C-Int are brought in close proximity upon X–Y interaction. This triggers the protein splicing reaction resulting in linkage of N- and C-GFP with a peptide bond. Interaction between X and Y is evaluated by measurement of fluorescence intensity from the matured GFP. The PRS-based approach requires no other agents including enzymes, substrates, and ATP. An advantage of the present method is that matured GFP accumulates in a targeted cell till degraded. Highly sensitive measurement is thus achievable for screening of protein–protein interactions.

2. Materials

2.1. DNA Plasmids

1. pET-15b (Novagen, Madison, WI).
2. A cDNA of a GFP (e.g., pAcGFP1 available from Clontech Laboratories, Palo Alto, CA) (*see* **Note 1**).
3. cDNAs of the N- and C-terminal *Ssp* DnaE intein (N-DnaE and C-DnaE) derived from *Synechocystis* sp. strain PCC6803.

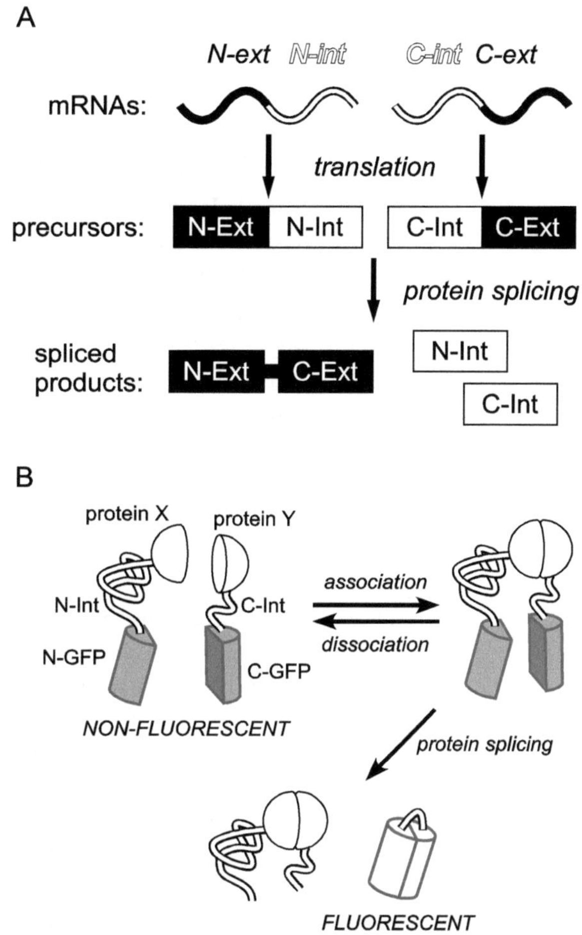

Fig. 15.1. Reconstitution of GFP fragments based on naturally split inteins. (**a**) Basic scheme for protein splicing reaction. Two mRNAs are translated into two precursor proteins. The precursor proteins are composed of intervening peptides (inteins) and flanking proteins (exteins). The exteins are excised from the inteins. Concomitantly, the exteins are ligated by the inteins with peptide bonds. (**b**) Schematic diagram of the present method for monitoring protein–protein interactions in bacteria. The N- and C-terminal fragments of GFP (N-GFP and C-GFP) are fused to the N- and C-terminal inteins (N-Int and C-Int), respectively. A protein of interest (protein X) and its target protein (protein Y) are linked to the opposite ends of the inteins. Interaction between protein X and Y triggers association of the inteins resulting in reconstitution of a full-length GFP by protein splicing reaction. The extent of the X–Y interaction is evaluated by measuring fluorescence intensities from the reconstituted GFP.

4. cDNAs of target proteins (*see* **Note 2**).
5. Forward and reverse oligonucleotide primers corresponding to each template cDNA.

2.2. Protein Expression and Fluorescence Measurements

1. An *E. coli* strain DH5α for constructions of plasmids.
2. An *E. coli* strain BL21(DE3)pLysS for expression of recombinant fusion proteins.

3. Luria–Bertani (LB) medium for culture of *E. coli*.
4. LB agar for culture of *E. coli* on dishes.
5. 100 mg/mL ampicillin sodium solution: Dissolve ampicillin sodium in deionized water and sterilize the solution by filtration. The stock solution is stored at –30°C.
6. 34 mg/mL chloramphenicol stock solution: Dissolve chloramphenicol in molecular biology-grade ethanol and store the solution at –30°C.
7. 100 mM isopropyl-β-D-thiogalactopyranoside (IPTG; CAS number: 367-93-1) stock solution: Dissolve IPTG in deionized water and sterilize the solution by filtration. The stock solution is stored at –30°C.

3. Methods

The present indicators consist of EGFP fragments and the naturally split *Ssp* DnaE intein. EGFP is dissected at the position between amino acid residues 157 and 158. The resulting N- and C-terminal halves of EGFP fragments are named N-EGFP and C-EGFP, respectively. For efficient protein splicing, two amino acid mutations (K157Y and Q158C) are introduced into N- and C-EGFP. Calmodulin (CaM) and M13 are chosen as a model of interacting proteins. CaM and M13 are connected to N-DnaE and C-DnaE with flexible linkers ASNNGNGRNG and GNNGGNNDV (*see* **Note 3**), respectively (**Fig. 15.2b**). CaM–M13 interactions trigger the protein splicing reaction resulting in reconstitution of full-length EGFP. Evaluation of CaM–M13 interactions is performed by observation of fluorescence intensity from the reconstituted EGFP with a cooled CCD imaging system. The present method has an advantage that spliced GFP accumulates in a targeted cell until it is degraded. Therefore, this approach allows highly sensitive sensing of protein–protein interactions in *E. coli* (*see* **Note 4**).

3.1. Construction of DNA Plasmids

An *E. coli* strain DH5α as a bacterial host is used for construction of all the plasmids. pET-15b is an expression vector that allows expression of cDNA inserted into its multiple cloning site. Cells transformed with pET-15b vector are cultured in LB media containing 100 μg/mL ampicillin, because the cells acquire tolerance to ampicillin.

1. Each cDNA fragment shown in **Fig. 15.2** is amplified with PCR to introduce restriction enzyme sites and some mutations.
2. The sequences of the cDNAs are confirmed by a genetic analyzer (e.g., ABI310 available from Applied Biosystems, Foster, CA).

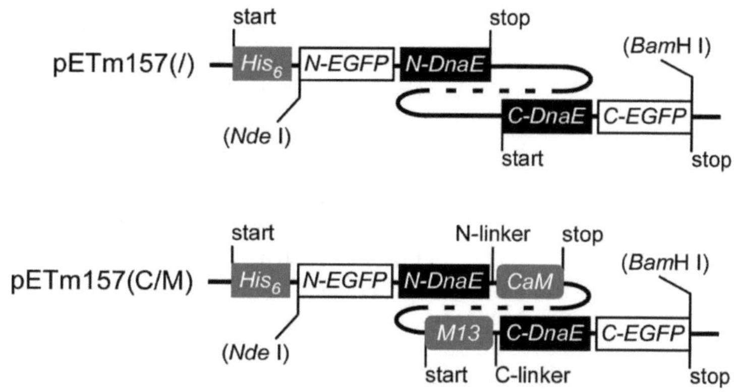

Fig. 15.2. Schematic structures of cDNA vector to be constructed. His_6 represents an epitope tag of polyhistidine. CaM and M13 are *Xenopus laevis* calmodulin and the CaM-binding protein derived from skeletal muscle myosin light-chain kinase. The proteins of interest are linked to *Ssp* DnaE intein fragments with peptide linkers: N-linker and C-linker are ASNNGNGRNG and GNNGGNNDV, respectively. The *dashed line* represents a Shine–Dalgarno sequence. The constructed cDNAs are cloned into the *Nde* I–*Bam*H I site of pET-15b.

3. The fragments of cDNAs are ligated and inserted into the *Nde* I–*Bam*H I site of pET-15b as illustrated in **Fig. 15.2** (*see* **Note 5**).

3.2. Protein Expression and Fluorescence Measurements

BL21(DE3)pLysS as a bacterial host is used for expressions of chimeric proteins.

BL21(DE3)pLysS is cultured in LB media containing 34 μg/mL chloramphenicol in order to retain pLysS in the host cells. pLysS is a pET-compatible plasmid that encodes T7 lysozyme to inhibit T7 RNA polymerase. The expressions of the recombinant fusion proteins are induced by the addition of IPTG (at a final concentration of 1.0 mM) to bacterial culture media.

1. Thaw tubes of competent cells on ice. Resuspend the competent cells gently.
2. Add 0.5 μg of purified plasmids directly to 40 μL of the cells. Mix the sample gently and return the tube to the ice.
3. Incubate the tube on ice for 5 min.
4. Heat the tube for 30 s in a 42°C water bath.
5. Place the tube on ice for 2 min.
6. Add 100 μL of LB medium or sterile deionized water to the tube and suspend the medium gently (*see* **Note 6**).
7. Spread the cell suspension on a LB agar dish containing 100 μg/mL ampicillin, 34 μg/mL chloramphenicol, and 1.0 mM IPTG.
8. Invert and incubate the dish for 12–16 h at 37°C.

9. Measure the fluorescence of transformed *E. coli* colonies by exposing the dish to a long-wavelength (470 nm) excitation with blue LED (e.g., LAS-1000plus, Fujifilm, Tokyo, Japan). The fluorescence is observed by a cooled charge-coupled device (CCD) with an emission filter (530DF30). The acquired images of the colonies are analyzed by an image processing software such as ImageGauge v.3.41 (Fujifilm) (*see* **Note 7**).

4. Notes

1. pAcGFP1 encodes a modified cDNA of *Aequorea coerulescens* GFP. We used pEGFP encoding the cDNA of enhanced GFP (EGFP) in the previous report (29). EGFP is a modified *Aequorea victoria* GFP.
2. Hereafter, we describe experimental procedures using calmodulin (CaM) and its target protein M13 as an interacting pair. Ternary complexes of CaM–M13–Ca^{2+} are formed at the physiological concentration ranges of Ca^{2+} in *E. coli* (28).
3. Flexible peptide linkers should be inserted between the protein of interest and the *Ssp* DnaE intein for efficient reconstitution of split GFP fragments by the protein splicing reaction. Preferable lengths of the peptide linker are 5–30 amino acids in the presented system. Here we used "GN" linkers that principally consist of flexible glycines and hydrophilic asparagines. Asparagines protect the linkers from aggregation in living cells. A "GS linker" is also applicable to this system. A "GS linker" is constituted of flexible glycines and soluble serines (e.g., GGGGSGGGGS (19, 21) and GGGSSGGG (18)).
4. Generally, it is desirable to validate protein–protein interaction with a couple of different methods in order to achieve higher accuracy in the analysis.
5. Each construct encodes a two-operon gene, since the constructs carry a cassette consisting of (translation termination codon)–(Shine–Dalgarno sequence)–(translation initiation codon) (30). We used "TGAGGAGGTTTAAAAT" as the Shine–Dalgarno sequence.
6. If selecting for kanamycin or streptomycin/spectinomycin resistance, shake or incubate the cells at 37°C for 30 min prior to plating on selective media.
7. An example of the result is shown in **Fig. 15.3**.

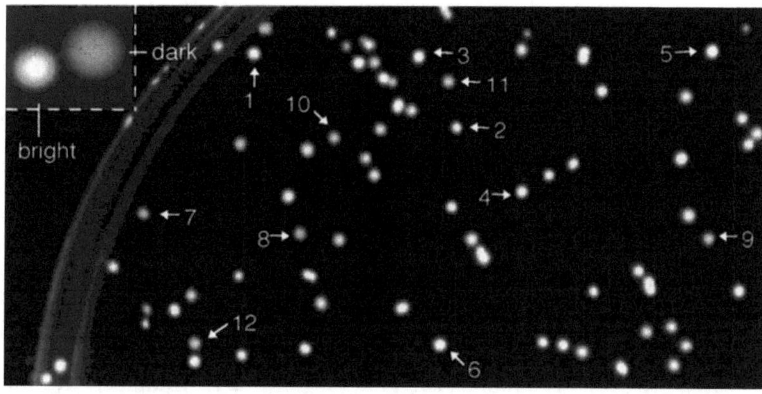

Fig. 15.3. A representative image of fluorescence from bacterial colonies on an LB agar plate. BL21(DE3)pLysS cells were transformed with a mixture of plasmids, pETm157(/) and pETm157(C/M). The LB agar plate contained 1.0 mM IPTG, 0.1 mM ampicillin, and 0.1 mM chloramphenicol. The transformed cells were cultured on the plate for 16 h at 37°C. Inset is a magnified image of bright and dark colonies on the plate. The cells of the bright colony carried pETm157(C/M), while the darker one carried pETm157(/) (14).

Acknowledgment

This work was financially supported by Core Research for Evolutional Science and Technology (CREST) of Japan Science and Technology Agency (JST) and Japan Society for the Promotion of Science (JSPS).

References

1. Blattner, F. R., Plunkett, G., Bloch, C. A., et al. (1997) The complete genome sequence of *Escherichia coli* K-12. *Science* 277, 1453–1474.
2. Goffeau, A., Barrell, B. G., Bussey, H., et al. (1996) Life with 6000 genes. *Science* 274, 546–567.
3. The C. *elegans* Sequencing Consortium (1998) Genome sequence of the nematode *C. elegans*: a platform for investigating biology. *Science* 282, 2012–2018.
4. Arabidopsis Genome, I. (2000) Analysis of the genome sequence of the flowering plant *Arabidopsis thaliana*. *Nature* 408, 796–815.
5. Adams, M. D., Celniker, S. E., Holt, R. A., et al. (2000) The genome sequence of *Drosophila melanogaster*. *Science* 287, 2185–2195.
6. International Human Genome Sequencing Consortium (2004) Finishing the euchromatic sequence of the human genome. *Nature* 431, 931–945.
7. Walhout, A. J. M., Vidal, M. (2001) Protein interaction maps for model organisms. *Nat Rev Mol Cell Biol* 2, 55–62.
8. Johnsson, N., Varshavsky, A. (1994) Split ubiquitin as a sensor of protein interactions in-vivo. *Proc Natl Acad Sci USA* 91, 10340–10344.
9. Stagljar, I., Korostensky, C., Johnsson, N., te Heesen, S. (1998) A genetic system based on split-ubiquitin for the analysis of interactions between membrane proteins in vivo. *Proc Natl Acad Sci USA* 95, 5187–5192.
10. Dunnwald, M., Varshavsky, A., Johnsson, N. (1999) Detection of transient in vivo interactions between substrate and transporter during protein translocation into the endoplasmic reticulum. *Mol Biol Cell* 10, 329–344.

11. Remy, I., Michnick, S. W. (1999) Clonal selection and in vivo quantitation of protein interactions with protein-fragment complementation assays. *Proc Natl Acad Sci USA* 96, 5394–5399.
12. Pelletier, J. N., Arndt, K. M., Pluckthun, A., Michnick, S. W. (1999) An in vivo library-versus-library selection of optimized protein-protein interactions. *Nat Biotechnol* 17, 683–690.
13. Rossi, F., Charlton, C. A., Blau, H. M. (1997) Monitoring protein–protein interactions in intact eukaryotic cells by beta-galactosidase complementation. *Proc Natl Acad Sci USA* 94, 8405–8410.
14. Wehrman, T., Kleaveland, B., Her, J. H., Balint, R. F., Blau, H. M. (2002) Protein–protein interactions monitored in mammalian cells via complementation of beta-lactamase enzyme fragments. *Proc Natl Acad Sci USA* 99, 3469–3474.
15. Galarneau, A., Primeau, M., Trudeau, L. E., Michnick, S. W. (2002) beta-Lactamase protein fragment complementation assays as in vivo and in vitro sensors of protein–protein interactions. *Nat Biotechnol* 20, 619–622.
16. Magliery, T. J., Wilson, C. G. M., Pan, W. L., et al. (2005) Detecting protein–protein interactions with a green fluorescent protein fragment reassembly trap: scope and mechanism. *J Am Chem Soc* 127, 146–157.
17. Nyfeler, B., Michnick, S. W., Hauri, H. P. (2005) Capturing protein interactions in the secretory pathway of living cells. *Proc Natl Acad Sci USA* 102, 6350–6355.
18. Luker, K. E., Smith, M. C. P., Luker, G. D., Gammon, S. T., Piwnica-Worms, H., Piwnica-Worms, D. P. (2004) Kinetics of regulated protein–protein interactions revealed with firefly luciferase complementation imaging in cells and living animals. *Proc Natl Acad Sci USA* 101, 12288–12293.
19. Kim, S. B., Kanno, A., Ozawa, T., Tao, H., Umezawa, Y. (2007) Nongenomic activity of ligands in the association of androgen receptor with Src *Acs. Chem Biol* 2, 484–492.
20. Kaihara, A., Kawai, Y., Sato, M., Ozawa, T., Umezawa, Y. (2003) Locating a protein–protein interaction in living cells via split *Renilla luciferase* complementation. *Anal Chem* 75, 4176–4181.
21. Paulmurugan, R., Gambhir, S. S. (2003) Monitoring protein–protein interactions using split synthetic *Renilla luciferase* protein-fragment-assisted complementation. *Anal Chem* 75, 1584–1589.
22. Remy, I., Michnick, S. W. (2006) A highly sensitive protein–protein interaction assay based on *Gaussia luciferase. Nat Methods* 3, 977–979.
23. Joung, J. K. (2001) Identifying and modifying protein-DNA and protein–protein interactions using a bacterial two-hybrid selection system. *J Cell Biochem* Supplement 37, 53–57.
24. Joung, J. K., Ramm, E. I., Pabo, C. O. (2000) A bacterial two-hybrid selection system for studying protein-DNA and protein–protein interactions *Proc Natl Acad Sci USA* 97, 7382–7387.
25. Shaywitz, A. J., Dove, S. L., Kornhauser, J. M., Hochschild, A., Greenberg, M. E. (2000) Magnitude of the CREB-dependent transcriptional response is determined by the strength of the interaction between the kinase-inducible domain of CREB and the KIX domain of CREB-binding protein. *Mol Cell Biol* 20, 9409–9422.
26. Michnick, S. W., Ear, P. H., Manderson, E. N., Remy, I., Stefan, E. (2007) Universal strategies in research and drug discovery based on protein-fragment complementation assays. *Nat Rev Drug Discov* 6, 569–582.
27. Kerppola, T. K. (2008) Bimolecular fluorescence complementation (BiFC) analysis as a probe of protein interactions in living cells. *Annu Rev Biophys Biomol Struct* 37, 465–487.
28. Ozawa, T., Nogami, S., Sato, M., Ohya, Y., Umezawa, Y. (2000) A fluorescent indicator for detecting protein–protein interactions in vivo based on protein splicing. *Anal Chem* 72, 5151–5157.
29. Ozawa, T., Takeuchi, M., Kaihara, A., Sato, M., Umezawa, Y. (2001) Protein splicing-based reconstitution of split green fluorescent protein for monitoring protein–protein interactions in bacteria: improved sensitivity and reduced screening time. *Anal Chem* 73, 5866–5874.
30. Wu, H., Xu, M. Q., Liu, X. Q. (1998) Protein trans-splicing and functional mini-inteins of a cyanobacterial dnaB intein. *Biochim Biophys Acta* 1387, 422–432.

Chapter 16

Enhancing the Solubility of Recombinant Proteins in *Escherichia coli* by Using Hexahistidine-Tagged Maltose-Binding Protein as a Fusion Partner

Ping Sun, Joseph E. Tropea, and David S. Waugh

Abstract

In the field of biotechnology, fusing recombinant proteins to highly soluble partners is a common practice for overcoming aggregation in *Escherichia coli*. *E. coli* maltose-binding protein (MBP) has been recognized as one of the most effective solubilizing agents, having frequently been observed to improve the yield, enhance the solubility, and promote the proper folding of its fusion partners. The use of a dual hexahistidine–maltose-binding protein affinity tag (His_6–MBP) has the additional advantage of allowing the fusion protein to be purified by immobilized metal affinity chromatography (IMAC) instead of or in addition to amylose affinity chromatography. This chapter describes a generic method for the overproduction of combinatorially tagged His_6–MBP fusion proteins in *E. coli*, with particular emphasis on the use of recombinational cloning to construct expression vectors. In addition, simple methods for evaluating the solubility of the fusion protein and the passenger protein after it is cleaved from the dual His_6–MBP tag are presented.

Key words: Maltose-binding protein, MBP, solubility enhancer, His tag, His_6–MBP, Gateway cloning, TEV protease, tobacco etch virus protease.

1. Introduction

For functional and structural studies, recombinant proteins usually need to be purified on a relatively large scale and under native (nondenaturing) conditions. Originally developed to facilitate the detection and purification of recombinant proteins, affinity tags and other types of genetically engineered fusion partners are widely used in the biotechnology industry (1). However,

T.C. Evans, M.-Q. Xu (eds.), *Heterologous Gene Expression in E. coli*, Methods in Molecular Biology 705,
DOI 10.1007/978-1-61737-967-3_16, © Springer Science+Business Media, LLC 2011

gradually it has become evident that certain tags have additional benefits, such as a positive impact on the yield of recombinant protein, the enhancement of solubility, and even the promotion of the proper folding of their fusion partners (2). *Escherichia coli* maltose-binding protein (MBP) is one of the most effective solubility enhancers (3, 4). Moreover, MBP is the only solubility-enhancing protein that is also a natural affinity tag. Unfortunately, however, amylose affinity chromatography has proven to be problematic at times, not only because of the relatively low binding capacity of amylose resin for MBP but also because of persistent contaminants that require additional chromatographic steps to remove (5–7). We have demonstrated that a hexahistidine tag (His_6) can be added to the N-terminus of MBP without interfering with its ability to promote the solubility of its fusion partners (8). Therefore, IMAC can be used to circumvent the problems associated with amylose affinity chromatography.

In this chapter, we illustrate the utility of a dual His_6–MBP tag as a solubility-enhancing fusion partner. In our laboratory, we routinely construct expression vectors by recombinational cloning, as described here. However, alternative methods for constructing MBP and His_6–MBP fusion vectors by conventional cloning techniques (i.e., using restriction enzymes and DNA ligase) or by ligation independent cloning (LIC) have been described elsewhere (9–13), as have methods for the large-scale production and purification of MBP and His_6–MBP fusion proteins (10, 14–16).

2. Materials

2.1. Recombinational Vector Construction

1. The Gateway destination vector pDEST–HisMBP, which can be obtained from AddGene (http://www.addgene.org) or the authors.
2. The Gateway donor vector pDONR221 (Invitrogen, Carlsbad, CA, USA).
3. Chemically competent DB3.1 or one shot CcdB survival competent cells (Invitrogen, Carlsbad, CA, USA) for propagating pDEST–HisMBP and pDONR221 plasmids.
4. Competent *gyrA+* cells (e.g., DH5α, MC1061, HB101) (*see* **Note 1**).
5. Gateway BP Clonase II (Invitrogen, Carlsbad, CA, USA) (*see* **Note 2**).
6. Gateway LR Clonase II (Invitrogen, Carlsbad, CA, USA) (*see* **Note 2**).

7. Reagents and thermostable DNA polymerase for PCR amplification (*see* **Note 3**).
8. Synthetic oligodeoxyribonucleotide primers for PCR amplification (*see* **Fig. 16.1**).
9. TE buffer (10 mM Tris-HCl [pH 8.0], 1 mM EDTA).
10. 1000-Fold antibiotic stock solutions: Kanamycin, 35 mg/ml in H_2O and filter sterilize. Ampicillin, 100 mg/ml in H_2O and filter sterilize. Store at –20°C.
11. LB medium and LB agar plates containing ampicillin (100 μg/ml). LB medium: dissolve 10 g bacto tryptone, 5 g bacto yeast extract, and 5 g NaCl in 1 l of H_2O and sterilize by autoclaving. For LB agar, also add 12 g of bactoagar before autoclaving. To prepare plates, allow medium to cool until flask or bottle can be held in hands without burning (~50°C), then add 1 ml 1000-fold ampicillin stock solution, mix by gentle swirling, and pour or pipet 30 ml into each sterile petri dish (100 mm diameter).
12. E-gels and an E-gel base (Qiagen, Valencia, CA, USA) for submarine gel electrophoresis of DNA (*see* **Note 4**).
13. QIAquick gel extraction kit (Qiagen, Valencia, CA, USA) for the extraction of DNA from agarose gels.
14. QIAprep Spin Miniprep Kit (Qiagen, Valencia, CA, USA) for small-scale plasmid DNA isolation (*see* **Note 5**).
15. A microcentrifuge capable of operating at 14,000 rpm.
16. An incubator set at 37°C.

2.2. Protein Expression

1. Competent BL21-Pro cells (B&D Clontech, Palo Alto, CA, USA) containing the TEV protease expression vector pRK603 (17). pRK603 plasmid can be obtained from AddGene (http://www.addgene.org) (*see* **Notes 6** and **7**).
2. Competent BL21(DE3) CodonPlus-RIL cells (Stratagene, La Jolla, CA) (*see* **Note 7**).
3. A derivative of pDEST–HisMBP that produces a His_6–MBP fusion protein with a TEV protease recognition site in the linker between MBP and the passenger protein (*see* **Section 3.1**).
4. LB agar plates and broth containing both ampicillin (100 μg/ml) and kanamycin (35 μg/ml). *See* **Section 2.1**, **Step 11** for LB broth and LB agar recipes.
5. 1000-Fold chloramphenicol stock solution: 30 mg/ml in ethanol and filter sterilize, store at –20°C.
6. Isopropyl-β-D-thiogalactopyranoside (IPTG), analytical grade. Prepare a 1000-fold stock solution of 1 M in H_2O and filter sterilize. Store at –20°C.

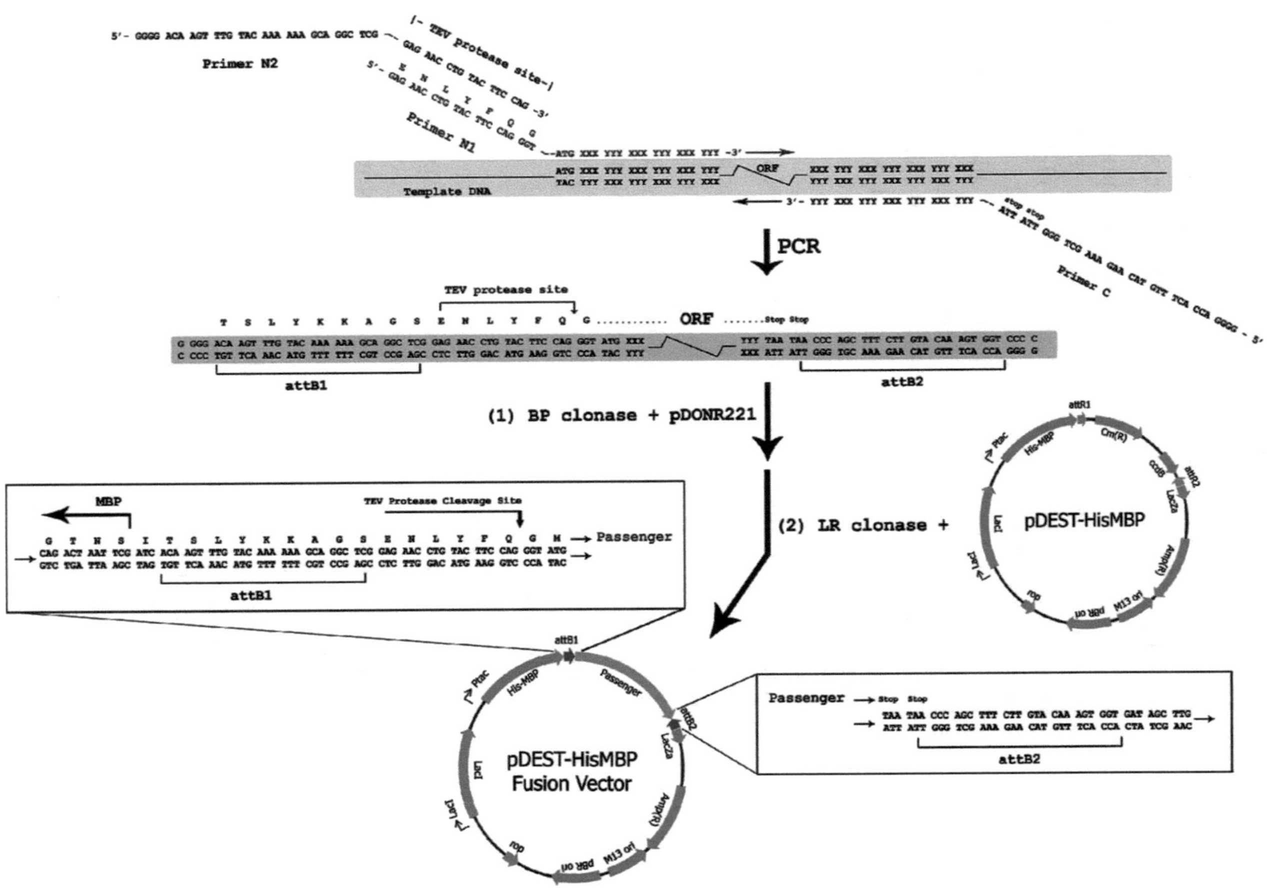

Fig 16.1. Construction of a pDEST–HisMBP fusion protein expression vector using PCR and Gateway cloning technology. The ORF of interest is amplified from the template DNA by PCR, using primers N1, N2, and C. Primers N1 and C are designed to base pair to the 5′- and 3′-ends of the coding region, respectively, and contain unpaired 5′-extensions as shown. Primer N2 base pairs with the sequence that is complementary to the unpaired extension of primer N1. The final PCR product is recombined with the pDONR221 vector to generate an entry clone, via the BP reaction. This entry clone is subsequently recombined with pDEST–HisMBP using LR Clonase to yield the final pDEST–HisMBP fusion vector. Abbreviations: Ptac, tac promoter; attR1/attR2 and attB1/attB2, recombination sites for Gateway cloning; Cm(R), chloramphenicol acetyl transferase (chloramphenicol resistance) gene; ccdB, gene encoding DNA gyrase poison CcdB; LacZα, gene encoding β-galactosidase; Amp(R), β-lactamase (ampicillin resistance) gene; M13 ori, origin of replication from bacteriophage M13; pBR ori, ColE1 origin of replication; rop, repressor of primer gene; lacI, gene encoding lactose repressor.

7. Anhydrotetracycline (ACROS Organics/Fisher Scientific, Springfield, NJ, USA). Prepare a 1000-fold stock solution by dissolving in 50% ethanol at 100 μg /ml. Store in a foil-covered tube at –20°C.
8. Temperature-controlled shaking incubator.
9. Sterile baffle-bottom flasks (Bellco Inc., Vineland, NJ, USA).
10. Cell lysis buffer: 20 mM Tris-HCl (pH 8.0), 1 mM EDTA.
11. Sonicator (with microtip).
12. 2x SDS-PAGE sample buffer (Invitrogen, Carlsbad, CA, USA) and 2-mercaptoethanol (Sigma-Aldrich, St. Louis, MO, USA).
13. SDS-PAGE gel, electrophoresis apparatus, and running buffer (*see* **Note 8**).
14. Gel stain (e.g., Gelcode® Blue from Pierce, Rockford, IL, USA, or PhastGel™ Blue R from GE Healthcare, Piscataway, NJ, USA).
15. Spectrophotometer.
16. 1.5 ml microcentrifuge tubes.

3. Methods

3.1. Construction of His_6–MBP Fusion Vectors by Recombinational Cloning

The Gateway recombinational cloning system is based on the site-specific recombination reactions that mediate the integration and excision of bacteriophage lambda into and from the *E. coli* chromosome, respectively. Please refer to the following manual by Invitrogen for detailed information: "Gateway Technology: A universal technology to clone DNA sequences for functional analysis and expression in multiple systems." Complete information about the Gateway system can be found at the following web site: http://www.invitrogen.com/site/us/en/home/Products-and-Services/Applications/Cloning/Gateway-Cloning.html.

3.1.1. pDEST–HisMBP

Currently there are no commercially available Gateway destination vectors for the production of His_6–MBP fusion proteins. However, pDEST–HisMBP can be obtained from the non-profit distributor of biological reagents AddGene, Inc. (http://www.addgene.org). A schematic diagram of pDEST–HisMBP is shown in **Fig. 16.1**. This plasmid was constructed by inserting an in-frame hexahistidine coding sequence between codons 3 and 4 of MBP in pKM596 (11).

The Gateway cloning cassette in pDEST–HisMBP carries a gene encoding the DNA gyrase poison CcdB, which provides a negative selection against non-recombined destination and donor vectors so that only the desired recombinant is obtained when the end products of the recombinational cloning reaction are transformed into *E. coli* and grown in the presence of ampicillin or kanamycin, respectively. pDEST–HisMBP and other vectors that carry the *ccdB* gene must be propagated in a host strain with a *gyrA* mutation (e.g., *E. coli* DB3.1) or in "CcdB Survival cells" that render the cells immune to the action of CcdB.

3.1.2. Gateway Cloning Protocol

To construct a His_6–MBP fusion protein expression vector, three primers (two forward primers and one reverse primer) are employed: Primer N1 (containing the TEV protease cleavage site as a 5′-unpaired extension), primer N2 (consisting of the TEV protease recognition site and an attB1 recombination site), and primer C (containing an attB2 recombination site as a 5′-unpaired extension) (**Fig. 16.1**). For the 3′-end of primer N1 and primer C, approximately 20–25 nucleotides that are complimentary to the ends of the open reading frame are required. The open reading frame of interest first is amplified using primers N1 and C. The first PCR amplicon then becomes the template for PCR with primers N2 and C. The PCR reaction is be performed in one step containing all three primers (*see* **Note 9**). However, to favor the accumulation of the desired product, the attB-containing primers are used at typical concentrations for PCR but the concentration of the gene-specific N-terminal primer N1 is 10- to 20-fold lower. The final PCR amplicon is inserted first into the donor vector pDONR221 by recombinational cloning with BP clonase and then into the destination vector pDEST–HisMBP in a second recombinational cloning reaction with LR clonase.

1. The PCR mix is prepared as follows (*see* **Note 10**): 1 μl template DNA (~10 ng/μl), 10 μl thermostable DNA polymerase 10X reaction buffer, 1.5 μl dNTP solution (10 mM each), 1.0 μl primer N1 (~30 ng), 3 μl primer N2 (~300 ng), 3.0 μl primer C (~300 ng), 1 μl thermostable DNA polymerase, 79.5 μl H_2O (to 100 μl total volume).
2. The reaction is placed in the PCR thermal cycler with the following program: initial melt for 5 min at 95°C; 35 cycles of 95°C for 30 s, 55°C for 30 s, and 68°C for 60 s (*see* **Notes 10** and **11**); hold at 4°C.
3. Purification of the PCR amplicon by agarose gel electrophoresis (*see* **Note 4**) is recommended to remove attB primer dimers.
4. To create the His_6–MBP fusion vector, the PCR product is recombined first into pDONR221 to yield an entry clone

intermediate (BP reaction), and then into pDEST–HisMBP (LR reaction; *see* **Note 12**).

a. Add to a microcentrifuge tube on ice: 50–100 ng of the PCR product in TE or H_2O, 200 ng of pDONR221 DNA, and enough TE to bring the total volume to 8 μl. Mix well.

b. Thaw BP Clonase II enzyme mix on ice (2 min) and then vortex briefly (2 s) twice.

c. Add 2 μl of BP Clonase II enzyme mix to the components in (a.) and vortex briefly twice.

d. Incubate the reaction at 25°C for at least 4 h (*see* **Note 13**).

e. Add to the reaction: 2 μl of the destination vector (pDEST–HisMBP) at a concentration of 50 ng/μl and 3 μl of LR Clonase II enzyme mix. The final reaction volume is 15 μl. Mix by vortexing briefly.

f. Incubate the reaction at room temperature for 3–4 h.

g. Add 2.5 μl of the proteinase K solution and incubate for 10 min at 37°C.

h. Transform 2 μl of the reaction into 50 μl of electrocompetent DH5α cells (*see* **Note 1**).

i. Pellet the cells by centrifugation, gently resuspend pellet in 100–200 μl of LB broth, and spread on an LB agar plate containing ampicillin (100 μg/ml), the selective marker for pDEST–HisMBP (*see* **Fig. 16.1**). Incubate the plate at 37°C overnight (*see* **Note 14**).

5. Plasmid DNA is isolated from saturated cultures started from individual ampicillin-resistant colonies and screened by PCR, using the gene-specific primers N1 and C, to confirm that the clones contain the expected gene. Alternatively, plasmids can be purified and screened by conventional restriction digests using appropriate enzymes. We routinely sequence clones that screen positive by either PCR or restriction digest to ensure that there are no PCR-induced mutations.

3.2. Protein Expression

To assess the yield and solubility of the fusion protein, the amount of total fusion protein produced in the crude cell extract is directly compared to the soluble fraction by visual inspection of a Coomassie blue-stained gel. A parallel in vivo cleavage experiment with TEV protease is run to determine if the fusion protein is a good substrate for the protease and whether or not the cleaved target protein remains soluble after it is released from His_6–MBP.

3.2.1. Selecting a Host Strain of *E. coli*

To achieve regulated expression of TEV protease, the in vivo processing experiment must be performed in a strain of *E. coli* that produces the Tet repressor, such as BL21-Pro or DH5α-Pro (B&D Clontech, Palo Alto, CA, USA). The Tet repressor blocks the synthesis of TEV protease mRNA and allows the enzyme to be regulated independently of the IPTG-inducible fusion protein. We have observed that delaying the induction of TEV protease until the fusion protein substrate has had time to accumulate in the cells often results in greater solubility of the passenger protein after cleavage (11, 17). Independent production of TEV protease from the expression vector pRK603 (17) is initiated by adding anhydrotetracycline to the cell culture, usually 2 h after induction of the fusion protein with IPTG. We prefer using BL21-Pro because of its robust growth characteristics and the fact that it lacks two proteases (Lon and OmpT) that are present in many *E. coli* K12 strains such as DH5α-Pro.

For the large-scale production of His_6–MBP fusion proteins, we prefer BL21(DE3) CodonPlus-RIL (Stratagene, La Jolla, CA), or Rosetta (DE3) (EMD, Madison, WI, USA) as host strains (*see* **Note 7**).

3.2.2. Protein Expression

1. Transform competent BL21-Pro or DH5a-Pro cells that already contain pRK603 with the His_6–MBP fusion protein expression vector and spread them on LB agar plates containing ampicillin (100 μg/ml) and kanamycin (35 μg/ml). Incubate the plate overnight at 37°C.
2. Inoculate 2–5 ml of LB medium containing ampicillin (100 μg/ml) and kanamycin (35 μg/ml) in a culture tube or shake flask with a single colony from the plate. Grow to saturation overnight at 37°C with shaking at 250 rpm.
3. The next morning, inoculate 50 ml of the same medium in a 250 ml baffled-bottom flask with 0.5 ml of the saturated overnight culture.
4. Grow the cells at 37°C with shaking to mid-log phase ($OD_{600\ nm}$ ~ 0.5).
5. Add IPTG (1 mM final concentration) and adjust the temperature to 30°C (*see* **Note 15**).
6. After 2 h, divide the culture into two separate flasks (20 ml in each). Label one flask "+" and the other "–".
7. Add anhydrotetracycline to the "+" flask (100 ng/ml final concentration).
8. After 2 h, measure the $OD_{600\ nm}$ of the cultures (dilute cells 1:10 in LB to obtain an accurate reading). An $OD_{600\ nm}$ of about 3–3.5 is normal, although lower densities are possible. If the density of either culture is much lower than this, it may be necessary to adjust the volume of the samples that are analyzed by SDS-PAGE.

Transfer 10 ml of each culture to a 15 ml conical centrifuge tube and pellet the cells by centrifugation (4000 × *g*) at 4°C.

9. Resuspend the cell pellets in 1 ml of lysis buffer and then transfer the suspensions to a 1.5 ml microcentrifuge tube.
10. Store the cell suspensions at –80°C overnight. Alternatively, the cells can be disrupted immediately by sonication (without freezing and thawing) and the procedure continued without interruption, as described below.

3.2.3. Sonication and Sample Preparation

1. Thaw the cell suspensions at room temperature, then place them on ice.
2. Lyse the cells by sonication (*see* **Note 16**).
3. Prepare samples of the total intracellular protein from the "+" and "–" cultures (T+ and T–, respectively) for SDS-PAGE by mixing 50 μl of each sonicated cell suspension from step 2 with 50 μl of 2x SDS-PAGE sample buffer containing 10% (v/v) 2-mercaptoethanol.
4. Pellet the insoluble cell debris (and proteins) by centrifuging the sonicated cell suspension from each culture at maximum speed in a microcentrifuge for 10 min at 4°C.
5. Prepare samples of the soluble intracellular protein from the "+" and "–" cultures (S+ and S–, respectively) for SDS-PAGE by mixing 50 μl of each supernatant from step 4 with 50 μl of 2x SDS-PAGE sample buffer containing 20% (v/v) 2-mercaptoethanol.

3.2.4. SDS-PAGE

We typically use pre-cast Tris-glycine or NuPAGE gradient gels for SDS-PAGE to assess the yield and solubility of MBP fusion proteins (*see* **Note 8**). Of course, the investigator is free to choose any appropriate SDS-PAGE formulation, depending on the protein size and laboratory preference.

1. Heat the T–, T+, S–, and S+ protein samples at 90°C for about 5 min and then spin them at maximum speed in a microcentrifuge for 5 min.
2. Assemble the gel in the electrophoresis apparatus, fill it with SDS-PAGE running buffer, load the samples (10 μl each), and carry out the electrophoretic separation according to standard lab practices. T and S samples from each culture ("+" and "–") are loaded in adjacent lanes to allow easy assessment of solubility. Molecular weight standards may also be loaded on the gel, if desired.
3. Stain the proteins in the gel with GelCode® Blue reagent, PhastGel™ Blue R, or a suitable alternative.

3.2.5. Interpreting the Results

The MBP fusion protein should be readily identifiable in the T– sample after the gel is stained since it will normally be the most abundant protein in the cells. Molecular weight standards can also be used to corroborate the identity of the fusion protein band. If the S– sample contains a similar amount of the fusion protein, this indicates that it is highly soluble in *E. coli.* If little or no fusion protein is observed in the S– sample, then this is an indicator of poor solubility. Of course, a range of intermediate states is also possible.

If the fusion protein is an efficient substrate for TEV protease, then little of it will be present in the T+ and S+ samples. Instead, one should observe a prominent band at ca. 42 kDa that corresponds to the His_6–MBP moiety and another prominent band migrating with the expected mobility of the passenger protein. If the fusion protein is a poor substrate for the protease, then the "+" samples will look similar to the "–" samples. If the passenger protein is soluble after it is released from His_6–MBP, then a similar amount will be present in the T+ and S+ lanes. At this point, some or all of the passenger protein may precipitate. If a substantial fraction of the passenger protein is insoluble, then troubleshooting may be necessary. Alternatively, an acceptable yield might still be obtained by scaling up cell production.

Two examples are illustrated in **Fig. 16.2**. In panel A, the second pair of lanes represent the total (T) and soluble (S) intracellular protein fractions from cells overproducing His_6–MBP fused to residues 2–191 of human dual specificity phosphatase 14 (DUSP14). Roughly equal amounts of the fusion protein are readily visible in these two lanes, indicating that it is highly soluble. In the third pair of lanes, the His_6–DUSP14(2–191) fusion protein was cleaved in vivo with TEV protease 2 h after its expression was induced by IPTG. In this case, most if not all of the cleaved DUSP14(2–191) is absent from the soluble fraction (S). Hence, this is an example of a protein that is rendered only temporarily soluble while it is fused to MBP and may not be properly folded. The first pair of lanes in panel A demonstrate that when DUSP14(2–191) is fused to a His tag alone (with no MBP), it is completely insoluble in *E. coli.* Panel B in **Fig. 16.2** shows exactly the same experiment, except that in this case the DUSP14 has been further truncated at its N-terminus (residues 18–191). As before, the His-tagged DUSP14(18–191) is totally insoluble whereas the His_6–MBP–DUSP14(18–191) fusion protein is highly soluble. The difference is that in this case nearly all of the DUSP14(18–191) is soluble after it has been cleaved from the His_6–MBP tag, suggesting that it has become properly folded as a result being fused to His_6–MBP. Indeed, DUSP14(18–191) produced in this manner readily hydrolyzes the phosphotyrosine mimetic *para*-nitrophenyl phosphate (data not shown).

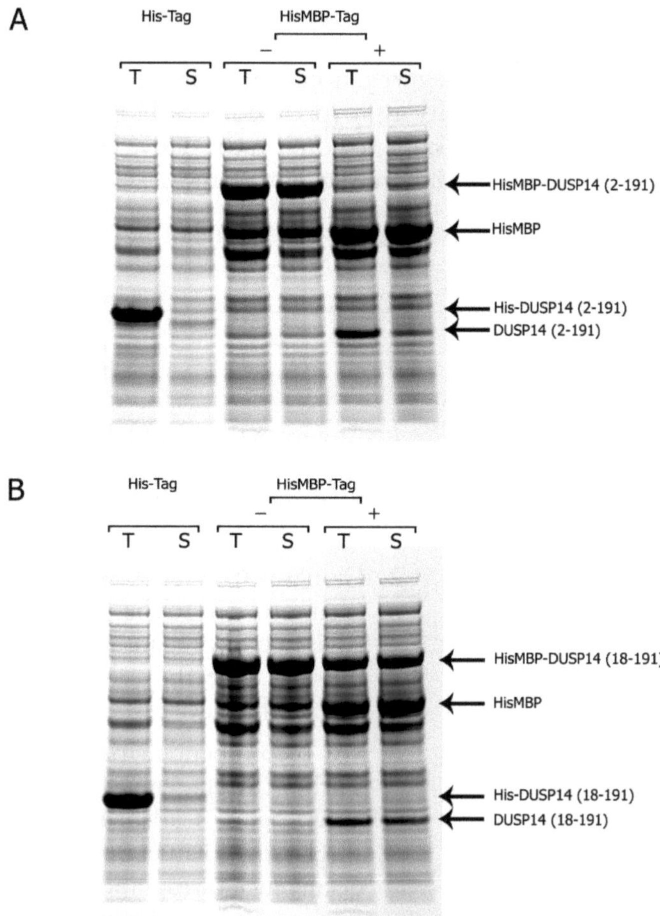

Fig 16.2. Overproduction and intracellular processing of pDEST–HisMBP fusion proteins by TEV protease. DUSP14(2–191) and DUSP14(18–191) were expressed from derivatives of pDEST–HisMBP in BL21-Pro cells that also contained the TEV protease expression vector pRK603 as described (*see* **Section 3.2**). "T" and "S" refer to the total and soluble fractions of the intracellular protein, respectively. All cultures were induced with IPTG to initiate the production of the His_6–MBP fusion proteins. Samples marked "+" were induced with anhydrotetracycline to initiate the production of TEV protease 2 h after the addition of IPTG, whereas samples marked "–" were not induced with anhydrotetracycline. The results of this experiment are discussed in **Section 3.2.5**.

Normally, once it has been established that an aggregation-prone passenger protein can be rendered soluble by fusing it to His_6–MBP, that the fusion protein can be cleaved by TEV protease, and that the passenger protein remains soluble after it is released from His_6–MBP, then the passenger protein is ready to be purified on a large scale. Detailed instructions for purification using a generic IMAC-based protocol have been described elsewhere (15). However, one should be aware that occasionally a passenger protein may accumulate in a soluble but inactive form. Exactly how and why this occurs is unclear. Fusion to MBP may enable certain proteins to evolve into kinetically trapped folding intermediates that are no longer susceptible to aggregation. Therefore, it is recommended that a biological activity assay (preferably) or biophysical techniques be used at an early stage to confirm the native conformation of the passenger protein before starting a large-scale purification.

4. Notes

1. Any gyrA+ strain of *E. coli* can be used. We prefer competent DH5α cells (Invitrogen, Carlsbad, CA, USA) because they are easy to use and have high transformation efficiencies.
2. Clonase enzyme mixes should be thawed quickly on ice and then returned to the –80°C freezer as soon as possible. It is advisable to prepare multiple aliquots of the enzyme mixes the first time they are thawed in order to avoid repeated freeze–thaw cycles.
3. We recommend a proofreading polymerase such as Pfu Turbo (Stratagene, La Jolla, CA, USA), Platinum Pfx (Invitrogen, Carlsbad, CA, USA), or Deep Vent (New England Biolabs, Beverly, MA, USA) to minimize the occurrence of mutations during PCR.
4. We typically purify fragments by electrophoresis using precast E-gels purchased from Invitrogen. However, suitable equipment and reagents for horizontal agarose gel electrophoresis can be purchased from a wide variety of scientific supply companies. DNA fragments are extracted from slices of the ethidium bromide-stained gel using a QIAquick gel extraction kit (Qiagen, Valencia, CA, USA) in accordance with the instructions supplied with the product.
5. We prefer the QIAprep Spin miniprep kit (Qiagen, Valencia, CA, USA), but similar kits can be obtained from a wide variety of vendors.
6. We prefer to use electrocompetent cells because of the high transformation efficiency that can be achieved (18). Detailed protocols for the preparation of electrocompetent cells and electrotransformation procedures can be obtained from the electroporator manufacturers (e.g., Bio-Rad, BTX, Eppendorf). Briefly, the cells are grown in 1l of LB medium (with antibiotics, if appropriate) to mid-log phase ($OD_{600\ nm}$ ~0.5) and then chilled on ice. The cells

are pelleted at 4°C, resuspended in 1 l of ice-cold H_2O, and then pelleted again. After several such washes with H_2O, the cells are resuspended in 3–4 ml of 10% glycerol, divided into 50 μl aliquots, and then immediately frozen in a dry ice/ethanol bath. The electrocompetent cells are stored at –80°C. Immediately prior to electrotransformation, the cells are thawed on ice and mixed with 10–100 ng of DNA (e.g., a plasmid vector or a Gateway reaction). The mixture is placed into an ice-cold electroporation cuvette and electroporated according to the manufacturer's recommendations (e.g., a 1.5 kV pulse in a cuvette with a 1 mm gap). Immediately add 0.450 ml of SOC medium (18) to the cells and allow them to grow at 37°C with shaking (ca. 250 rpm) for 1 h. Then spread 5–200 μl of the cells on an LB agar plate containing the appropriate antibiotic(s).

7. If the open reading frame encoding the passenger protein contains codons that are rarely used in *E. coli* (http://www.doe-mbi.ucla.edu/cgi/cam/racc.html), this can adversely affect the yield of an MBP fusion protein. In such cases, it is advisable to introduce an additional plasmid into the host cells that carry the cognate tRNA genes for rare codons. The pRIL plasmid (Stratagene, La Jolla, CA, USA) is a derivative of the p15A replicon that carries the *E. coli argU*, *ileY*, and *leuW* genes, which encode the cognate tRNAs for AGG/AGA, AUA, and CUA codons, respectively. pRIL is selected for by resistance to chloramphenicol. In addition to the tRNA genes for AGG/AGA, AUA, and CUA codons, the pRARE accessory plasmid in the Rosetta™ host strain (EMD, Madison, WI, USA) also includes tRNAs for the rarely used CCC and GGA codons. Like pRIL, the pRARE plasmid is a chloramphenicol-resistant derivative of the p15A replicon. Both of these tRNA accessory plasmids are compatible with derivatives of pDEST–HisMBP. On the other hand, they are incompatible with the vector pRK603 that we use for intracellular processing experiments (*see* **Section 3.2.1**). Nevertheless, because pRK603 and the tRNA accessory plasmids have different antibiotic resistance markers, it is possible to force cells to maintain both plasmids by simultaneously selecting for kanamycin and chloramphenicol resistance. Alternatively, the kanamycin-resistant TEV protease expression vector pKM586, a pRK603 derivative with the replication machinery of a pSC101 replicon, which can be obtained from the authors, can be stably maintained in conjunction with p15A-type tRNA plasmids.

8. We find it convenient to use pre-cast gels for SDS-PAGE gels (e.g., 1.0 mm × 10 well, 10–20% Tris-glycine

gradient), running buffer, and electrophoresis supplies from Invitrogen (Carlsbad, CA, USA).

9. Alternatively, the PCR reaction can be performed in two separate steps, using primers N1 and C in the first step and primers N2 and C in the second step. The PCR amplicon from the first step is used as the template for the second PCR. All primers are used at the typical concentrations for PCR in the two-step protocol.
10. The PCR reaction can be modified in numerous ways to optimize results, depending on the nature of the template and primers. *See* (18) (Vol. 2, Chapter 8) for more information.
11. PCR cycle conditions can also be varied. For example, the extension time should be increased for especially long genes. A typical rule of thumb is to extend for 60 s/kb of DNA.
12. This "one-tube" Gateway protocol bypasses the isolation of an "entry clone" intermediate. However, the entry clone may be useful if the investigator intends to experiment with additional Gateway destination vectors, in which case the BP and LR reactions can be performed sequentially in separate steps; detailed instructions are included with the Gateway PCR kit. Alternatively, entry clones can easily be regenerated from expression clones via the BP reaction, as described in the instruction manual.
13. At this point, we remove a 5 μl aliquot from the reaction and add it to 0.5 μl of proteinase K solution. After 10 min at 37°C, we transform 2 μl into 50 μl of competent DH5α cells (*see* **Note 1**) and spread 100–200 μl on an LB agar plate containing kanamycin (35 μg/ml), the selective marker for pDONR221. From the number of colonies obtained, it is possible to gauge the success of the BP reaction. Additionally, entry clones can be recovered from these colonies in the event that no transformants are obtained after the subsequent LR reaction.
14. If very few or no ampicillin-resistant transformants are obtained after the LR reaction, the efficiency of the process can be improved by incubating the BP reaction overnight.
15. The optimum temperature for TEV protease activity is 30°C. At 37°C, the protease does not fold properly in *E. coli* and little processing will occur. Reducing the temperature also improves the solubility of some MBP fusion proteins.
16. We routinely break cells in a 1.5 ml microcentrifuge tube on ice with two or three 30 s pulses using a VCX600

sonicator (Sonics & Materials, Newtown, CT, USA) with a microtip at 38% power. The cells are cooled on ice between pulses.

Acknowledgments

This project was supported by the Intramural Research Program of the NIH, National Cancer Institute, Center for Cancer Research. The content of this publication does not necessarily reflect the views or policies of the Department of Health and Human Services, nor does the mention of trade names, commercial products, or organizations imply endorsement by the US government.

References

1. Waugh, D. S. (2005) Making the most of affinity tags. *Trends Biotechnol* 23, 316–320.
2. Esposito, D., Chatterjee, D. K. (2006) Enhancement of soluble protein expression through the use of fusion tags. *Curr Opin Biotechnol* 17, 353–358.
3. Kapust, R. B., Waugh, D. S. (1999) *Escherichia coli* maltose-binding protein is uncommonly effective at promoting the solubility of polypeptides to which it is fused. *Protein Sci* 8, 1668–1674.
4. Fox, J. D., Routzahn, K. M., Bucher, M. H., Waugh, D. S. (2003) Maltodextrin-binding proteins from diverse bacteria and archaea are potent solubility enhancers. *FEBS Lett* 537, 53–57.
5. Lichty, J. J., Malecki, J. L., Agnew, H. D., Michelson-Horowitz, D. J., Tan, S. (2005) Comparison of affinity tags for protein purification. *Protein Expr Purif* 41, 98–105.
6. Pryor, K. D., Leiting, B. (1997) High-level expression of soluble protein in *Escherichia coli* using a His6-tag and maltose-binding-protein double-affinity fusion system. *Protein Expr Purif* 10, 309–319.
7. Routzahn, K. M., Waugh, D. S. (2002) Differential effects of supplementary affinity tags on the solubility of MBP fusion proteins. *J Struct Funct Genomics* 2, 83–92.
8. Nallamsetty, S., Austin, B. P., Penrose, K. J., Waugh, D. S. (2005) Gateway vectors for the production of combinatorially-tagged His6-MBP fusion proteins in the cytoplasm and periplasm of *Escherichia coli*. *Protein Sci* 14, 2964–2971.
9. Alexandrov, A., Dutta, K., Pascal, S. M. (2001) MBP fusion protein with a viral protease cleavage site: one-step cleavage/purification of insoluble proteins. *Biotechniques* 30, 1194–1198.
10. Riggs, P. (2001) Expression and purification of maltose-binding protein fusions. *Curr Protoc Mol Biol* Chapter 16 , Unit16 6.
11. Fox, J. D., Waugh, D. S. (2003) Maltose-binding protein as a solubility enhancer. *Methods Mol Biol* 205, 99–117.
12. Cabrita, L. D., Dai, W., Bottomley, S. P. (2006) A family of *E. coli* expression vectors for laboratory scale and high throughput soluble protein production. *BMC Biotechnol* 6, 12.
13. Donnelly, M. I., Zhou, M., Millard, C. S., Clancy, S., Stols, L., Eschenfeldt, W. H., Collart, F. R., Joachimiak, A. (2006) An expression vector tailored for large-scale, high-throughput purification of recombinant proteins. *Protein Expr Purif* 47, 446–454.
14. Jeon, W. B., Aceti, D. J., Bingman, C. A., Vojtik, F. C., Olson, A. C., Ellefson, J. M., McCombs, J. E., Sreenath, H. K., Blommel, P. G., Seder, K. D., Burns, B. T., Geetha, H. V., Harms, A. C., Sabat, G., Sussman, M. R., Fox, B. G., Phillips, G. N., Jr. (2005) High-throughput purification and quality assurance of *Arabidopsis thaliana* proteins for eukaryotic structural genomics. *J Struct Funct Genomics* 6, 143–147.
15. Nallamsetty, S., Waugh, D. S. (2007) A generic protocol for the expression and purification of recombinant proteins in

Escherichia coli using a combinatorial His6-maltose binding protein fusion tag. *Nat Protoc* 2, 383–391.
16. Tropea, J. E., Cherry, S., Nallamsetty, S., Bignon, C., Waugh, D. S. (2007) A generic method for the production of recombinant proteins in *Escherichia coli* using a dual hexahistidine-maltose-binding protein affinity tag. *Methods Mol Biol* 363, 1–19.
17. Kapust, R. B., Waugh, D. S. (2000) Controlled intracellular processing of fusion proteins by TEV protease. *Protein Expr Purif* 19, 312–318.
18. Sambrook, J., Russell, D. W. (2001) *Molecular Cloning: A Laboratory Manual.* Cold Spring Harbor Laboratory Press, Cold Spring Harbor, NY.

Chapter 17

Introducing Predetermined Mutations Throughout a Target Gene Using TDEM (Transposon-Directed Base-Exchange Mutagenesis)

Yun Cheol Kim

Abstract

Transposon-directed base-exchange mutagenesis (TDEM) is an efficient and controllable method for introducing a mutation(s) into a gene. Each round of TDEM removes a predetermined number of bases (up to 11 base pairs) from a randomly selected site within the target gene and replaces them with any length of DNA of predetermined sequence. Therefore, the number of bases to be deleted and inserted can be precisely regulated. Because each round of TDEM generates mutation(s) at a single site, the number of mutations introduced can be determined by the number of cycles of TDEM. Furthermore, using a novel frame-checking procedure, non-functional mutants containing a frameshift or stop codon can be minimized. Thus, TDEM can be used to introduce a limited and predetermined change at each round of mutagenesis, thereby providing a useful tool for studying protein structure and function.

Key words: Transposon, mutagenesis, mutation insert, amino acid scanning mutagenesis.

1. Introduction

Current mutagenesis methods can be divided into two groups: controlled or random mutagenesis (1). PCR-based site-direct mutagenesis (2) is the method of choice for controlled mutagenesis. It has been widely used for alanine (3, 4) or cysteine scanning mutagenesis (5, 6) to investigate protein function or structure, respectively. Mutator bacterial strains (7), gene shuffling (8), and error-prone PCR (epPCR) (9) have been widely used for random mutagenesis to generate mutant libraries containing on average one to four mutations per kilobase of DNA (1, 10, 11).

T.C. Evans, M.-Q. Xu (eds.), *Heterologous Gene Expression in E. coli*, Methods in Molecular Biology 705,
DOI 10.1007/978-1-61737-967-3_17, © Springer Science+Business Media, LLC 2011

Although the nucleotide changes can be regulated using site-directed mutagenesis, it can be costly and labor-intensive as different mutations require the use of separate oligonucleotides, PCR, and cloning. While random mutagenesis methods are simple tools for generating mutant libraries, the rate and the sequence of mutations cannot be controlled so that the size and the complexity of the mutant library following each round of mutagenesis are unknown.

TDEM uses the random integration property of a transposon (12, 13) so that the mutation can be placed at a randomly selected site within the target gene. In addition, the bases to be inserted and the number of bases to be deleted can be separately determined by the researcher. Therefore, TDEM combines random mutagenesis with controllable site-directed mutagenesis.

2. Materials

2.1. Target Gene of Interest (GOI) Preparation

1. Sequence analysis software: Gene Construction Kit™ (Textco Biosoftware, West Lebanon, NH), MacVector (Accelrys, San Diego, CA), or other software having restriction enzyme mapping capability.
2. DNase free water: Commercial double distilled water (ddH_2O) (Invitrogen, Carlsbad, CA) or lab-made autoclaved double distilled water.
3. GsF primer: 5′-GGTCTCNATGGNN **NNNNNNNNN NNNNNNNN**-3′ (*see* **Note 1** and **Fig. 17.1**).
4. GsR primer: 5′-GGTCTCNGGTTNN **NNNNNNNNN NNNNNNNN**-3′ (*see* **Note 1** and **Fig. 17.1**).
5. Pfu DNA polymerase (Stratagene, La Jolla, CA).
6. 10 mM dNTP (Invitrogen).
7. PCR machine (MJ research, Waltham, MA, or comparable).
8. Agarose gel electrophoresis apparatus (Mupid-ExU (Eurogentec, San Diego CA) or comparable horizontal electrophoresis chamber).
9. Power supply (BioRad, Hercules, CA).
10. Ethidium bromide: Prepare 10 mg/ml solution of ethidium bromide (Sigma, St. Louis, MO) in water and store in brown bottle at room temperature.
11. Seakeme agarose: Appropriate amount of Seakeme agarose (Sigma) is added to 1X TAE buffer and melted with a microwave oven for 3 min. When the temperature of the

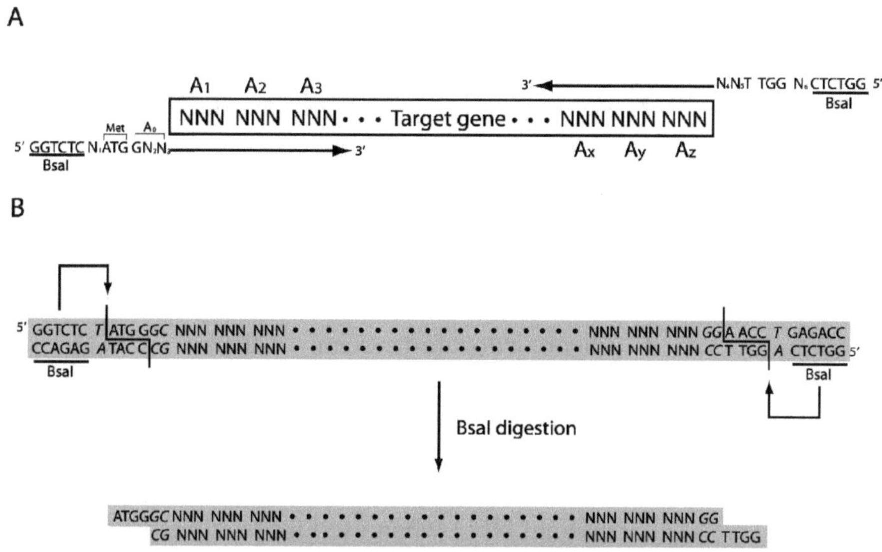

Fig. 17.1. Amplification of the gene of interest using primers containing *Bsa*I sequences. (**a**) Primer sequences for amplification of the target gene. The extra bases in the primer are depicted as codons and *Bsa*I sites. Within the primer N1 and N2 can be any bases and N2, N3, N4, and N5 should be chosen so as to not introduce stop codons. A_1–A_Z represent the amino acids based on the codons of the GOI. (**b**) The amplified target gene containing *Bsa*I sites at each ends. *Bsa*I digestion generates cohesive ends with different overhangs so that the digested DNA cannot be self-ligated.

agarose solution drops to about 60°C, add ethidium bromide solution to a concentration of 0.1 μg/ml and then pour into a gel casting chamber with a comb and let it cool to solidify.

12. 50X TAE buffer: mix 242 g of Tris base (Sigma), 57.1 ml glacial acetic acid (Fisher), 100 ml of 0.5 M EDTA, pH 8.0, and add water to 1 l.
13. DNA molecular weight markers: 1 kb ladder plus (Invitrogen, Carlsbad, CA) or 1 kb ladder (NEB) (*see* **Note 2**).
14. 10X DNA loading buffer (10 ml): Dissolve 0.025 g of bromophenol blue (Sigma) and 4 g sucrose (Sigma) in 10 ml 1X TAE.
15. Transilluminator (UVP, Upland, CA).
16. Razor blades (VWR, West Chester, PA): Individually wrapped disposable razor blades are required for cutting DNA bands from the agarose.
17. QIAquick gel extraction kit (Qiagen, Valencia, CA).
18. T4 DNA ligase (New England Biolabs (NEB), Ipswich, MA).
19. T4 polynucleotide kinase (NEB).
20. Water baths set at 37 and 70°C.

2.2. Preparation of Vector, pCompact-Kana

1. pCompact-Kana vector (available upon request from the author).
2. VectF primer: 5′-CCCGGACAGCAAGCGAACCGGAATTGC-3′.
3. VectR primer: 5′-CATGTGAGCAAAAGGCCAGCAAAAGGCC-3′.
4. Calf intestinal alkaline phosphatase (CIP) (NEB).
5. 10X NEBuffer 2 restriction enzyme buffer (NEB).
6. Phenol:chloroform:isoamyl alcohol (25:24:1) buffered with Tris pH 8.0 (Sigma).
7. PelletPaint (Novagen, San Diego, CA).
8. 100% ethanol (molecular biology grade) (Sigma): We observed that low-grade ethanol results in denaturation of plasmid DNA when used for precipitation.
9. Sodium acetate: Dissolve 40.8 g of sodium acetate (Sigma) in 80 ml of ddH_2O. Adjust the pH to 5.2 with glacial acetic acid. Adjust the volume to 100 ml with ddH_2O. Sterilize by autoclaving.
10. 70% ethanol: 35 ml of 100% ethanol (molecular biology grade) with 15 ml of autoclaved ddH_2O.
11. Microcentrifuge with refrigeration.

2.3. Cloning into pCompact-Kana

1. Top10 chemical-competent cell (Invitrogen) or comparable chemical-competent *E. coli*. The *E. coli* strain should be carefully chosen to ensure that the *E. coli* is sensitive to ampicillin, kanamycin, chloramphenicol, and zeocin.
2. 37°C shaking incubator for liquid LB culture.
3. 37°C static incubator for LB plate culture.
4. Kanamycin: Dissolve 0.5 g kanamycin (Sigma) in 10 ml of ddH_2O to make 1000X stock solution and store at –20°C.
5. QIAprep spin miniprep kit (Qiagen).
6. Plasmid maxiprep kit (Qiagen).
7. Luria–Broth (LB): dissolve 10 g of trypton, 5 g of yeast extract, and 10 g of sodium chloride in 1 l of water and autoclave to sterilize.
8. Bactoagar: Add 15 g of bactoagar (BD Difco, Franklin Lakes, NJ) to 1 l of LB and 5 μl of antifoam (Sigma) and autoclave. When cooled to 60°C, add appropriate antibiotics and pour in Petri dishes to solidify.
9. SOC: Dissolve 20 g of tryptone (BD Difco), 5 g of yeast extract (BD Difco), 0.5 g of NaCl (Sigma), and 2.5 ml of 1 M KCl (Sigma) in 1 l of ddH_2O, adjust pH to 7.0 with

10 N NaOH, autoclave to sterilize, and add 20 ml of sterile 1 M glucose (Sigma) immediately before use.

2.4. Transposition into the GOI

1. Transposon: M1-CamR or compatible transposons (Finnzyme, Espoo, Finland).
2. HyperMu MuA transposase (Epicentre, Madison, WI).
3. *Bsa*I (NEB).
4. Chloramphenicol (Sigma): Dissolve 50 mg in 10 ml of ddH_2O to make 1000X stock solution and store at –20°C.
5. Top10 electro-competent cell (Invitrogen) or comparable *E. coli* strain electro-competent cell.

2.5. Inserting Mutation Insert (MI)

1. Zeocin: Commercial zeocin (Invitrogen) solution is 2000X stock.
2. *Not*I (NEB).
3. Mutation insert (available upon request from the author).

2.6. Digestion of Right and Left Arms of the MI

1. *Bpm*I (NEB).
2. *Bsg*I (NEB).

2.7. Removing Unwanted Frameshift Mutants by Frame Checking

1. Plasmid pFrameCheck (available upon request from the author).
2. Carbenicillin: Dissolve 0.5 g carbenicillin (Sigma) in 10 ml of ddH_2O to make 1000X stock solution and store at –20°C.

3. Methods

TDEM is an approach for generating amino acid substitutions or deletions based on the use of Mu transposons and a DNA cassette termed the mutation insert (MI). The transposon is used to generate breaks at a single random position in a target DNA sequence into which the MI is inserted. The MI is used to delete a given number of base pairs from the target DNA and to subsequently replace them with an MI-derived sequence. Therefore, the MI carries all the information required to both remove the predetermined number of base pairs from the target gene and replace them with a defined sequence. Using different MIs, it is possible to efficiently introduce an unlimited number of different bases into the target gene and up to 11 bases within the target gene can be removed.

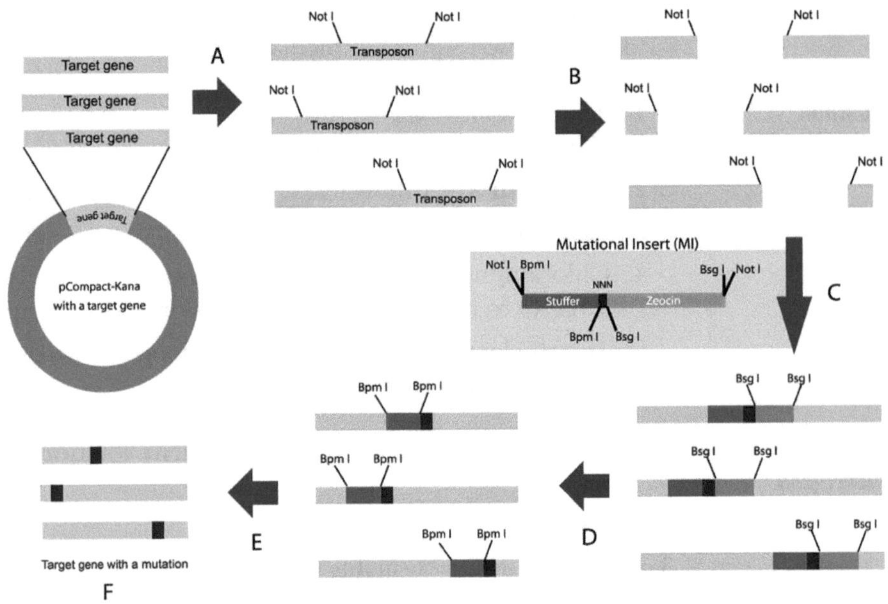

Fig. 17.2. Schematic diagram showing introduction of a mutation using transposon. (**a**) Transposition with MuA transposase and Mu-transposon (Tn) into a target gene. The Tn contains an antibiotic resistance gene for selection. (**b**) *NotI* digestion to remove transposon. The linear plasmid is then purified using agarose gel electrophoresis. (**c**) Cloning of the mutational insert (MI) into the target gene. The MI contains the zeocin resistance gene. (**d**) *BsgI* digestion, followed by the production of blunt ends and self-religation. This step removes the stuffer from the MI. (**e**) *BpmI* digestion, followed by the production of blunt ends and self-religation. This step replaces the bases of the target gene with predetermined bases. (**f**) The final gene contains a mutation at a random site.

The procedure for TDEM and the structure of the mutation insert (MI) are depicted in **Fig. 17.2**. Briefly, transposition into random sites within the target gene is carried out using the Mu transposon containing *Not*I sites at its ends (*see* **Fig. 17.2a**). After transposition, the vector is digested with *Not*I to open it at the transposon integration site and to remove the transposon (*see* **Fig. 17.2b**). Subsequently, the mutation insert (MI), containing a predetermined sequence (referred to as incorporating base (IB)) between *Bsg*I and *Bpm*I sites, is ligated into the linearized vector (*see* **Fig. 17.2c**). The plasmids are then digested with *Bsg*I to remove the right-hand region of the stuffer (*see* **Fig. 17.2d**). After self-ligation, the plasmids are again digested with *Bpm*I to remove a predetermined number of bases from the target and the left-hand region of the MI (*see* **Fig. 17.2e**) resulting in the mutant library containing predetermined bases at random positions within the target (*see* **Fig. 17.2f**).

As shown in **Fig. 17.3**, the MI consists of three components: (1) distal *Bsg*I and *Bpm*I sites to remove bases from the target gene, (2) nucleotides to be inserted into the target gene (IB) between the proximal *Bsg*I and *Bpm*I sites, and (3) a zeocin resistance gene (*zeo*^{r}). *Bpm*I and *Bsg*I, type IIs restriction

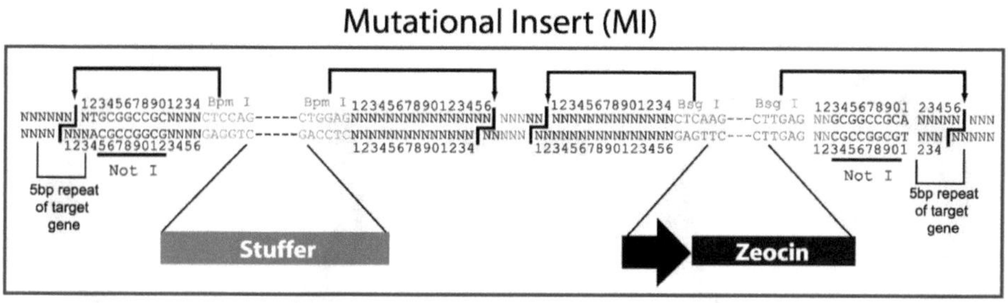

Fig. 17.3. Structure of the mutation insert (MI). Mutation insert (MI) is the 1.2 kb DNA fragment containing the zeocin resistance gene. *Bsg*I digestion removes the 5 bp repeat of the target sequence and *Bpm*I digestion removes up to 11 bp from a target gene at the site of transposition. After *Bsg*I and *Bpm*I digestion and self-ligation, predetermined bases are incorporated into the site from which the bases were removed.

endonucleases, cut 16 bp from their recognition sequence. Therefore, by adjusting the relative position of the distal restriction enzyme sites, the number of bases to be deleted from the target can be altered. On the other hand, the sequence to be inserted is determined by the IB located between the proximal *Bpm*I and *Bsg*I sites. The IB can be of any length. Thus, the number of bases to be deleted and the length of the DNA to be inserted can be separately determined so that many types of mutant libraries can be generated. Lastly, the *zeo*^{r} gene confers resistance to zeocin so that only plasmids containing the MI are recovered following selection with zeocin and kanamycin.

3.1. Preparation of Target Gene of Interest

1. Amplify the gene of interest (GOI) with GsF and GsR primers containing *Bsa*I sites (*see* **Fig. 17.1**).
2. Purify the PCR products from agarose gel using QIAquick gel extraction kit (Qiagen).
3. Treat the purified PCR products with polynucleotide kinase (PNK) (NEB) at 37°C for 30 min. The reaction mixture would be the following:
   Up to 200 ng purified DNA.
   5 μl 10X T4 DNA ligase buffer (NEB).
   Add water to 49 μl.
   1 μl PNK.
   50 μl total volume (*see* **Note 3**).
4. Incubate the reaction mixture at 70°C for 20 min to inactivate the PNK.
5. Store the reaction mixture at –20°C for later use.

3.2. Preparation of Vector, pCompact-Kana

1. Amplify pCompact-Kana, which has only an origin of replication and kanamycin resistance gene (*see* **Fig. 17.4**). pCompact-Kana (1.7 kb) can be easily amplified by PCR to

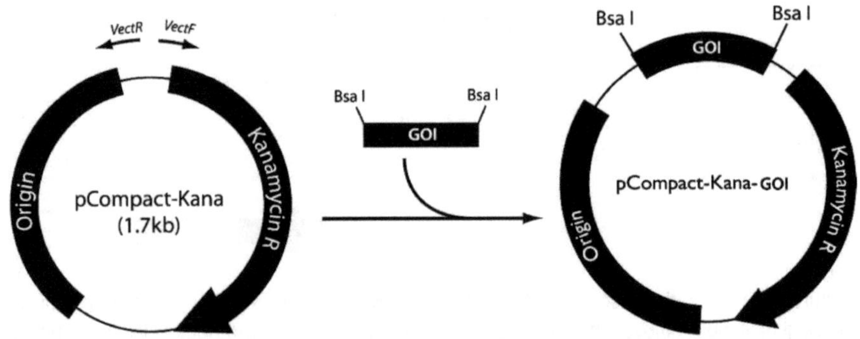

Fig. 17.4. Cloning of GOI into pCompact-Kana. The plasmid, pCompact-Kana, has only an origin of replication from pUC19 and a kanamycin resistance gene, so it can be easily amplified by PCR. The locations of the primers are depicted as *arrows*, VectF and VectR. The *Bsa*I site containing GOI can be cloned into the PCR-amplified pCompact-Kana. Use of this vector maximizes the possibility of introducing the transposon into the target gene. Plasmids with insertions into an essential part of the vector (origin and kanamycin resistance gene) will not grow.

generate a blunt-ended vector for cloning due to its small size.

PCR mixture:

10 μl 10X cloned *Pfu* buffer l.

3 μl VectF primer (10 pmol/μl stock).

3 μl VectR primer (10 pmol/μl stock).

3 μl dNTPs (10 mM stock).

2 μl *Pfu* DNA polymerase.

10–100 pg supercoiled pCompact-Kana as template DNA.

Water to 100 μl.

PCR cycle conditions:

Initial denaturation: 95°C for 2 min.
1 cycle

Denaturation: 95°C for 30 s.

Annealing: 55°C for 30 s.

Extension: 72°C for 4 min.
30 cycles

Final extension: 72°C for 10 min (*see* **Note 4**).

2. Analyze the amplified pCompact-Kana DNA using agarose gel electrophoresis and purify the amplified DNA from the gel using QIAquick gel extraction kit.
3. Treat 100–200 ng of the purified linear pCompact-Kana with 1 μl of calf intestinal alkaline phosphatase (CIP) in the 1X NEBuffer 2 restriction enzyme buffer at 37°C for 20 min in order to remove phosphate groups at the 5′-ends (*see* **Note 5**).

3.3. Cloning the GOI into pCompact-Kana

1. Set up a ligation mixture as follows:
   100 ng GOI (prepared in **Section 3.1**).
   50 ng linear pCompact-Kana (prepared in **Section 3.2**).
   2 μl 10X T4 DNA ligase buffer.
   Add water up to 19 μl.
   1 μl T4 DNA ligase.
2. Incubate at room temperature for 20 min.
3. Transform into Top10 (or any other comparable *E. coli* strain) chemical-competent cells using 2 μl of above reaction mixture and spread the appropriate amount of the transformation mixture onto an LB agar plate containing 50 μg/ml of kanamycin. Incubate at 37°C overnight (12–16 h).
4. Pick colonies and inoculate into 3 ml of LB broth containing 50 μg/ml of kanamycin and grow at 37°C overnight.
5. Prepare the plasmids from the cultured broth using QIAprep spin miniprep kit.
6. Identify clones containing the GOI using restriction enzyme digestion or PCR screening.
7. Sequence several clones to identify a correct clone and prepare a sufficient amount of the plasmid using plasmid midi or maxiprep kit.
8. (Optional) Remove unwanted restriction enzyme sites (*Bpm*I, *Bsg*I, *Bsa*I, and *Not*I) from the target DNA. *Bpm*I, *Bsg*I, *Bsa*I, and *Not*I restriction enzymes are used throughout the TDEM procedure. Therefore, the target gene should be free of these restriction enzyme sites. Commercially available software can be used to determine the position of these restriction enzyme sites in the target gene. When removing these enzyme sites from the target gene, the altered bases should not change the amino acid sequence and the resulting codons should be codons preferred by the organisms in which phenotypic screening is performed (*see* **Note 6**).

3.4. Transposition into the GOI

1. Set up the transposition reaction mixture as follows:
   60 fmol target DNA (GOI in pCompact-Kana).
   DNase-free ddH_2O up to 8 μl
   10 μl 10X reaction buffer for HyperMu MuA transposase.
   1 μl (20 ng, 25 fmol) transposon (M1-CamR).
   1 μl HyperMu MuA transposase.
   20 μl total reaction volume.
2. Mix the above reaction mixture and incubate at 37°C for 2 h.

3. Add an equal volume of phenol:chloroform:isoamyl alcohol (PCI) to the transposition reaction mixture and vortex.
4. Centrifuge at maximum revolutions per minute for 1 min to separate the aqueous layer from the phenol–chloroform layer.
5. Carefully transfer the upper aqueous layer to a new tube.
6. Add 1 μl of PelletPaint and mix well (*see* **Note 7**).
7. Add 2 μl (1/10 of the reaction volume) of 3 M sodium acetate (pH 5.2) and mix well.
8. Add 40 μl (twice the reaction volume) of 100% ethanol and mix well.
9. Centrifuge at maximum revolutions per minute at 4°C for 5 min.
10. Decant the solution and wash the pellet with 70% ethanol followed by drying the pellet.
11. Resuspend the pellet with 3 μl of ddH_2O and perform electro-transformation.
12. Add 1 ml of SOC and shake at 37°C for 1 h. Then spread 50 μl of the transformation mixture onto the LB agar plate containing 50 μg/ml of kanamycin and 7 μg/ml of chloramphenicol. Inoculate the rest of the transformation mixture into 100–200 ml of LB containing 50 μg/ml of kanamycin and 7 μg/ml of chloramphenicol and culture the broth at 37°C shaking at 250 rpm for 20 h.
13. Count the colonies after 20 h in order to estimate the total number of transposition events (*see* **Note 8**).
14. After obtaining a sufficient number of transposition events, harvest the library plasmid using plasmid maxiprep kit. The total number of transposition events should be recorded for later reference.
15. Although transposition into the vector can impair replication of the plasmid, the plasmid mixture will still contain plasmids with transposons in the vector sequence (**Fig. 17.5**). The following procedure is used to eliminate these plasmids:

    Analyze the parental target DNA and transposed DNA prepared above step 14 by agarose gel and identify the plasmid containing a transposon based on the DNA size (*see* **Note 9**).

    Purify the supercoiled plasmid containing the transposon from the gel using QIAquick gel extraction kit (Qiagen).

    Digest the purified plasmid with *Bsa*I in order to isolate plasmids containing the transposon in the target sequence. As illustrated in **Fig. 17.5**, the digested DNA will generate four distinct bands based on the integration site.

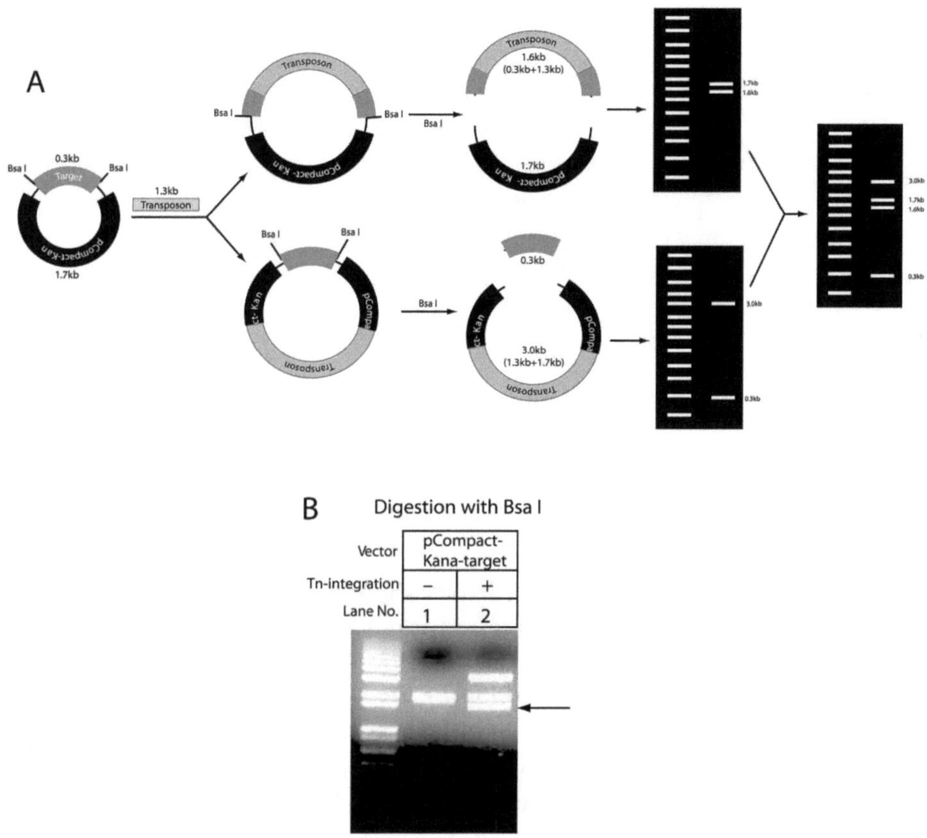

Fig. 17.5. The Mu transposon attacks both the target gene and the vector sequence. (**a**) Schematic diagram of transposon integration into either the vector or target gene and the fragments generated following *Bsa*I digestion. The virtual gel pictures are a schematic representation of the four distinctive bands seen following transposon integration and digestion with *Bsa*I. (**b**) Following transposition, the supercoiled plasmids were harvested and analyzed by agarose gel after digestion with *Bsa*I. The marks indicate the fragment of the target gene with the transposon integrated. L, 1 kb DNA ladder; lane 1, pCompact-Kana-target; lane 2, pCompact-Kana-target with transposon.

Purify the target gene containing the transposon and vector which is free of the target gene from the gel using QIAquick gel extraction kit.

Ligate the purified DNA bands using T4 DNA ligase (NEB) as follows:

50 μl purified target containing transposon.

50 μl purified vector only DNA.

20 μl 10X T4 DNA ligase buffer.

75 μl ddH$_2$O.

5 μl T4 DNA ligase.

200 μl total volume.

Incubate at room temperature for 30 min followed by extraction with PCI.

Repeat steps 3–14 to precipitate ligated DNA and electro-transform into Top10 cells (*see* **Note 10**).

3.5. Replace the Transposon with the Mutation Insert

1. The purified plasmid mixture is a library with the transposon randomly located throughout the target gene. In order to open up the site at which the transposon is located, digest 5 μg of the plasmid DNA library with *Not*I as follows (**Fig. 17.2b**). *Not*I sites are located at both ends of the transposon.

   5 μg plasmid DNA (up to 43 μl).

   5 μl 10X NEBuffer 3.

   Add ddH_2O to 48 μl.

   2 μl *Not*I.

   Total volume 50 μl.
2. Incubate at 37°C for 1 h.
3. Load the digested DNA on an agarose gel to analyze and purify the plasmids digested at random sites within the target gene.
4. Two DNA bands should be seen. The larger band is the plasmid digested at random sites within the target gene and the smaller band is the 1.2 kb transposon. Excise the larger DNA band (1.7 + the size of the target gene in kilobase).
5. After purifying the DNA from the gel using QIAquick gel extraction kit, set up a ligation mixture as follows:

   100 ng purified plasmid DNA (up to 50 μl).

   1 μl 300 ng/μl mutation insert.

   6 μl 10X T4 DNA ligase buffer.

   Add ddH_2O to 48 μl.

   2 μl T4 DNA ligase.
6. Incubate at room temperature for 1 h.
7. Extract the reaction mixture with PCI followed by ethanol precipitation as detailed in steps 3–10 of unit 3.4.
8. Resuspend the pellet with 3 μl of water and perform electro-transformation.
9. Add 1 ml of SOC and shake at 37°C for 1 h. Then spread 5 μl of the transformation mixture onto an LB agar plate containing 25 μg/ml of zeocin and 50 μg/ml of kanamycin. Inoculate the rest of the transformation mixture into 100–200 ml of LB containing 25 μg/ml of zeocin and 50 μg/ml of kanamycin and culture at 37°C shaking at 250 rpm for 16 h.
10. Estimate the total number of transformants based on the number of colonies in the plate (*see* **Note 11**).
11. After obtaining a sufficient number of transformants, harvest the plasmid library using Plasmid maxiprep kit.

3.6. Digestion of Right and Left Arms of the MI

1. Purify the supercoiled plasmid from the agarose gel using the procedure described in step 15 of **Section 3.4** (*see* **Note 12**).
2. Digest 5 μg of the purified supercoiled plasmid DNA with *Bsg*I as follows in order to remove the right part of the MI.

   5 μg plasmid DNA (up to 43 μl).

   5 μl 10X NEBuffer 3 buffer.

   Add ddH_2O to 48 μl.

   2 μl *Bsg*I.

   50 μl total volume.
3. Incubate at 37°C for 1 h.
4. Extract with PCI once (*see* steps 3–5 of **Section 3.4**), add an equal volume of water and 10X DNA loading buffer (*see* **Note 13**).
5. Load the digested DNA on an agarose gel to analyze and purify the *Bsg*I-digested plasmid. Two DNA bands should be seen. The larger band is the plasmid with the target gene and the smaller band is the 0.3 kb left part of the MI. Excise the larger DNA band (2.5 added to the size of the target gene in kb).
6. After purifying the DNA from the gel using QIAquick gel extraction kit, blunt the DNA ends on ice as follows:

   3 μg purified plasmid DNA (up to 50 μl).

   3 μl 10 mM dNTP.

   6 μl 10X T4 DNA polymerase buffer.

   Add ddH_2O to 48 μl.

   2 μl T4 DNA polymerase.
7. Incubate at 12°C for 25 min (*see* **Note 14**).
8. Extract with PCI and precipitate with ethanol (*see* steps 3–10 of **Section 3.4**).
9. Resuspend the pellet with 30 μl of water and set up a self-ligation mixture as follows:

   3 μg purified plasmid DNA with blunt ends (up to 50 μl).

   20 μl 10X T4 DNA ligase buffer.

   Add ddH_2O to 196 μl.

   4 μl T4 DNA ligase.
10. Purify the ligation mixture by repeating steps 3–10 of unit 3.4.
11. Resuspend the pellet with 3 μl of water and perform electro-transformation into Top10 electro-competent cell.

12. Add 1 ml of SOC and shake at 37°C for 1 h. Then spread 5 μl of the transformation mixture onto an LB agar plate containing 50 μg/ml of kanamycin. Inoculate the rest of the transformation mixture into 100–200 ml of LB containing 50 μg/ml of kanamycin and culture at 37°C with shaking at 250 rpm for 16 h. At this step the LB does not contain zeocin because the zeocin resistance gene is removed following digestion with *Bsg*I.
13. Estimate the total number of transformants based on the number of colonies on the plate (*see* **Note 11**).
14. After obtaining a sufficient number of transformants, harvest the library plasmid using plasmid midi or maxiprep kit. The plasmids now contain only the right side of the MI randomly inserted throughout the target gene.
15. In order to remove the left side of the MI and insert the novel bases, repeat steps 1–14 using *Bpm*I instead of *Bsg*I. After *Bpm*I digestion, the plasmids will contain the desired mutations within the target gene (*see* **Fig. 17.2f**).

3.7. Removing Unwanted Frameshift Mutants by Frame Checking

Although the above plasmid contains the desired mutation throughout the target gene, more than 50% of the plasmid will contain undesired mutations due to the incorrect digestion of *Bpm*I and *Bsg*I. These enzymes are type IIs restriction enzymes, which show relatively high non-specific digestion. If a stringent screening system is to be used, the plasmid library can be directly screened. However, when a more defined mutation library is desired, the following steps can be used to remove unwanted mutations. In this frame-checking procedure, the gene mutated by TDEM is cloned between the gIII signal sequence of the coat protein of phage fd and the beta-lactamase gene (ampicillin-resistant gene). If the mutated genes contain unwanted frame shifts or stop codons within the gene, the signal sequence will not be physically linked to active beta-lactamase resulting in cell death following ampicillin selection. However, if the mutated genes do not contain a frameshift or stop codons, the signal sequence will be attached to an active GOI – beta-lactamase fusion protein and beta-lactamase will be transported to the periplasmic space resulting in ampicillin resistance (*see* **Fig. 17.6**).

1. The mutated target gene was harvested from pCompact-Kana-GOI by digesting with *Bsa*I as follows:

   10 μg plasmid library (up to 87 μl).

   10 μl 10X NEBuffer 3.

   1 μl 100X BSA (NEB).

   Add ddH_2O up to 98 μl.

   2 μl *Bsa*I.
2. Incubate at 37°C for 1 h.

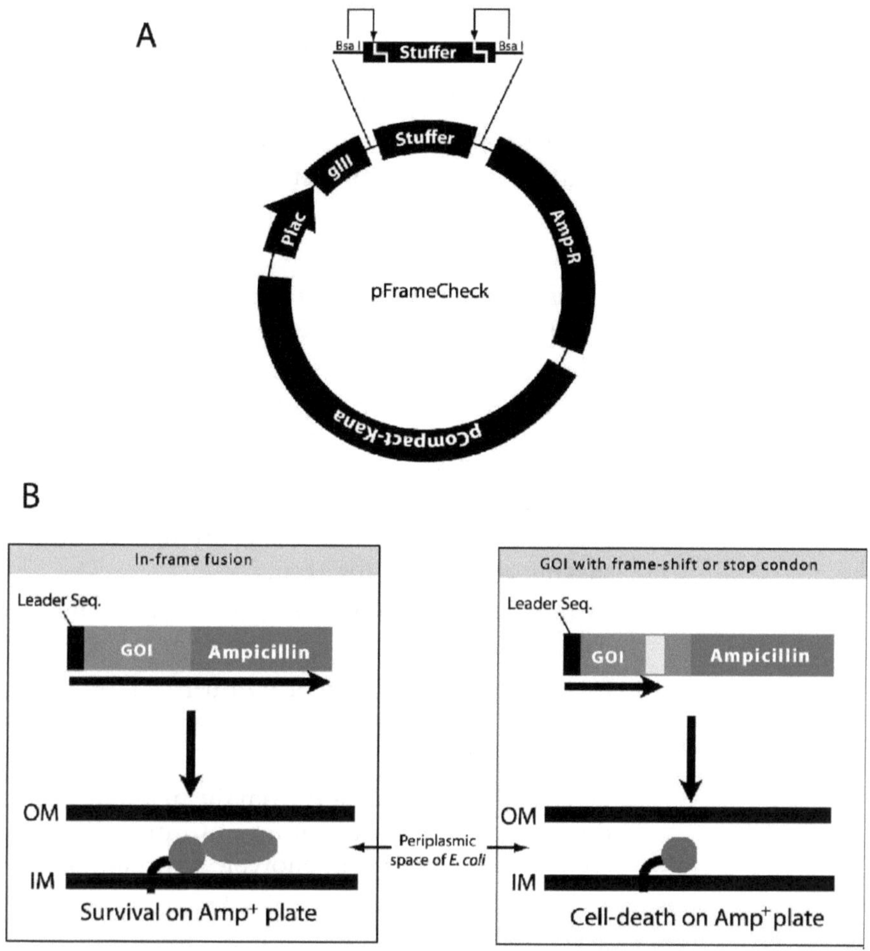

Fig. 17.6. Elimination of frameshift mutants. In the process of TDEM, additional bases can be deleted or added resulting in a frameshift mutation instead of the specific replacement of bases. These frameshift mutants can be eliminated by subcloning the mutated gene between the gIII signal sequence of the coat protein of phage fd and the ampicillin resistance gene. (**a**) The pFrameCheck vector. When the mutated target gene within pCompact-Kana is digested using *Bsa*I, the overhangs are compatible with pFrameCheck and the open reading frame will be maintained so that the gIII signal sequence is fused to beta-lactamase through the mutated target gene. (**b**) Only colonies having the gene of interest (GOI) in frame with the gIII signal sequence and beta-lactamase can transport the beta-lactamase to the periplasmic space resulting in resistance to ampicillin or carbenicillin.

3. Analyze the digested DNA by agarose gel and harvest the digested GOI and pCompact-Kana vector from the gel using QIAquick gel extraction kit. *Bsa*I digestion will yield two bands. Save the linearized parental pCompact-Kana for later use. The band containing the GOI will be cloned into pFrameCheck vector.
4. Ligate the GOI with the *Bsa*I-digested-pFrameCheck vector as follows (*see* **Note 15**):
   2 μg *Bsa*I-digested pFrameCheck.

3 μg gel purified mutated GOI.

10 μl 10X T4 DNA ligase buffer.

Add ddH_2O up to 96 μl.

4 μl T4 DNA ligase.

5. Incubate at room temperature for 1 h.
6. Extract with PCI and precipitate with ethanol (*see* steps 3–10 of **Section 3.4**).
7. Resuspend the pellet with 3 μl of water and electro-transform into Top10 electro-competent cells.
8. Add 1 ml of SOC and shake at 37°C for 1 h. Then spread 5 μl of the transformation mixture onto an LB agar plate containing 30 μg/ml of carbenicillin and 50 μg/ml of kanamycin. Inoculate the rest of the transformation mixture into 100–200 ml of LB containing 30 μg/ml of carbenicillin and 50 μg/ml of kanamycin and culture at 37°C shaking at 250 rpm for 16 h.
9. Estimate the total transformants based on the number of colonies on the plate.
10. After obtaining a sufficient number of transformants, harvest the library plasmid using plasmid midi or maxiprep kit (*see* **Note 16**).
11. Harvest the mutated GOI from the pFrameCheck by digestion with *Bsa*I and clone it into pComact-Kana or another compatible vector for phenotypic screening or expression.

4. Notes

1. The underlined sequences in the GsF and GsR primer sequences are *Bsa*I recognition sites, and the bold letter N's represent the gene-specific sequences. Five bases at the ends of the primers are for the cohesive ends so that the sequences will be compatible with vectors used at later steps during the procedure. Although the same restriction sites are at each end of the GOI, different four bases overhang are generated facilitating directional cloning at various steps in the procedure (*see* **Fig. 17.1**).
2. 1 kb plus ladder (Invitrogen) is convenient for evaluating a wide range of DNA fragments and the 1 kb ladder from NEB is optimal for estimating DNA quantity as the manufacturer provides information on the quantity of DNA within each band of the ladder.

3. Because PNK is active in the ligase buffer, T4 DNA ligase buffer can be used for convenience. In this case, PNK-treated DNA can be used directly for ligation without buffer exchange after heat inactivation of the PNK.
4. Use of this vector maximizes the possibility of introducing the transposon into the target gene. Plasmids with insertions into an essential part of the vector (origin and kanamycin resistance gene) will not grow.
5. Although the use of non-phosphorylated primers for PCR generates plasmids lacking 5′-phosphate, additional CIP treatment can further minimize the possibility of vector self-ligation.
6. Restriction enzyme site removal can be achieved either using PCR with mutation containing primers or commercial kits such as QuickChange® Site Directed Mutagenesis Kit (Stratagene).
7. Addition of PelletPaint is very helpful for recovering DNA present at low concentrations.
8. The transposon will also integrate into vector regions outside of the GOI, and Mu transposase possesses intrinsic site preference depending on the base composition of the target DNA. Therefore, in order to obtain all possible mutants, the initial number of transposition events should exceed the number of bases of the GOI by many fold (practically, 50- to 100-fold). For example, more than 50,000 transposition events are required for a GOI of 1 kb. Therefore, it may be necessary to scale up the transposition reaction to obtain a sufficient number of transposition events. The scale of the transposition reaction can be estimated from the initial transposition experiment.
9. Because transposition into a target DNA can occur several times, multiple DNA bands may be observed in the gel. In order to harvest plasmids containing only one transposon in the vector, agarose gel analysis and gel purification are required.
10. The total number of transformants should exceed the number of original transposition events in order to recover all of the initial transposition events. After this procedure, all of the plasmids will have a single transposon in the target gene.
11. The total number of transformants should exceed the number of the previous step in order not to loss possible mutants. It may be necessary to scale up the ligation reaction in order to obtain a sufficient number of transformants.

12. Because multiple copies of the MI can be ligated into the vector, *Bsg*I digestion of the purified plasmid without gel purification will result in a very complicated pattern in the agarose gel. In order to avoid this, only monomeric supercoiled plasmid should be purified before digesting with *Bsg*I.
13. The addition of water after PCI extraction is required to dilute residual organic solvents in the DNA solution such as phenol or chloroform which will interfere with loading onto the agarose gel.
14. PCR machine can be used for this low temperature incubation.
15. Prior to this ligation, sufficient amounts of the pFrameCheck vector should be prepared by digesting with *Bsa*I and gel-purifying linearized pFrameCheck. Because the undigested pFrameCheck contains a LacZα fragment between the two *Bsa*I sites, *Bsa*I digestion of the pFrameCheck will generate 2 and 0.3 kb fragments. The 2 kb DNA is the linearized pFrameCheck with the compatible ends with the GOI.
16. About 95–98% of the plasmids recovered following this frame-checking procedure do not contain either a frameshift or stop codon within the target gene.

Acknowledgments

The author would like to thank professor Sherie L. Morrison for her support, advice, and encouragement. This work was supported by a seed grant from the UCLA AIDS Institute (NIH grant AI28697) and by grant AI29470 from the National Institute of Health.

References

1. Sen, S., Venkata Dasu, V., Mandal, B. (2007) Developments in directed evolution for improving enzyme functions. *Appl Biochem Biotechnol* 143, 212–223.
2. Sarkar, G., Sommer, S. S. (1990) The "megaprimer" method of site-directed mutagenesis. *Biotechniques* 8, 404–407.
3. Cunningham, B. C., Wells, J. A. (1989) High-resolution epitope mapping of hGH-receptor interactions by alanine-scanning mutagenesis. *Science (New York, N Y)* 244, 1081–1085.
4. Miyamoto, M., Han, G. D., Kimura, T., Furuichi, Y., Komiyama, T. (2005) Alanine-scanning mutagenesis of HM-1 killer toxin and the essential residues for killing activity. *J Biochem* 137, 517–522.

5. Mitchell, S. M., Lee, E., Garcia, M. L., Stephan, M. M. (2004) Structure and function of extracellular loop 4 of the serotonin transporter as revealed by cysteine-scanning mutagenesis. *J Biol Chem* 279, 24089–24099.
6. Sahin-Toth, M., Kaback, H. R. (1993) Cysteine scanning mutagenesis of putative transmembrane helices IX and X in the lactose permease of *Escherichia coli*. *Protein Sci* 2, 1024–1033.
7. Low, N. M., Holliger, P. H., Winter, G. (1996) Mimicking somatic hypermutation: affinity maturation of antibodies displayed on bacteriophage using a bacterial mutator strain. *J Mol Biol* 260, 359–368.
8. Stemmer, W. P. (1994) Rapid evolution of a protein in vitro by DNA shuffling. *Nature* 370, 389–391.
9. Cadwell, R. C., Joyce, G. F. (1992) Randomization of genes by PCR mutagenesis. *PCR Methods Appl* 2, 28–33.
10. Bornscheuer, U. T., Altenbuchner, J., Meyer, H. H. (1998) Directed evolution of an esterase for the stereoselective resolution of a key intermediate in the synthesis of epothilones. *Biotechnol Bioeng* 58, 554–559.
11. Fujii, R., Kitaoka, M., Hayashi, K. (2004) One-step random mutagenesis by error-prone rolling circle amplification. *Nucleic Acids Res* 32, e145.
12. Haapa, S., Taira, S., Heikkinen, E., Savilahti, H. (1999) An efficient and accurate integration of mini-Mu transposons in vitro: a general methodology for functional genetic analysis and molecular biology applications. *Nucleic Acids Res* 27, 2777–2784.
13. Biery, M. C., Stewart, F. J., Stellwagen, A. E., Raleigh, E. A., Craig, N. L. (2000) A simple in vitro Tn7-based transposition system with low target site selectivity for genome and gene analysis. *Nucleic Acids Res* 28, 1067–1077.

Chapter 18

Fluorescent Site-Specific Labeling of *Escherichia coli* Expressed Proteins with Sfp Phosphopantetheinyl Transferase

Aihua Zhang, Luo Sun, John Buswell, Nancy Considine, Inca Ghosh, Anastasiya Masharina, Christopher Noren, and Ming-Qun Xu

Abstract

Fluorescent tagging of proteins has become a critical step in optical analysis of protein function in vitro and in living cells. Here we describe a two-tag system for expression and isolation of a protein of interest from *Escherichia coli* and subsequent site-specific fluorescent labeling with Sfp phosphopantetheinyl transferase (Sfp synthase). In the example presented, adenoviral protein E3-14.7 K (E14.7) was expressed as a tripartite fusion protein with a fluorophore-targeting peptide tag and a chitin-binding domain. This system allows for rapid isolation of the recombinant fusion protein from crude bacterial cell lysate via a single chitin column. Sfp synthase-mediated labeling with fluorophore conjugated to coenzyme A-SH (CoA-SH) resulted in covalent attachment of a fluorescent dye to a specific residue of the peptide tag via a phosphopantetheinyl linker. The fluorescently labeled E14.7 fusion protein was analyzed with a fluorescence imager and subsequently transfected into mammalian cells for imaging with a fluorescence microscope.

Key words: Protein labeling, Sfp synthase, ACP synthase, ACP tag, ybbR, S6 tag, fluorescent labeling.

1. Introduction

Optical analysis of proteins is an important pathway toward understanding of protein dynamics and interaction networks in cells. This approach has been facilitated by fluorescently tagging proteins using various chemical modification and protein

T.C. Evans, M.-Q. Xu (eds.), *Heterologous Gene Expression in E. coli*, Methods in Molecular Biology 705,
DOI 10.1007/978-1-61737-967-3_18,

fusion techniques (1, 2). The advantages and limitations of different approaches are assessed in accordance with the needs of the research project. Traditional chemical modification cannot achieve specific labeling in complex samples and requires purification of the target protein. Assays with fluorophore-linked antibodies are widely used for characterization of proteins, but can rarely address protein labeling and detection inside living cells. The use of autofluorescent proteins, such as green fluorescent protein (GFP) from the jellyfish *Aequorea victoria*, has been one of the most enabling technologies in optical visualization of proteins in living cells and animals (3). Various techniques have been developed for tagging proteins with small molecule fluorophores for in vitro analysis of protein function and for the study of protein dynamics in living cells (4–6). Orthogonal protein labeling with fluorophores with different spectra allows for simultaneous visualization and characterization of multiple protein targets (7, 8). In addition, fluorescent labeling can generate imaging agents for monitoring protein–protein interactions on cell surface, protein uptake, and trafficking in living cells (9–11). Thus, development of robust methods of specifically derivatizing a protein of interest with a fluorophore can facilitate monitoring of protein localization, interactions, and transportation. Of special interest is the use of small tags for labeling since they are less likely to affect the structure and function of a protein (the fusion partner). It has been shown that small protein tags, for example, acyl carrier protein (ACP, 77 amino acid residues), peptide carrier protein (PCP), and peptide substrates (ybbR or S6 tag) can be fluorescently labeled by Sfp synthase or ACP synthase (7, 10, 12). These enzymes catalyze the transfer of a fluorophore or biotin conjugated to coenzyme A-SH (CoA-SH), to the side chain of a specific serine residue in the target sequence.

This work presents an example for the production of fluorochrome-labeled proteins by expressing and isolating a chimeric protein consisting of adenoviral protein E3-14.7 K, a small 12-residue peptide tag S6 and a C-terminal 6 kDa chitin-binding domain from *E. coli* cells (**Fig. 18.1**). E3-14.7 K has been implicated in regulation of TNF-mediated apoptosis and other receptor trafficking events; however, the molecular mechanisms have not been fully understood. Fluorescent labeling of E3-14.7 may be utilized for investigation of its role in inhibition of receptor internalization and its interactions with other components (13). Site-specific enzymatic labeling with Sfp synthase was employed to label the fusion protein with various fluorescent substrates. We demonstrate in vitro analysis of the labeled proteins with a fluorescent imager, protein transfection, and imaging of live mammalian cells by fluorescence confocal microscopy.

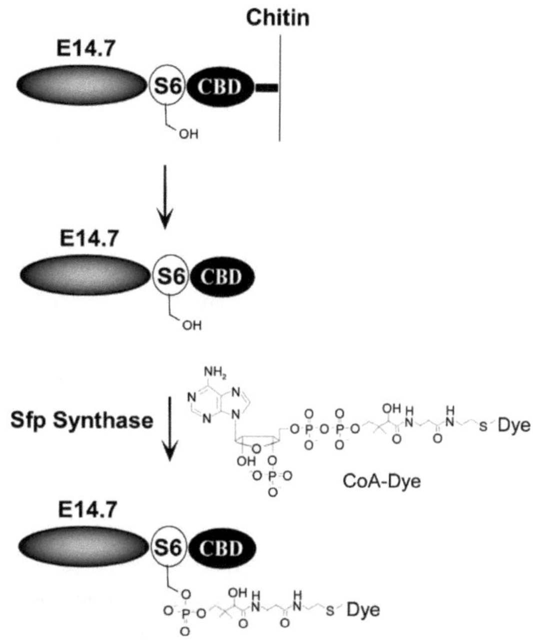

Fig. 18.1. Schematic of protein purification and labeling by Sfp phosphopantetheinyl transferase (Sfp synthase). Adenoviral protein E3-14.7 K (E14.7) was expressed as a fusion protein comprising E14.7, S6 peptide, and an elutable chitin-binding domain (CBD). This elutable CBD tag allows for binding of the fusion protein to a chitin resin in a buffer containing 2 M NaCl and protein elution by a shift of salt concentrations from 2 to 0–0.5 M NaCl (14, 15). The purified fusion protein was labeled by Sfp phosphopantetheinyl transferase (Sfp synthase) and CoA-small molecule conjugates by the transfer of the dye to a specific serine residue (*underlined*) in the S6 tag (GDSLSWLLRLLN) (7).

2. Materials

2.1. Cell Culture and Protein Production

1. DMEM–F-12 medium (Invitrogen, Carlsbad, CA)
2. Phosphate-buffered saline (Cell Signaling Technology, Danvers, MA)
3. SlowFade Antifade Kit (Invitrogen)
4. *E. coli* T7 Express (New England Biolabs, Ipswich, MA)
5. Kanamycin (Sigma, St. Louis, MO)
6. Ampicillin (American Pharmaceutical Partners, Schaumburg, IL)
7. Luria–Bertani (LB) media
8. Isopropyl thiogalactosidase (IPTG; American Bioanalytical, Natick MA)
9. LB agar (Difco, Lawrence, KS)
10. Ni-NTA resin (Qiagen, Valencia, CA)

11. Imidazole (Sigma)
12. NaCl (Sigma)
13. Tris-HCl (Sigma)
14. Ethylenediaminetetraacetic acid (EDTA; Sigma)
15. Chitin beads (New England Biolabs)
16. Sfp synthase storage buffer: 10 mM Tris-HCl, pH 7.5, 1 mM EDTA, and 10% (v/v) glycerol
17. Cell lysis and protein-binding buffer: 20 mM Tris-HCl, pH 8.5, 2 M NaCl
18. Elution buffer: 20 mM Tris-HCl, pH 8.5, 50 mM NaCl

2.2. Reagents for Sfp Labeling

1. pET29-Sfp (contact Christopher T. Walsh at Harvard University)
2. Coenzyme A trilithium salt (Sigma)
3. Maleimide-activated DyLight 488, DyLight 549, and DyLight 649 (Thermo Scientific, Waltham, MA)
4. Tris-buffered saline (TBS): 50 mM Tris-HCl, pH 7.5, 0.15 M NaCl
5. Glycerol solution: 10% (v/v) glycerol
6. Sodium phosphate buffer: prepare a solution of 100 mM sodium phosphate (pH 7.0) by mixing the stock solutions of 0.5 M NaH_2PO_4 (monobasic) and 0.5 M Na_2HPO_4 (dibasic) (Sigma)
7. Acetonitrile (Sigma)
8. Trifluoroacetic acid (TFA) solution: 0.1% (v/v) trifluoroacetic acid (Sigma)
9. Dimethyl sulfoxide (DMSO; SIGMA)
10. $MgCl_2$ solution: 10 mM $MgCl_2$
11. HEPES buffer: 50 mM 4-(2-hydroxyethyl)-1-piperazineethanesulfonic acid, pH 7.5
12. Coomassie blue (Brilliant Blue R; Sigma)
13. Coomassie blue staining solution: 0.1% (w/v) Brilliant Blue R, 10% (v/v) glacial acetic acid, 50% (v/v) methanol
14. Destain solution: 10% (v/v) glacial acetic acid, 50% (v/v) methanol

2.3. Equipment

1. Innova 4230 floor incubator/shaker (New Brunswick Scientific, Edison, NJ)
2. Beckman J2-21 centrifuge (Beckman, Ramsey, MN)
3. Cell disruptor (Active Ultrasonics, La Chaux-de-Fonds, Switzerland)

4. Centriprep YM-10 concentrator (Millipore, Billerica, MA)
5. HPLC system (Beckman)
6. Protein peptide preparative HPLC column (Vydac, Deerfield, IL)
7. Voyager matrix-assisted laser desorption/ionization time of flight (MALDI-TOF) system (Applied Biosystems, Foster City, CA)
8. 10–20% Novex Tris-Glycine Mini gel (Invitrogen)
9. XCell Surelock Mini-Cell (Invitrogen)
10. TE22 Mini Tank Transfer Unit (Hoefer Scientific Instruments, Horsham, PA)
11. PowerPac power supply (Bio-Rad, Hercules, CA)
12. Power Supply (VWR, West Chester, PA)
13. Typhoon Imager 9400 (GE Healthcare, Piscataway, NJ)

2.4. Protein Electrophoresis and Blotting

1. 3X Sodium dodecyl sulfate-polyacrylamide gel electrophoresis (SDS-PAGE) reducing sample buffer: 3 mM NaCl, 70 mM Tris-HCl, pH 6.8 at 25°C, 1 mM EDTA, 2% (w/v) SDS, 0.01 % (w/v) phenol red, 10% (v/v) glycerol, 40 mM dithiothreitol
2. SDS-PAGE Tris-glycine buffer: 0.1% SDS, 0.19 M glycine, 0.025 M Tris, pH 8.3.
3. Nitrocellulose membrane (Protran; Whatman, Florham Park, NJ)
4. Prestained Protein Marker (New England BioLabs)
5. Broad Range Protein Marker (New England BioLabs)
6. Coomassie blue staining solution
7. Destain solution

2.5. Protein Transfection and Cell Imaging

1. Chamber slide: LAB-TEK II chambered coverglass (Nalge Nunc International)
2. COS-7: African Green Monkey SV40-transfected kidney fibroblast cell line
3. Serum medium: Dulbecco's modified Eagle's medium (DMEM) with 10% fetal bovine serum and L-glutamine
4. Cell culture incubator: Forma Series II 3110 Water Jacketed CO_2 incubator (Thermal Fisher Scientific), 5% CO_2 and 37°
5. TransPass P Transfection Reagent (New England Biolabs).
6. DAPI: 4′,6-diamidino-2-phenylindole, a fluorescent stain that binds to DNA (Invitrogen)
7. Fluorescence microscopy: Carl Zeiss LSM 510 meta Laser Scanning Microscope (Carl Zeiss)

3. Methods

To generate protein imaging agents by site-specific fluorescent labeling, small peptide tags, for example, the 12-residue S6 tag and 77-residue ACP tag, are preferred to larger protein tags as they presumably have less effect on the folding and function of a target protein, such as molecular interactions and translocation. Sfp-catalyzed labeling was mapped to a specific serine residue (underlined) in the ACP tag (Ser35), S6 (GDSLSWLLRLLN) or ybbR (DSLEFIASKLA) peptides (7, 10, 12). This experimental procedure includes the following steps (1) expression and purification of Sfp synthase, (2) synthesis of CoA-fluorochrome conjugates, (3) construction of a gene of interest fused to a targeting polypeptide and an affinity tag, (4) purification of the fusion protein, (5) labeling the purified protein, and (6) imaging the labeled protein by a fluorescent imager and microscope. The labeling reaction has been previously described (6). To facilitate rapid isolation of recombinant proteins from *E. coli* host cells, affinity tags such as an elutable CBD or six-histidine tag can be selected as a fusion partner. A mutant of the 6 kDa *Bacillus circulans* CBD carrying Trp687Phe mutation permits reversible binding to chitin resin by a shift in salt concentrations (e.g., 2 M NaCl and 100 mM NaCl) thereby facilitating rapid isolation of the fusion protein from the cell lysate via a single column step (14).

3.1. Cloning and Purification of E14.7 K Fusion Protein

1. Clone the coding region of adenoviral E3-14.7 K (E14.7) in-frame to S6 tag and CBD in a pTXB1 derivative (15). The resulting pE14.7S6CBD expresses a 23 kDa fusion protein E14.7-S6-CBD under the control of the IPTG-inducible T7/*lac* promoter.
2. Grow *E. coli* strain T7 Express harboring pE14.7S6CBD at 37°C in 1 l of LB medium containing 100 μg/ml of ampicillin to an A_{600} of 0.5.
3. Induce protein expression with 0.3 mM isopropyl-β-D-thiogalactoside (IPTG) at 16°C overnight.
4. Collect the cells by centrifugation at 3000×*g* for 30 min.
5. Resuspend the cells in 100 ml of cell lysis buffer containing 20 mM Tris-HCl, pH 8.0, 2 M NaCl.
6. Remove the debris by centrifugation at 16,000×*g* for 30 min.
7. Load the clarified supernatant onto a column at 4°C with a 10 ml bed volume of chitin beads equilibrated in the cell lysis buffer.
8. Wash the column with 40 column volumes of the cell lysis buffer.

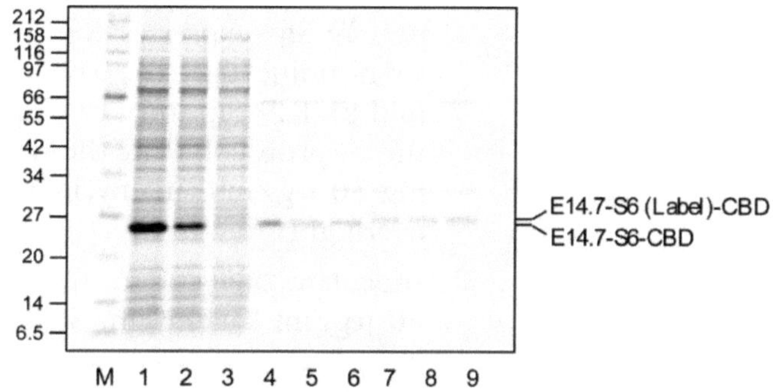

Fig. 18.2. SDS-PAGE analysis of protein purification and protein labeling. Chitin chromatography and Sfp-mediated protein labeling reactions with different fluorescent substrates were visualized on a 10–20% Tris-glycine gel after electrophoretic separation of proteins under denaturing conditions. Adenoviral protein E3-14.7 K (E14.7) was expressed in *E. coli* as a fusion protein E14.7-S6-CBD. Protein labeling in the presence of Sfp synthase and CoA-fluorophore conjugate resulted in the covalent linkage of the fluorophore to the side chain of a specific serine residue in the S6 tag, yielding E14.7-S6 (label)-CBD product. *Lane 1*, crude cell extract showing the expressed fusion protein; *lane 2*, clarified cell extract after sonication and centrifugation; *lane 3*, clarified cell extract after passage over a chitin resin, resulting in the binding of E14.7-S6-CBD; *lane 4*: protein fraction eluted from the chitin column; *lane 5*, incubation of E14.7-S6-CBD without Sfp synthase and fluorescent substrate; *lane 6*, labeling reaction without fluorescent substrate; *lane 7*, labeling with CoA-DyLight 488; *lane 8*, labeling with CoA-Dylight 549; *lane 9*, labeling with CoA-Dylight 649.

9. Elute the fusion protein with 20 mM Tris-HCl, pH 8.0, 0.1 M NaCl. Collect ten 3 ml fractions.
10. Analyze the elution fractions by 12% Tris-glycine denaturing gels (**Fig. 18.2**).
11. Determine the protein concentrations by Bradford assay (16).

3.2. Synthesis of CoA-Fluorophore Conjugates

1. Synthesize CoA-small molecule substrate by conjugation of CoA-SH to maleimide-activated fluorophores including DyLight 488, DyLight 549, and DyLight 649. We use the synthesis of CoA-DyLight 488 as an example.
2. To a solution of maleimide DyLight 488 (1.0 mg, 1.2 μmol) in 0.25 ml dimethyl sulfoxide (DMSO), add CoA trilithium salt (1.6 mg, 2.1 μmol) in 0.75 ml sodium phosphate 100 mM, pH 7.0, and stir the reaction mixture at room temperature for 1 h in the dark.
3. Directly purify the reaction mixture by preparative HPLC using a reverse-phase protein peptide C18 column with a gradient of 0–50% (vol/vol) acetonitrile over 30 min.
4. Lyophilize the purified compound and confirm its identity by MALDI-TOF (negative mode): 1564.48.

3.3. Expression and Purification of Sfp Synthase

1. pET29-Sfp expresses Sfp protein fused to a C-terminal six-histidine tag (6). Transform the Sfp expression plasmid pET29-Sfp into T7 Express cells following manufacturer's protocol. Plate the cells on an LB agar plate containing 50 μg/ml kanamycin and incubate the plate at 37°C overnight.
2. Inoculate one colony to 1 l of LB medium containing 50 μg/ml kanamycin. Shake at 200 r.p.m. at 37°C on a New Brunswick Innova 4230 floor incubator/shaker until the OD_{600} nm reaches 0.5.
3. Add IPTG to the LB culture to a final concentration of 0.3 mM.
4. Continue to shake overnight at 200 r.p.m. at 16°C and then harvest the cells by centrifugation in 1000 ml centrifuge bottle at 5000 r.p.m. for 30 min.
5. Resuspend the cells in 50 ml of a solution of 5 mM imidazole, 0.5 M NaCl, and 20 mM Tris-HCl, pH 7.9.
6. Lyse the cells by sonication in an ice-chilled water bath and prepare the clarified cell extract by centrifugation at 12,000×*g* for 30 min at 4°C.
7. Purify the Sfp protein by chromatography using 10 ml bed volume of N-NTA resin in a 15 mm diameter column according to the manufacturer's instructions.
8. Analyze the fractions of the eluant from the Ni-NTA column by SDS-PAGE and pool the fractions that contain Sfp synthase (26 kDa) and are >90% pure.
9. Dialyze the protein twice against 10 mM Tris-HCl, pH 7.5, 1 mM EDTA, and 10% glycerol.
10. Aliquot the protein solution into Eppendorf tubes and store the tubes at –80°C. The purification procedure typically yields approx. 50 mg of pure (>90%) Sfp synthase per liter of culture. The enzyme is very stable and can be stored at the above conditions for at least 12 months without significant loss of activity.

3.4. Sfp-Mediated Protein Labeling

1. We usually set up a reaction of 50 μl final volume containing 10 mM $MgCl_2$ and 50 mM HEPES, pH 7.5.
2. Add 1 μM Sfp synthase, 5 μM S6-tagged protein E14.7-S6-CBD, and 10 μM substrate (CoA-DyLight 488, CoA-DyLight 549, or CoA-DyLight 649).
3. Incubate the reaction mixture at room temperature for 30 min in dark.
4. Store the samples in dark at –20°C.

3.5. Gel-Shift Analysis

1. Add 25 μl 3X SDS-PAGE loading buffer (NEB) to 50 μl of the labeling mixture.
2. Heat the samples at 95°C for 5 min and load on a 10–20% SDS-PAGE gel.
3. Load approximately 1 μg of protein substrate (E14.7-S6-CBD) per lane on a polyacrylamide denaturing gel.
4. Perform electrophoretic separation at 140 V with a Cell Surelock Mini-Cell system.
5. Stain the gel with Coomassie blue.
6. Identify the labeled protein product based on the mobility shift of the labeled species by comparing to the mobility of the unlabeled protein in the control reaction (without Sfp synthase) (**Fig. 18.2**). Estimate the labeling efficiency based on the ratio of the labeled product to the unlabeled protein present in the labeling reaction.

3.6. Detection of Fluorescent Proteins

1. For in-gel detection by a fluorescent imager, subject 100–250 ng of protein substrate to electrophoretic separation by SDS-PAGE. Otherwise, electroblot the protein bands onto 0.45 μm nitrocellulose membrane with a Mini Trans-blot cell. Keep the gel or blot in dark.
2. Scan the gel or blot with Typhoon Imager 9400 following the manufacturer's instructions. Position the blot, with the protein-side down, onto the clean glass platen.
3. Select the instrument settings for imaging fluorescent proteins in gels or blots (**Fig. 18.3**). Focal Plane: platen; PMT voltage: 300 V; pixel size: 100 μm.
4. Choose the filter settings for the following fluorescent substrates:

   CoA-DyLight 488: 488/526 nm, excitation/emission filter set;

   CoA-DyLight 549: 532/580 nm excitation/emission filter set;

   CoA-DyLight 649: 633/670 nm excitation/emission filter set.

3.7. Transfection and Cell Imaging

1. Set up cells to be transfected so they are approximately 10×10^4–40×10^4 cells per well (approximately 70–80% confluent) at the time of transfection.
2. Add 0.4 μg in 2 μl of protein to a sterile tube containing the 50 μl of serum-free medium.
3. Add 1 μl of TransPass P Transfection Reagent (mix well before use).

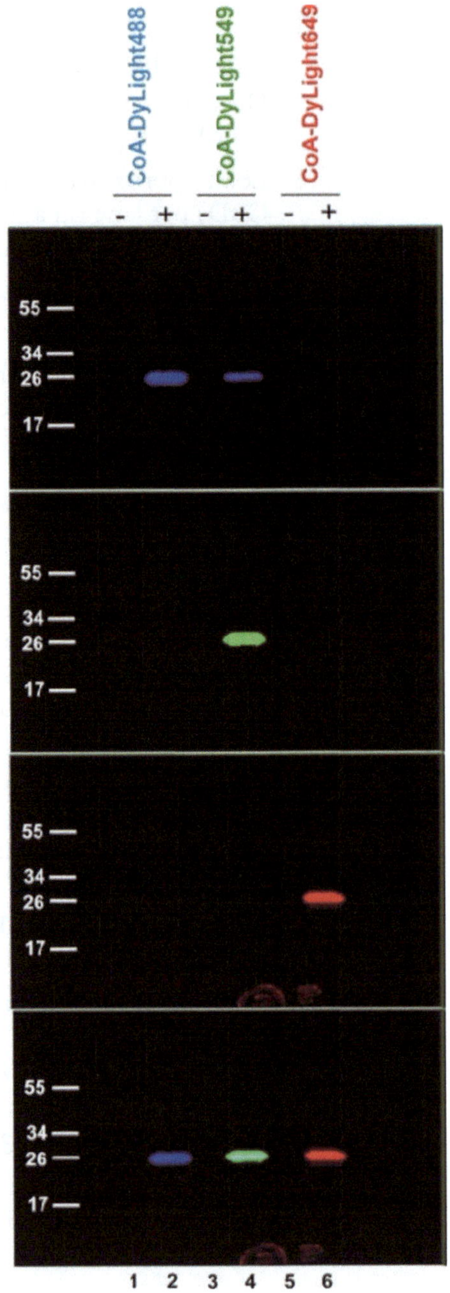

Fig. 18.3. Fluorescent in-gel detection of Sfp-mediated labeling of E14.7 fusion protein. E14.7-S6-CBD fusion protein was labeled at 25°C for 30 min with Sfp synthase in the presence of CoA-DyLight 488, CoA-DyLight 549, or CoA-DyLight 649 conjugate. The proteins were separated in 10–20% Tris-glycine gel under denaturing conditions. The gel was imaged with Typhoon 9400 at 300 V PMT with 488/526 nm excitation/emission filter set (*blue, top panel*), 532/580 nm excitation/emission filter set (*green*) or 633/670 nm excitation/emission filter set (*red*). A merged image of the three channels is shown (*bottom panel*).

4. Gently mix the transfection complex mixture by flicking the tube.
5. Incubate at room temperature for 20 min.
6. Remove serum-containing growth media from cells by aspirating, wash cells with serum-free medium and add 1 ml of serum-free medium to each well.
7. Add the transfection complex mixture to cells.
8. Return plate to incubator and incubate for 2–5 h.
9. Add 1 ml of complete media (containing 10% serum) to each well.
10. Replace media on the following day and continue incubation.
11. Wash cells with serum-free medium before imaging to remove any untransfected protein.
12. Cover the cells with fresh prewarmed serum medium and immediately image the cells using appropriate filter sets (**Fig. 18.4**).

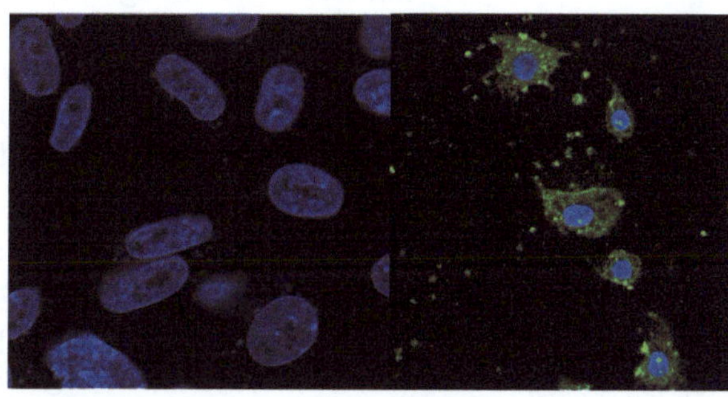

Fig. 18.4. Protein transfection and live cell imaging by fluorescence microscopy (17). Cos-7 cells transfected with DyLight 488 labeled E14.7-S6-CBD (*right panel*) were photographed 24 h post-transfection by a fluorescent confocal microscope. Nuclear staining was with DAPI. Cos-7 cells without the labeled protein were also photographed (*left panel*) showing only DAPI nuclear staining. The micrographs shown are the merged images of the FITC and DAPI channels.

4. Notes

1. In general, the 8 kDa ACP tag or short peptides S6 or ybbR can be efficiently labeled by Sfp enzyme; however, certain sequences and tags may have adverse effect on Sfp-mediated labeling reaction when it is present adjacent to the targeting peptide.

2. To assure consistent and high-level protein expression, inoculate medium with freshly grown colonies or cell culture.
3. Maleimide-activated molecules can be conjugated to CoA-SH to generate CoA-small molecule substrates for enzymatic labeling by Sfp synthase.
4. ACP tag can be also labeled using ACP synthase with CoA-small molecule substrates.
5. Enzymatic labeling reaction can be carried out in clarified cell lysate or using purified protein sample.
6. Run prestained protein markers to monitor electrophoresis and transfer on the nitrocellulose membrane.
7. Prior to protein transfer to nitrocellulose soak the nitrocellulose membrane in water for 1 min and then in transfer buffer. Make sure there are no bubbles between a gel and nitrocellulose membrane. Squeeze out the air bubbles using a pencil or glass pipette as a roller.
8. Both SDS-PAGE gels and protein blots can be scanned by a fluorescence imager. It is also possible to spot the protein samples directly onto a nitrocellulose membrane. Careful handling is required for laying down the gel or membrane (with the protein-side facing down) to avoid air bubbles which affect the quality of fluorescence imaging and quantification.
9. In a preliminary test, it is advisable to examine a labeling reaction containing 100 ng of purified protein. A titration of 1–100 ng of protein sample may be performed to further characterize the labeling efficiency and sensitivity.
10. Store labeled proteins at –20 or –80°C in dark to avoid photobleaching and protein degradation.
11. A labeling reaction may be validated by a shift in mobility of the labeled protein by Coomassie blue-stained SDS-PAGE. An unlabeled control (without enzyme) should be performed and analyzed.
12. Prepare a stock solution of substrate (in 0.25 or 1 mM final concentration) by dissolving the substrate in water or dimethyl sulfoxide (DMSO) and store the aliquots at –20°C (or at –80°C) in dark.
13. The transfection complex mixture is composed of protein and TransPass P Transfection Reagent in serum-free medium. For example, at the incubation step (in 6-well format), transfection complex mixture consisting of 2 μg protein and 3 μl TransPass P Transfection Reagent in 200 μl serum-free medium is added to a well containing cells in a 1 ml volume.

Acknowledgments

The authors thank New England Biolabs and Donald G. Comb, Jim Ellard, Richard Roberts, Ted Davis, Salvatore Russello, Thomas C. Evans, Jr., Williams Jack, Jack Benner, Sriharsa Pradhan, Christopher Provost for their encouragement and valuable suggestions. We thank the Organic Division for peptide and fluorescent substrate synthesis. We also thank Drs. Christopher Walsh and Jun Yin for kindly providing pET29-Sfp for expression and purification of Sfp synthase.

References

1. Wang, H., Chen, X. (2008) Site-specifically modified fusion proteins for molecular imaging. *Front Biosci* 13, 1716–1732.
2. Chen I., Ting A. Y. (2005) Site-specific labeling of proteins with small molecules in live cells. *Curr Opin Biotechnol* 16, 35–40.
3. Tsien, R. Y. (1998) The green fluorescent protein. *Annu Rev Biochem* 67, 509–544.
4. Griffin B. A., Adams, S. R., Tsien, R. Y. (1998) Specific covalent labeling of recombinant protein molecules inside live cells. *Science* 10, 269–272.
5. Keppler, A., Gendreizig, S., Gronemeyer, T., Pick, H., Vogel, H., Johnsson, K. (2002) A general method for the covalent labeling of fusion proteins with small molecules *in vivo*. *Nat Biotechnol* 21, 86–89.
6. Yin, J., Lin, A. J., Golan, D. E., Walsh, C. T. (2006) Site-specific protein labeling by Sfp phosphopantetheinyl transferase. *Nat Protoc* 1, 280–285.
7. Zhou, Z., Cironi, P., Lin, A. J., Xu, Y., Hrvatin, S., Golan, D. E., et al. (2007) Genetically encoded short peptide tags for Orthogonal protein labeling by Sfp and Acps phosphopantetheinyl transferases. *ACS Chem Biol* 2, 337–346.
8. Gautier, A., Juillerat, A., Heinis, C., Corrêa, I. R., Kindermann, M., Beaufils, F., Johnsson, K. (2008) An engineered protein tag for multiprotein labeling in living cells. *Chem Biol* 15, 128–136.
9. Meyer, B. H., Segura, J., Martinez, K. L., Hovius, R., George, N., Johnsson, K., Vogel, H. (2006) FRET imaging reveals that functional neurokinin-1 receptors are monomeric and reside in membrane microdomains of live cells. *Proc Natl Acad Sci USA* 103, 2138–2143.
10. Yin, J., Lin, A. J., Buckett, P. D., Wessling-Resnick, M., Golan, D. E., Walsh, C. T. (2005) Single-cell FRET imaging of transferrin receptor trafficking dynamics by Sfp-catalyzed, site-specific protein labeling. *Chem Biol* 12, 999–1006.
11. Jacquier, V., Prummer, M., Segura, J.-M., Pick, H., Vogel, H. (2006) Visualizing odorant receptor trafficking in living cells down to the single-molecule level. *Proc Natl Acad Sci USA* 103, 14325–14330.
12. Yin, J., Straight, P. D., McLoughlin, S. M., Zhou, Z., Lin, A. J., Golan, D. E., et al. (2005) Genetically encoded short peptide tag for versatile protein labeling by Sfp phosphopantetheinyl transferase. *Proc Natl Acad Sci USA* 102, 15815–15820.
13. Schneider-Brachert, W., Tchikov, V., Merkel, O., Jakob, M., Hallas, C., et al. (2006) Inhibition of TNF receptor 1 internalization by adenovirus 14.7 K as a novel immune escape mechanism. *J Clin Invest* 116, 2901–2913.
14. Ferrandon, S., Sterzenbach, T., Mersha, F., Xu, M.-Q. (2003) A single surface tryptophan in the chitin-binding domain from *Bacillus circulans* chitinase A1 plays a pivotal role in binding chitin. *Biochim Biophys Acta* 1621, 31–40.
15. Evans, T. C., Jr., Benner, J., Xu, M.-Q. (1998) Semisynthesis of cytotoxic proteins using a modified protein splicing element. *Protein Sci* 7, 2256–2264.
16. Bradford, M. M. (1976) A rapid and sensitive method for the quantitation of microgram quantities of protein utilizing the principle of protein-dye binding. *Anal Biochem* 72, 248–254.
17. Spector, D., Goldman, R. (eds.) (2004). *Live Cell Imaging: A Laboratory Manual.* Cold Spring Harbor Laboratory Press, Cold Spring Harbor, NY.

Index

T.C. Evans, M.-Q. Xu (eds.), *Heterologous Gene Expression in E. coli*, Methods in Molecular Biology 705, DOI 10.1007/978-1-61737-967-3, © Springer Science+Business Media, LLC 2011

Index

MIX
Papier aus verantwortungsvollen Quellen
Paper from responsible sources
FSC® C105338

If you have any concerns about our products,
you can contact us on
ProductSafety@springernature.com

In case Publisher is established outside the EU,
the EU authorized representative is:
Springer Nature Customer Service Center GmbH
Europaplatz 3, 69115 Heidelberg, Germany

Printed by Libri Plureos GmbH
in Hamburg, Germany